TK
7871.15.P5
A48
1995
AGC-5079

Plastics for Electronics

Electronic Packaging and Interconnection Series
Charles M. Harper, Series Advisor

CLASSON • *Surface Mount Technology for Concurrent Engineering and Manufacturing*
GINSBERG AND SCHNORR • *Multichip Modules and Related Technologies*
HARPER • *Electronic Packaging and Interconnection Handbook*
HARPER AND MILLER • *Electronic Packaging, Microelectronics, and Interconnection Dictionary*
HARPER AND SAMPSON • *Electronic Materials and Processes Handbook, 2/e*
LICARI • *Multichip Module Design, Fabrication, and Testing*

Related Books of Interest

BOSWELL • *Subcontracting Electronics*
BOSWELL AND WICKAM • *Surface Mount Guidelines for Process Control, Quality, and Reliability*
BYERS • *Printed Circuit Board Design with Microcomputers*
CAPILLO • *Surface Mount Technology*
CHEN • *Computer Engineering Handbook*
COOMBS • *Printed Circuits Handbook, 3/e*
DI GIACOMO • *Digital Bus Handbook*
DI GIACOMO • *VLSI Handbook*
FINK AND CHRISTIANSEN • *Electronics Engineers' Handbook*
GINSBERG • *Printed Circuits Design*
JURAN AND GRYNA • *Juran's Quality Control Handbook*
MANKO • *Solders and Soldering, 3/e*
RAO • *Multilevel Interconnect Technology*
SZE • *VLSI Technology*
VAN ZANT • *Microchip Fabrication*

To order or receive additional information on these or any other McGraw-Hill titles, in the United States please call 1-800-822-8158. In other countries, contact your local McGraw-Hill representative.   BC14BCZ

Plastics for Electronics

Materials, Properties, and
Design Applications

William M. Alvino

McGraw-Hill, Inc.
New York   San Francisco   Washington, D.C.   Auckland   Bogotá
Caracas   Lisbon   London   Madrid   Mexico City   Milan
Montreal   New Delhi   San Juan   Singapore
Sydney   Tokyo   Toronto

Library of Congress Cataloging-in-Publication Data

Alvino, William M.
  Plastics for electronics : materials, properties, and design applications / William M. Alvino.
      p.   cm.—(Electronic packaging and interconnection series)
  Includes bibliographical references and index.
  ISBN 0-07-001435-3
  1. Electronics—Materials. 2. Plastics. 3. Electronic packaging—Materials. I. Title. II. Series.
  TK7871.15.P5A48   1994
  620.1'92'0246213—dc20   94-38303
                                                                        CIP

Copyright © 1995 by McGraw-Hill, Inc. All rights reserved. Printed in the United States of America. Except as permitted under the United States Copyright Act of 1976, no part of this publication may be reproduced or distributed in any form or by any means, or stored in a data base or retrieval system, without the prior written permission of the publisher.

1 2 3 4 5 6 7 8 9 0   DOC/DOC   9 0 9 8 7 6 5 4

ISBN 0-07-001435-3

The sponsoring editor for this book was Stephen S. Chapman, the editing supervisor was Stephen M. Smith, and the production supervisor was Pamela A. Pelton. It was set in Century Schoolbook by McGraw-Hill's Professional Book Group composition unit.

Printed and bound by R. R. Donnelley & Sons Company.

Information contained in this work has been obtained by McGraw-Hill, Inc., from sources believed to be reliable. However, neither McGraw-Hill nor its authors guarantee the accuracy or completeness of any information published herein and neither McGraw-Hill nor its authors shall be responsible for any errors, omissions, or damages arising out of use of this information. This work is published with the understanding that McGraw-Hill and its authors are supplying information, but are not attempting to render engineering or other professional services. If such services are required, the assistance of an appropriate professional should be sought.

To Barbara

Contents

Preface xi
Acknowledgments xiii

Part 1 Fundamentals of Plastics for Electronics

Chapter 1. Introduction to the Use of Plastics in the Electrical Industry   3

General Comments   3
The Role of Plastics in Electronics   4
Extent of Plastics in the Electronics Industry: Statistics   6
Growth   8
A Resource   10
References   12

Chapter 2. Fundamentals   13

Polymer Definition   13
Types of Polymers   13
Structure and Properties   18
The Physical State of a Polymer   24
Synthesis   26
Bulk Polymerization   26
Solution Polymerization   26
Emulsion Polymerization   27
Suspension Polymerization   27
Terminology   27
References   35

Chapter 3. Thermoplastics   37

Acrylics   52
Acetals   54
Cellulosics   55
Fluoroplastics   55
Ionomer   57
Ketone Plastics   58

viii   Contents

    Liquid-Crystal Polymers 60
    Nylon 64
    High-Temperature Nylon 65
    Polyamide-Imides 66
    Polyimides 67
    Polyetherimide 69
    Polyarylates and Polyesters 70
    Polycarbonates 72
    Polyolefins 72
    Polyolefin Copolymers 74
    Polyphenylene Oxide 75
    Polyphenylene Sulfide 76
    Styrenics 77
    Polysulfones 77
    Vinyls 78
    Cross-linked Thermoplastics 79
    References 81

Chapter 4. Thermosets 85

    Allyl Resins 87
    Bismaleimides 91
    Epoxy Resins 91
    Phenolic Resins 94
    Polyesters 95
    Polyurethanes 97
    Silicones 98
    Cyanate Ester Resins 99
    Benzocyclobutenes 103
    Polyxylylene 104
    Urea and Melamine Resins 105
    Silicon-Carbon Thermosets 106
    References 108

Chapter 5. Elastomers 109

    Elastomer Types 110
    Elastomer Properties 117
    References 127

Chapter 6. Alloys and Blends 129

    Chemistry and Properties 129
    Preparation and Processing 131
    Other Types 132
    Blend Properties and Characterization 133
    References 137

Chapter 7. Processing of Plastics and Elastomers — 139
- Processing Methods — 140
- References — 152

Chapter 8. High-Performance Polymers — 153
- Thermal Stability — 154
- Polyimides — 155
- Condensation and Addition Polyimide Hybrids — 164
- Polybenzimidazoles — 164
- Other High-Performance Polymers — 166
- Cyanate Esters — 166
- Silicon-Carbon Resins — 166
- Polyphenylene Rigid-Rod Polymers — 167
- Heterocyclic Polymers — 167
- High-Performance Fibers — 171
- Advanced High-Performance Thermoplastics — 171
- Summary — 173
- References — 173

Chapter 9. Organic Coatings — 177
- Overview — 177
- Coating Types — 179
- Other Coating Processes and Applications — 198
- References — 202

Part 2 Design Considerations for Plastics

Chapter 10. Plastic Properties and Testing — 205
- Mechanical Properties — 215
- Electrical Properties — 217
- Thermal Properties — 222
- Environmental Properties — 227
- Thermoplastics — 228
- Thermosets — 233
- Elastomers — 235
- Aging of Polymers — 237
- Nuclear Radiation Effects — 241
- References — 257

Chapter 11. Design Considerations — 259
- A Brief Review — 259
- Engineering Properties — 260
- Environmental Properties — 262
- Assembly — 264

Contents

Machining, Finishing, and Decorating	266
References	269

Part 3 Utilization of Plastics in Electronics

Chapter 12. Forms, Applications, and Uses — 273

Testing and Characterization	273
Films and Sheets	274
Flexible Circuits	277
Coatings and Encapsulants	282
Adhesives	282
Tapes	296
Sleeving and Tubing	299
Reinforced Electrical Insulation Materials	303
References	314

Chapter 13. Plastics in Microelectronics — 317

Resists	318
Other Developments	323
Materials for Electronic Packages	326
References	338

Part 4 Reference Sources on Plastic Materials

Chapter 14. Information Sources — 343

Technical Societies	343
Plastics Information Databases	347
Conferences	347
Publications	347
Newspapers	350
Trade Associations and Professional Organizations	350
Trade Publications	351
Business Consulting Groups and Publishers	352
Government Publications	354
Plastics Suppliers	355
References	355

Index   357

Preface

We live in a world where materials play such a significant role that they impact our lives on a daily basis. Technological advances in every discipline are related in some way to a material of construction that is used in the design, development, and manufacture of a device or its components.

A class of materials that has broad application in many fields is plastics. One such field is the electrical/electronics industry, and plastics are used in many things from large power cables to microelectronic devices. The electronics engineer using plastic materials often has limited knowledge of the types available; their properties, strengths, and weaknesses; how they are made, used, and measured; and the interdependent relationships that exist between their structure, properties, and performance. This book is intended primarily to fill in that engineer's gap of knowledge concerning plastics in electronics.

While the contents of this book is limited to plastics and elastomers, it nevertheless is broad enough in scope to address significant aspects of polymer materials in the electrical/electronic industry. This book will be useful to people engaged in industries that utilize polymers in general and plastics in particular, and to the teacher and student engaged in materials science studies.

Plastics for Electronics is divided into four parts. Part 1 is composed of Chaps. 1 to 9 and contains basic information on polymer materials used in electronics. The first chapter deals with the demographics of plastics in the electrical/electronics industry. Chapter 2 deals with plastic fundamentals, while Chaps. 3 to 6 cover the different types of polymers. Chapter 7 reviews the processing of polymers. Chapter 8 describes those polymers that have special properties that set them apart from the ordinary plastics and enable them to be classified as high-performance polymers. Chapter 9 surveys plastics as organic coatings in the electronics industry. Part 2 deals with design considerations of plastics, and includes two chapters. Chapter 10 covers plastic properties and testing, while Chap. 11 deals with factors affecting the selection of plastics. Part 3 deals with the utilization of plastics in electronics. Chapter 12 describes the various and sundry applications of plastics in the electronics industry. Chapter 13 provides the reader with current and next-generation uses of plastics in electronics. Part 4, Chap. 14, provides information on where to go to get more information on polymers in general and plastics in particular.

The author hopes you will find this book to be an excellent resource on plastics information.

William M. Alvino

Acknowledgments

The author would like to express his appreciation to the following: Charles A. Harper for his encouragement, discussions, constructive comments, and careful reading of the drafts; Ronald N. Sampson for encouragement and helpful discussions; Terry McCarron for initial manuscript typing and Barbara J. Jackson for performing the lion's share of typing, editing, retyping, etc.; Maurice Hanes for supplying the comprehensive information on photoresists; and the Westinghouse Science and Technology Center for providing the opportunity to do both fundamental and applied research work in the fields of polymer science, electrical insulation, and microelectronics to acquire the knowledge and skills necessary to complete this work.

Plastics for Electronics

Part 1

Fundamentals of Plastics for Electronics

Chapter 1

Introduction to the Use of Plastics in the Electrical Industry

General Comments

Plastics are ubiquitous. Since their beginnings, circa 1840, when Goodyear discovered the vulcanization of rubber, these materials have established themselves as key substances in almost every area of our lives. Indeed, they have spanned almost 10 decades of growth and use in applications too numerous to mention, so that one might refer to this period as the "plastics age." Plastics are no longer considered just useful materials for the fabrication of toys. Plastics represent a very large and versatile group of materials because of their unique combination of properties, i.e., excellent electrical, chemical, mechanical properties, combined with light weight, low cost, and ease of processing and fabrication. While plastics are not the panacea of industry's materials problems, they do offer such a unique combination of properties that they have become one of the important classes of materials and have found very widespread use in the electronics industry. Beginning in the early 1900s with the use of phenolic resins in electrical housing and wiring platforms to the use of plastics today in microelectronic devices, plastics are used as wire and cable insulation, electrical connectors, circuit boards, encapsulants, coatings, flexible circuitry, multilayer circuitry, photoresists, adhesives, interlayer dielectrics, and the list goes on. Their use in this industry has spawned the development of many support organizations whose function is to set standards and specifications that govern the property measurement, use, and dissemination of information about these materials.[1,2] Some of these organizations are

4 Fundamentals of Plastics for Electronics

TABLE 1.1 Organizations in the Electrical and Electronics Industries

Symbol	Name
ASTM	American Society for Testing and Materials
NEMA	National Electrical Manufacturer's Association
IEEE	Institute of Electrical and Electronics Engineers
ICEA	Insulated Cable Engineers Association
AEIC	Association of Edison Illuminating Companies
ANSI	American National Standards Institute
UL	Underwriters Laboratories
IEC	International Electrotechnical Commission
EPRI	Electrical Power Research Institute
SAE	Society of Automotive Engineers
Agencies	MIL-1 Specifications on Insulation
Military	MIL-W Specifications on Insulated Wire
EASA	Electrical Apparatus Service Association
EIA	Electronic Industries Association
IEPS	International Electronics Packaging Society
IPC	Institute for Interconnecting and Packaging Electronic Circuits
SPE	Society of Plastics Engineers
SPI	Society of the Plastics Industry
ISHM	International Society for Hybrid Microelectronics

listed in Table 1.1. The extent of this list is but one testament to the importance of these materials in the electronics industry.

A distinction should be made regarding the use of the terms *electrical* and *electronic*. While both terms can be used to describe the electrical industry in the broad sense, the term *electrical* is used in a more traditional way to refer to the insulating properties of plastics, as in wire and cable insulation, switches, connectors, etc. The term *electronic* refers to the use of plastics in the semiconductor industry, e.g., as photoresists, intermetallic dielectrics, and packaging. Typical applications in both areas are shown in Table 1.2.

The Role of Plastics in Electronics

The materials used in the packaging of electronic devices, whether they are used as insulation in large power cables or in microelectronic devices, play a key role in the operation and life of the product. While the main function of the packaged electronic assembly is the conduc-

TABLE 1.2 Plastics Applications in the Electrical and Electronic Industries

Application	Electrical	Electronic
Housing and wiring platforms	X	
Wire and cable insulation	X	
Magnet wire	X	
Power cable insulation	X	
Building wire	X	
Lamp cord	X	
Connectors	X	
Circuit boards	X	X
Encapsulants	X	X
Resists		X
Conformal coatings		X
Packaging		X
Flexible circuits	X	
Terminal blocks	X	
Switches	X	
Adhesives	X	X
Bobbins	X	
Cloth, fabric, paper insulation	X	X
Tapes	X	
Heat-shrinkable tubing	X	
Sleeving	X	
Relays	X	
Sockets	X	

tion of signals through the device, of equal and often greater importance is the material that provides the electrical insulation. The primary function of the insulation is to prevent the loss of the signal currents and to confine them to the desired paths. In addition to the fact that the plastic materials used in the electronic industry must have excellent dielectric properties, they must combine this with superior chemical, mechanical, thermal, and environmental properties to meet the requirements of the electronics industry in the various and sundry applications.

Plastics are indeed used in many applications and serve a variety of functions in the electronics industry, as we have seen in Table 1.2. Hence, it is critical that these materials have a unique balance of properties to meet the intended application requirements.

Extent of Plastics in the Electronics Industry: Statistics[3]

Of the 25 largest manufacturing industries in 1992, the electronics parts, semiconductors, and plastics materials rank 14, 15, and 16. Their values in industry shipments in 1987 dollars are $32.9, $28.8, and $27.5 billion, respectively. The major markets and distribution of all resins (thermoplastics and thermosets) are shown in Figs. 1.1 and 1.2. In terms of pounds of plastic used in the electrical and electronics industries, the percentages shown in the figures represent 209 million lb for all thermoset plastics and 2.6 billion lb for all thermoplastics. In the reinforced plastics (composites) area, which consists of a reinforcing fiber and a resin (thermoplastic and thermoset), an additional 260 million lb of plastic resin is used by the electronics industry. The U.S. composites shipments for all market areas are shown in Table 1.3. The reinforcing fiber is glass, aramid, carbon, graphite, Teflon, boron, and polybenzoxazole (developmental) while the resins used include unsaturated polyesters, vinyl esters, epoxy, silicone, cyanate ester, bismaleimides, and various thermoplastics. Further, the integrated-circuit industry contributes additional usage of plastics to the numbers men-

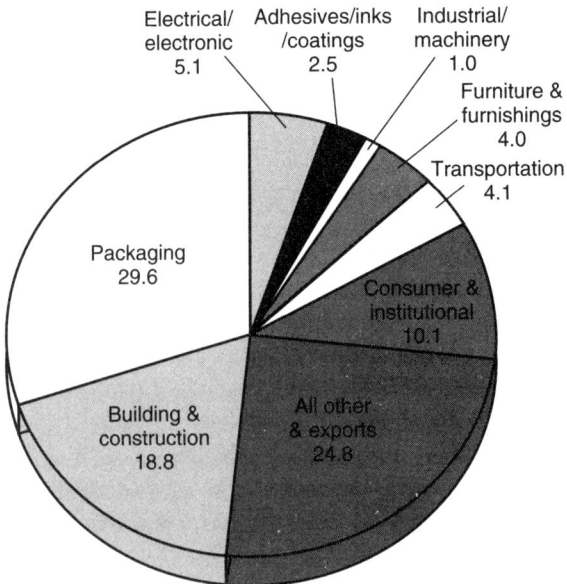

Figure 1.1 The 1991 percentage distribution of plastic resin sales and captive use by major market: all resins. (Source: Ref. 3. Reprinted with permission from the Society of the Plastics Industry.)

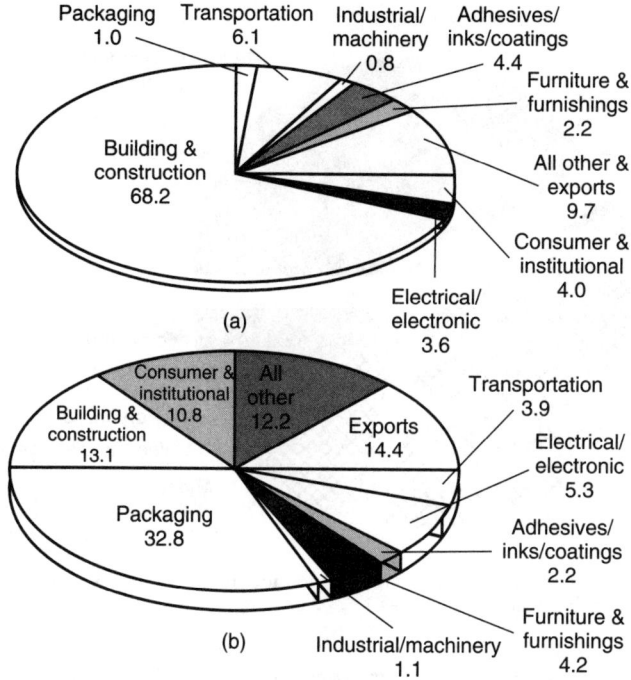

Figure 1.2 The 1991 percentage distribution of plastic resin sales and captive use by major market: (a) thermoset resins, (b) thermoplastic resins. (Source: Ref. 3. Reprinted with permission from the Society of the Plastics Industry.)

TABLE 1.3 U.S. Composites Shipments: 1991–1993 (10^6 lb)

Markets	1991	1992		1993 (Forecast)	
Aircraft, aerospace, military	38.7	32.3	(−16.5)	31.8	(−1.5)
Appliance and business equipment	135.2	143.2	(+5.9)	150.7	(+5.2)
Construction	420.0	483.0	(+15.0)	512.0	(+6.0)
Consumer products	148.7	162.2	(+9.1)	171.9	(+6.0)
Corrosion-resistant equipment	355.0	332.3	(−6.4)	337.3	(+1.5)
Electrical and electronic	231.1	260.0	(+12.5)	273.8	(+5.3)
Marine	275.0	304.4	(+10.7)	321.0	(+5.5)
Transportation	682.2	750.0	(+9.9)	800.0	(+6.7)
Other	73.8	83.4	(+13.0)	88.3	(+5.9)
Total	2359.7	2550.8	(+8.1)	2686.8	(+5.4)

SOURCE: Ref. 4. Reprinted with permission from *Plastics World*, Cahners Publications.

8 Fundamentals of Plastics for Electronics

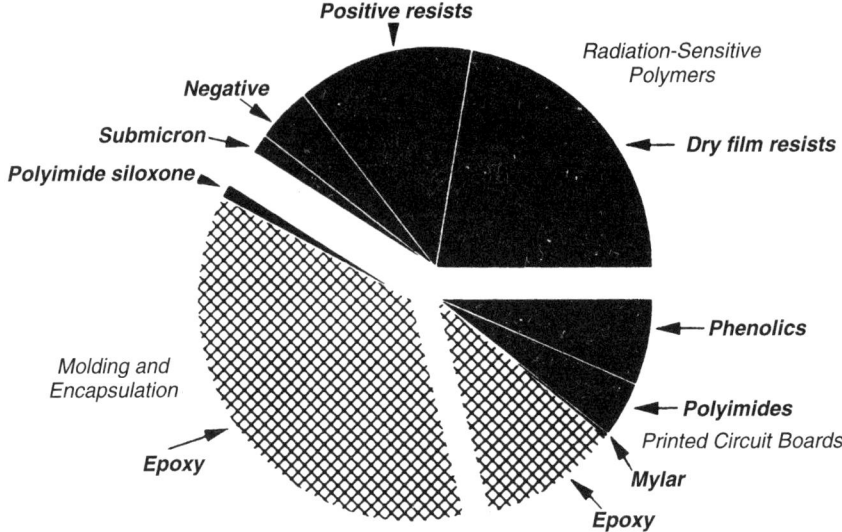

Figure 1.3 Polymers for electronic applications: worldwide usage. (Source: Ref. 5. Reprinted with permission from Academic Press.)

tioned previously. A whole new industry sprang up after the invention of the transistor by researchers at AT&T Bell Laboratories and with it a tremendous growth in the use of plastics to serve this industry. The integrated-circuit industry has grown to a multibillion-dollar industry worldwide. The major uses of plastics in this industry are shown in Fig. 1.3. There has indeed been tremendous growth in the electronics industry as a whole and in the microelectronics area in particular. With this growth came increased demand for materials that could fulfill the stringent requirements of the electronics industry. The plastics industry responded with the development of totally new plastics as well as the improvement of existing resins. The demand and use for plastic materials to serve in both traditional electrical roles and in advanced electronic packaging will continue to grow well into the next century. New and exciting technologies and developments in piezoelectric and pyroelectric polymer areas, optical waveguides, nonlinear optical materials, molecular electronics, optical storage multichip module (MCM), and other related electronic areas abound and create new opportunities for the use of polymers.

Growth

Plastic products are expected to grow 31 percent over the 1991 estimate of $66.3 billion to roughly $87 billion in 1996.[6] This forecast is predi-

TABLE 1.4 Outlook for Plastic Product Consumption by Key Industries

Industry	Percentage of total consumption		Annual growth of real consumption, 1991–1996, %
	1991	1996	
Electronics	21.1	26.0	10.1
High technology	10.7	15.1	13.1
Health care	9.9	9.1	3.8
Construction	9.6	8.7	3.6
Transportation	5.7	5.5	4.9
Motor vehicles	5.3	5.1	4.8
Food products	6.0	5.2	2.6
Dishware	3.5	3.1	3.1
Furnishings and appliances	3.3	3.0	3.4
All others	40.9	39.4	4.8

SOURCE: Ref. 6. Reprinted with permission from Advanstar Publications.

cated on the continuing popularity of plastic materials and the trend toward material substitution. The most important customer for the manufactured plastics products is the electronics industry (driven primarily by computers and semiconductors). By 1996 it is estimated that electronics will account for almost 26 percent of the dollars spent on plastics products. This compares to 21 percent spent in 1991 and 11 percent spent in 1980. The outlook for plastics consumption in key industries is shown in Table 1.4.

In the integrated-circuit area, the two primary driving forces are increased performance and reduced cost. The increase in the level of integration in a chip is the most effective way of meeting the cost and performance requirements. The progress of the industry is shown in Fig. 1.4.[5] Along with the higher level of integration came a need for denser interconnections on the package which affected the printed-circuit technology. This evolution is depicted in Figs. 1.5 and 1.6.[5] This will place great demands on the power-handling ability, thermal stability, and cooling, and new plastic materials will be required that have lower dielectric constants to accommodate the decreased interconnection spacing and thermal loads imposed by the new designs as well as improved mechanical, dimensional, and chemical stability. The increasing need for high-temperature materials in electrical, electronics, and aerospace industries, which together account for 50 percent of the high-temperature polymers market, should spur the development of these materials even further in the 1990s. The high-temperature polymer business is about $2 billion per year and has had a growth of 15 percent per year since 1984 and should increase by about 8 percent per year in the 1990s, according to Kline & Co. (Fairfield, New Jersey).

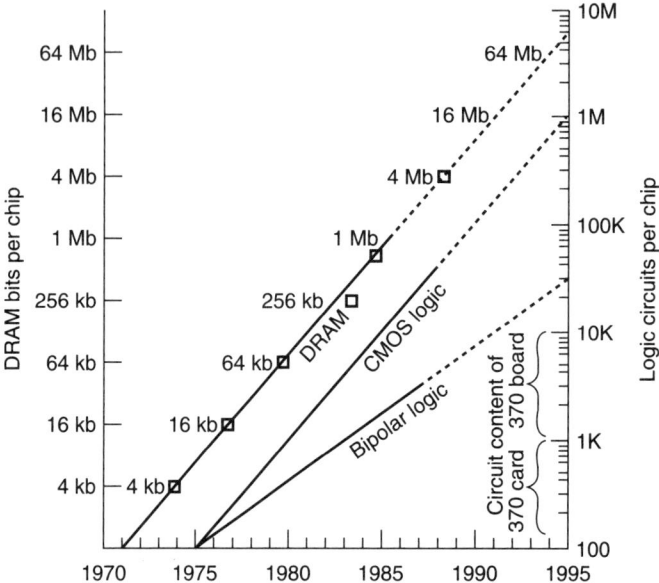

Figure 1.4 Progress in the level of chip integration. (Source: Ref. 5. Reprinted with permission from Van Nostrand Reinhold.)

Materials such as polyketones, polysulfones, polybenzimidazoles, polyimides, polyphenylene sulfide, polyetherimide and liquid-crystal polymers, and other newer high-temperature polymers will see more use in the electronics industry.[7]

A Resource

The electronics engineer using plastic materials often has only limited knowledge of the materials available, their properties, applications, their strengths and weaknesses, and the interdependent relationships that exist between their structure, properties, and performance. This kind of information is needed in the overall design scheme in order to use a particular plastic for a given application. This book hopefully will provide that information. This book is about plastics in electronics. It is about what they are, how they are used, and what their properties are. It is about where they are used and about how one can get more information on plastics in electronics. And last but not least, this book is a resource for both the plastics and electronic practitioner and those who are not.

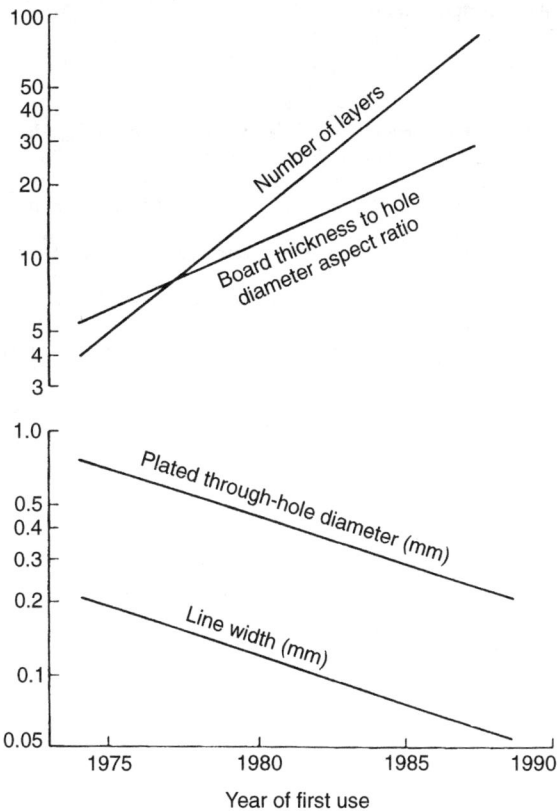

Figure 1.5 Evolution of printed-circuit board technology. (Source: Ref. 5. Reprinted with permission from Van Nostrand Reinhold.)

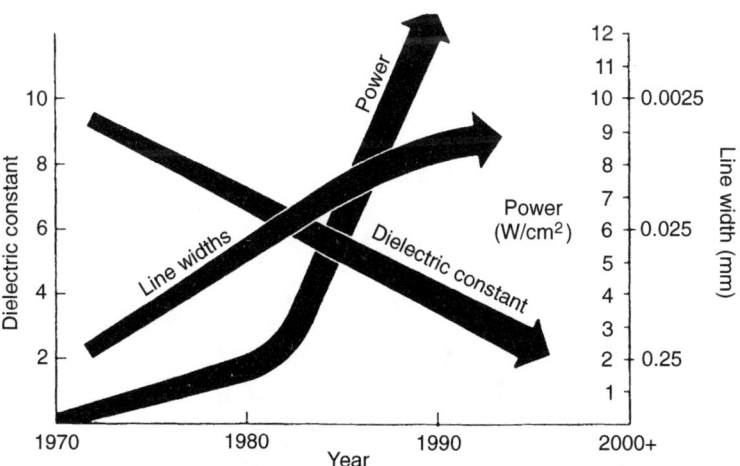

Figure 1.6 Trends in packaging. (Source: Ref. 5. Reprinted with permission from Van Nostrand Reinhold.)

References

1. H. Mark, N. M. Bikales, C. G. Overberger, and G. Menges, eds., *Encyclopedia of Polymer Science and Engineering,* 2d ed., vol. 5, Wiley, New York, 1990, p. 433.
2. R. Greene, ed., *Modern Plastics Encyclopedia,* McGraw-Hill, New York, 1991.
3. *Facts and Figures of the U.S. Plastics Industry,* Society of the Plastics Industry, Washington, DC, 1992.
4. J. Callari and C. Kirkland, "New Applications Fuel Reinforced Plastics Gains," *Plastics World,* March 1993, p. 16.
5. C. P. Wong, ed., *Polymers for Electronic and Photonic Applications,* Academic Press, New York, 1993.
6. Plastics Site Selector Supplement to *Advanced Composites,* Advanstar Communications, 1992, p. 2.
7. *Chemical Week* Markets Newsletter, February 10, 1993, p. 39.

Chapter 2

Fundamentals

Polymer Definition

Polymers are macromolecules (large molecules) formed by the linking together of large numbers of small molecules, called *monomers*. The process involved in the joining of these monomers is called *polymerization*. Plastics are a group of synthetic polymers made up of chains of atoms or molecules. The long molecular chains contain various combinations of oxygen, hydrogen, nitrogen, carbon, silicon, chlorine, fluorine, and sulfur. As more repeating units are added, the plastic's molecular weight increases and can reach into the millions,[1] but typically most polymers for practical applications have molecular weights from 5000 to 200,000. There are exceptions. The unusual properties of polymers are based on their great length and internal structure.

Types of Polymers

Polymers are classified as natural or synthetic. The majority of commercial applications of plastics are derived from synthetic polymers. However, some natural polymers are cotton, cellulose, wool, silk, natural rubber, and starch. The synthetic polymers can be differentiated by the way in which their monomers are joined, i.e., via addition or condensation polymerization. In addition, the molecular chains are linked by the successive addition of one monomer to another. Typical polymers are the polyolefins, polystyrenes, acrylics, vinyls, and fluoroplastics.

Condensation polymers are prepared by the reaction of two different molecules, both having two reactive end groups. Molecular weight is built up by the linking together of these end groups and the elimination of a small molecule from the reaction medium to attain high molecular weight. Examples include nylons, polyesters, phenolics, and polyimides.

14   Fundamentals of Plastics for Electronics

Addition and condensation polymerization are two different processes and result in very different product structures. Addition is usually a fast chain reaction occurring via ionic or free-radical intermediates, where the monomer units add rapidly to the growing chain, one after another. Interaction between the chains either does not occur or results in the termination of growth of one or both chains. Condensation polymerization is normally a stepwise slow reaction where any size molecule may interact and lead to larger molecules capable of further reaction to yield still larger molecules. Structurally the polymers from the two processes are different; the addition polymer does not normally include in the main chain a grouping which can be attacked and degraded by solvents. On the other hand, a condensation polymer usually contains such a grouping. Figures 2.1 and 2.2 show typical condensation and addition polymers.

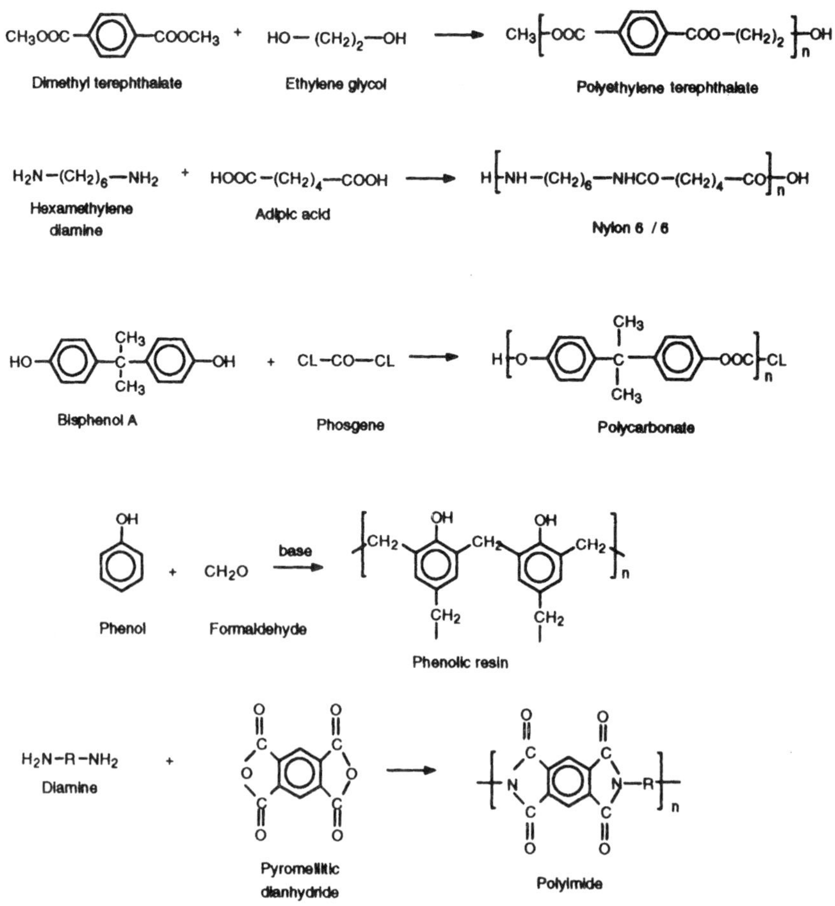

Figure 2.1   Condensation polymers.

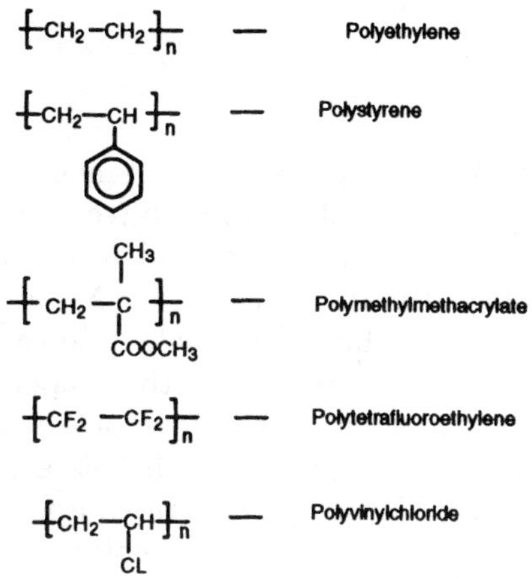

$+CH_2-CH_2+_n$ — Polyethylene

$+CH_2-CH+_n$ — Polystyrene
 |
 ⏣

$+CH_2-\underset{COOCH_3}{\overset{CH_3}{C}}+_n$ — Polymethylmethacrylate

$+CF_2-CF_2+_n$ — Polytetrafluoroethylene

$+CH_2-\underset{CL}{CH}+_n$ — Polyvinylchloride

Figure 2.2 Addition polymers.

All polymers can be classified in the manner described above, but they can also be further subclassified to more accurately define their structural and compositional characteristics. Polymers can be linear, branched, crystalline, amorphous, liquid-crystalline copolymers, elastomers, and alloys. All these except the elastomers can be divided into two major groups: thermoplastics and thermosets. Both types of plastics are fluid enough to be formed and molded at some stage in their conversion to finished product. Thermoplastics solidify by cooling and can be remelted. Thermoset resins undergo cross-linking to form a three-dimensional network and, unlike thermoplastics, cannot be remelted and reshaped. With few exceptions, polymers are not used in their natural state and are usually mixed with other materials to yield a compounded polymer, which may be in the form of a pellet, granule, powder, or liquid. The properties of a polymer can be varied, and the process that is used to polymerize the monomer with one or more different monomers is called *copolymerization*. These polymers are called *copolymers* or *terpolymers,* depending on whether two or three comonomers are used during the copolymerization. Many commercially important polymers derive their useful properties from the fact that the polymer chain comprises more than one kind of monomer. These are known as *copolymers* in contrast to *homopolymers* which contain only a single type of monomer. Copolymerization, e.g., may significantly change toughness, solvent resistance, and stability.[2] Vinyl chloride–vinyl acetate and vinylidene chloride–vinyl chloride plastics are such copolymers.

The way in which the monomer segments are distributed along the polymer chain affects the properties of the composition, and this can vary as indicated in the structures on p. 16, in which M_1 and M_2 represent different monomers.

$M_1M_2M_1M_2M_1M_2$ Alternating copolymer
$M_1M_2M_1M_1M_1M_2M_2M_1M_2M_2M_2M_1$ Random copolymer
$M_1M_1M_1M_1M_2M_2M_2M_2$ Block copolymer

The complexity of copolymerization increases rapidly as the number of monomers increases. Three- and four-component systems have been examined. Such systems are of real as well as theoretical interest because multicomponent polymers exist as commercial entities. Among plastics the acrylonitrile butadiene styrene (ABS) resins belong in the three-component category. Another technique used to vary the properties of polymers is to mechanically blend one polymer with another to form an alloy. The properties of these alloys generally fall between those of the starting polymers. Alloys are discussed later in Chap. 6. Elastomers differ significantly from plastics. While both are polymers,

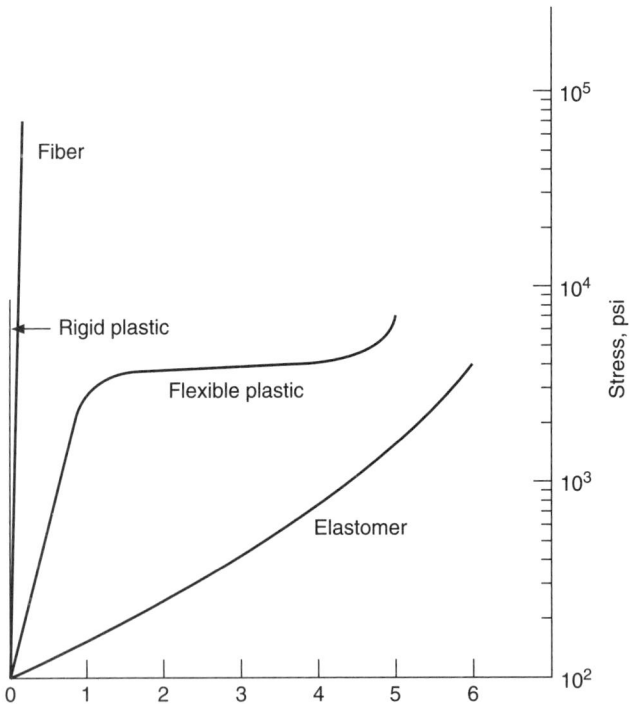

Figure 2.3 Stress-strain plots for a typical elastomer, plastic, and fiber. (Source: Ref. 1. Reprinted with permission from McGraw-Hill.)

elastomers easily undergo very large reversible elongations at relatively low stresses. In order for this to happen, the polymer must be completely amorphous with a low glass transition temperature T_g and low secondary forces, so as to obtain high mobility of the polymer chains. Some degree of cross-linking is needed so that the deformation is rapidly reversible. Figure 2.3 and Table 2.1 illustrate the differences between plastics, fibers, and elastomers by way of a strain-strain plot. Typical properties of these material classes are shown in Table 2.2.

TABLE 2.1 Classification of Polymers

Property	Elastomers	Plastics	Fibers
Initial modulus of elasticity, (lb/in^2)	15–150	1500–15,000	150,000–1,500,000
Upper limit of extensibility, percent	100–1000	20–100	<10
Character of stress deformation	Almost completely and instantaneously elastic	Partly reversible elasticity; little delayed elasticity; some permanent set	Some instantaneously reversible elasticity; some delayed elasticity; some permanent set
Effect of temperature upon mechanical properties	Elastic modulus increases with temperature (within a limited range)	Marked temperature dependency (over a wide range)	Little temperature dependency from −50°C to +150°C
Crystallization tendency	Low (unstressed)	Moderate to high	Very high
Molecular cohesion, cal/mol	1000–2000	2000–5000	5000–10,000
Representative polymers	Polysulfide Natural rubber Polychloroprene Polybutadiene Polyisobutylene	Phenolics Poly(vinyl chloride) Amino resins Polystyrene Poly(vinyl acetate)	Polyamides Poly(vinylidene chloride) Silk Cellulose Poly(ethylene terephthalate)

SOURCE: Ref. 20.

TABLE 2.2 Comparison of Fibers, Plastics, and Elastomers

	Tensile strength at break, lb/in^2	Elongation at break, %
Fibers		
Cellulose acetate	20,000–24,000	25–28
Nylon 66	65,000–123,000	19–32
Polyethylene, low-density	12,000–27,000	25–50
Polyethylene, high-density	49,000–86,000	10–20
Polytetrafluoroethylene	50,000	13
Poly(vinyl chloride)	48,000–54,000	14–20
Cotton	30,000–120,000	5–8
Wool	22,000	10–50
Silk	50,000	25
Plastics		
Cellulose acetate	3,000–11,000	4–55
Nylon 66	7,000–10,900	90
Polyethylene, low-density	1,000–2,300	90–650
Polyethylene, high-density	3,100–5,500	15–100
Polytetrafluoroethylene	2,000–4,500	200–400
Poly(vinyl chloride)	5,000–9,000	2–40
Filled polyester	3,000–50,000	0.5–2.0
Unfilled phenolic	7,000–8,000	1.0–1.5
Elastomers		
Natural rubber	3,300–4,500	470–600
SBR (styrene-butadiene rubber)	4,000	580
Neoprene	3,500	580

SOURCE: Ref. 3.

Structure and Properties

In addition to the broad categories of thermoplastics and thermosets, polymeric materials can be classified in terms of their structure, i.e., linear, branched, cross-linked, amorphous, crystalline, and liquid crystalline. Recall that a polymer molecule consists of monomer molecules that have been linked together in one continuous length. Such a polymer is termed a *linear polymer.* Branched polymers are those which have side branches of linked monomer molecules protruding from various points along the main polymer chain. Cross-linked polymers are those in which adjacent molecules are linked together, resulting in a complex interconnected network. Figure 2.4 is a schematic illustration of these structures.

In some thermoplastics, the chemical structure is such that the polymer chains will fold on themselves and pack together in an organized manner; see Fig. 2.5. The resulting organized regions show the behav-

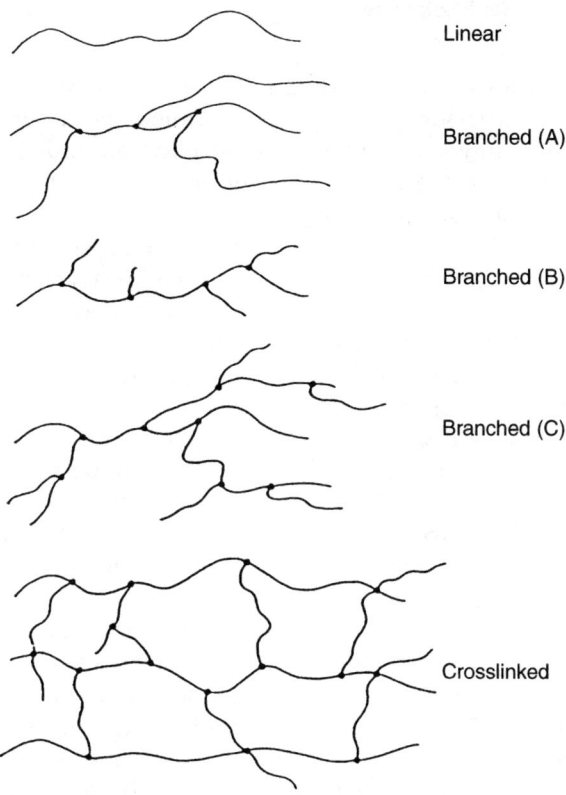

Figure 2.4 Structures of polymer molecules. (Source: Ref. 1. Reprinted with permission from McGraw-Hill.)

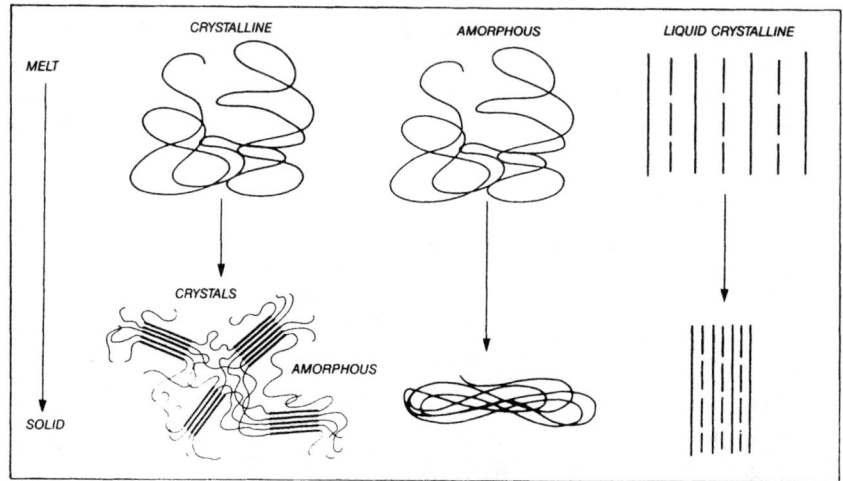

Figure 2.5 Two-dimensional representation of crystalline, amorphous, and liquid-crystalline structures. (Source: Ref. 4. Reprinted with permission.)

ior characteristics of crystals. Plastics that have these regions are called *crystalline*. Plastics without these regions are called *amorphous*. All the crystalline plastics have amorphous regions between and connecting the crystalline regions. For this reason, the crystalline plastics are often referred to as *semicrystalline* in the literature.

Liquid-crystalline polymers are best thought of as being a separate and unique class of plastics. The molecules are stiff, rodlike structures which are organized in large parallel arrays or domains in both the melted and solid states. These large, ordered domains provide liquid-crystalline polymers with unique characteristics compared to those of the crystalline and amorphous polymers. Many of the mechanical and physical property differences between plastics can be attributed to their structure. As a generalization, the ordering of crystalline and liquid-crystalline thermoplastics makes them stiffer, stronger, and less impact-resistant than their amorphous counterparts. Also, crystalline and liquid-crystalline materials have a higher resistance to creep, heat, and chemicals. However, crystalline materials are more difficult to process because they have higher melt temperatures and tend to shrink and warp more than amorphous polymers. Amorphous polymers soften gradually and continuously as heat is applied, and in the molding process they do not flow as easily as melted crystalline polymers do. Liquid-crystalline polymers have the high melting temperature of crystalline plastic, but soften gradually and continuously as amorphous polymers do. They have the lowest viscosity, warpage, and shrinkage of the thermoplastics.

Some crystalline and amorphous plastics are listed in Table 2.3. And some generalized properties of these films are shown in Table 2.4.

The properties of polymers are determined by the size of the macromolecule, which includes the elements of the molecular structure. These elements include composition, molecular weight, molecular shape, molecular conformation, and molecular aggregation. All these

TABLE 2.3 **Crystalline and Amorphous Polymers**

Typical crystalline thermoplastic resins	Typical amorphous thermoplastic resins
Acetal	Polystyrene
Nylon	ABS
Polyethylene	SAN
Polypropylene	Polycarbonate
Polyester	Polyvinyl chloride (PVC)

TABLE 2.4 Variation of Polymer Properties

Property	Crystalline	Amorphous	Liquid crystalline
Specific gravity	Higher	Lower	Higher
Tensile strength	Higher	Lower	Highest
Tensile modulus	Higher	Lower	Highest
Ductility, elongation	Lower	Higher	Lowest
Resistance to creep	Higher	Lower	High
Max. usage temperature	Higher	Lower	High
Shrinkage and warpage	Higher	Lower	Lowest
Flow	Higher	Lower	Highest
Chemical resistance	Higher	Lower	Highest

elements are responsible for the observed properties of plastics, such as ductility, brittleness, density, melting point, electrical characteristics, flow, elasticity, strength, thermal stability, and radiation stability.

One of the most important characteristics of a polymer is its molecular weight because the properties of polymers are a consequence of their high molecular weight. Strength does not usually develop in polymers until a minimum molecular weight (5000 to 10,000) is attained. The molecular weight of most polymers ranges from 10,000 to 1,000,000. The molecular weights of some typical polymers are shown in Table 2.5. The molecular weights of gelled or cross-linked plastics such as epoxy phenolic, cyanate ester, and bismaleimide resins are essentially infinite. Above this size (10,000 molecular weight) there is a rapid increase in mechanical properties and a leveling off as the molecular weight increases further. See Fig. 2.6. In most instances there is some molecular weight range for which a given polymer property will be optimum for a particular application. This critical range will vary from one polymer to another. Not all polymers are homogeneous and composed of molecules of different sizes. To fully characterize the size of a polymer, one should know its molecular weight and its molecular weight distribution. Both properties significantly affect processing and strength. The molecular weight of a polymer is expressed as an average due to the heterogeneous nature of the molecules that comprise the polymer. There are two ways of calculating the size of the molecules in the polymer: the number average molecular weight M_n (based on the number of molecules present) and the weight average M_w based on the total weight of the molecules of each size. These relationships are defined in Eqs. (1) and (2).

TABLE 2.5 Degree of Polymerization and Molecular Weight of Polymers

Material	Molecular weight range	Degree of polymerization
Native cellulose in cotton or wood	1,200,000–1,500,000	7,400–9,250
Cellulose in bleached cotton linters	400,000–500,000	2,500–3,100
Cellulose in purified wood pulp	300,000–450,000	1,850–2,750
Regenerated cellulose in rayon	75,000–100,000	460–620
Regenerated cellulose in staple fiber	60,000–75,000	370–460
Regenerated cellulose in cellophane	50,000–60,000	310–370
Native rubber in Hevea latex	140,000–210,000	2,050–3,100
Rubber after being milled in air	55,000–70,000	810–1,030
Cellulose nitrate used for molding	400,000–700,000	1,480–2,600
Cellulose nitrate used for extrusion	150,000–300,000	550–1,100
Cellulose nitrate used for coatings	50,000–100,000	185–370
Polystyrene for injection molding	120,000–180,000	1,150–1,730
Polystyrene for coatings	80,000–120,000	770–1,150
Poly(vinyl chloride)	250,000	4,000
Polyisobutylene	120,000–200,000	2,150–3,500
Poly(hexamethylene adipamide)	16,000–32,000	140–280
Polymethylmethacrylate	500,000–1,000,000	5,000–10,000

SOURCE: Adapted from H. F. Mark, "Molecular Structure and Mechanical Behavior of High Polymers," table 1, p. 8, in S. B. Twiss, ed., *Advancing Fronts in Chemistry*, vol. 1: *High Polymers*, Reinhold Publishing Corp., New York, 1945.

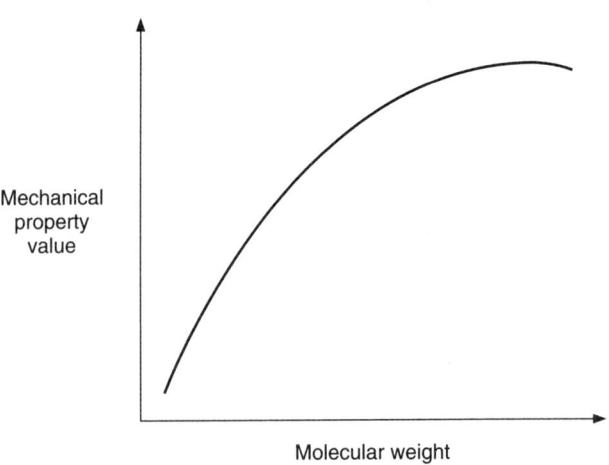

Figure 2.6 Dependence of physical properties on molecular weight.

$$M_n = \frac{\text{total weight of all molecules}}{\text{total number of molecules}} \qquad (1)$$

$$M_w = \frac{\Sigma \, (\text{weight of all molecules of each size} \times \text{its molecular weight})}{\text{total weight of all molecules}} \qquad (2)$$

The significance of these definitions can be shown by an analogy, in which piles of stones are compared with samples of polyethylene and each stone within a pile represents a molecule. For example, two piles of stones might have the same weight, but contain stones of different sizes, e.g.,[5]

PILE A

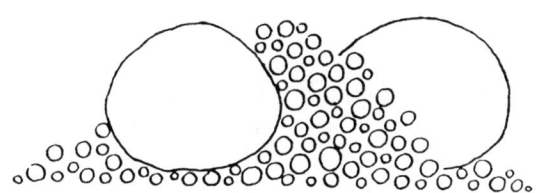

PILE B

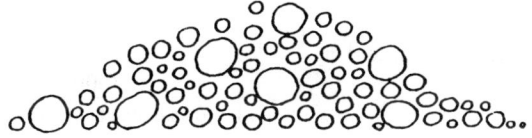

Pile A	Pile B
500 stones @ 1 lb = 500 lb	400 stones @ 1 lb = 400 lb
2 stones @ 250 lb = 500 lb	100 stones @ 6 lb = 600 lb
Total weight = 1000 lb	Total weight = 1000 lb

The average weight of the stones in a given pile (corresponding to the number average molecular weight M_n) is the total weight divided by the number of stones:

$$M_n = \frac{1000 \text{ lb}}{502 \text{ stones}} = 1.99 \qquad M_n = \frac{1000 \text{ lb}}{500 \text{ stones}} = 2.0$$

Although the size distribution of the stones in one pile is vastly different from that in the other pile, this important fact is not shown by the average figure, which corresponds to the number average molecular weight. The term *molecular weight,* when applied to polyethylene, usually refers to the number average molecular weight and, as in the case of the stones, does not provide a measure of the size distribution of the molecules within the sample.

To return to the stone analogy, the equivalent of weight average molecular weights would be calculated as follows:[5]

$$500 \times 1 = 500 \qquad 400 \times 1 = 400$$
total wt. of   wt. of individual   total wt. of   wt. of individual
1-lb stones   1-lb stones   1-lb stones   1-lb stones

$$500 \times 250 = 125{,}000 \qquad 600 \times 6 = 3600$$
total wt. of   wt. of individual   total wt. of   wt. of individual
250-lb stones   250-lb stones   6-lb stones   6-lb stones

$$\overline{125{,}500} \qquad\qquad \overline{4000}$$

$$M_w = \frac{125{,}000}{1000 \text{ (total wt. of all stones)}} = 125.5$$

$$M_w = \frac{4000}{1000 \text{ (total wt. of all stones)}} = 4.0$$

The ratio of the weight average molecular weight to the number average molecular weight gives a good indication of the molecular weight distribution:

$$\frac{M_w}{M_n} = \frac{125.5}{1.99} = 63 \text{ (broad)} \qquad \frac{M_w}{M_n} = \frac{4.0}{2.0} = 2 \text{ (narrow)}$$

The Physical State of a Polymer

Plastics in general are hard and brittle, but they (thermoplastics) soften when heated and can be elongated to a large extent, much as an elastomer is stretched at room temperature. If the temperature is increased, the polymer is converted to a viscous melt if it is a thermoplastic. Cross-linked thermoset polymers cannot flow but continue to soften until degradation occurs. This transformation of a plastic from a hard (glassy) state to a rubberlike state is called the *glass transition*. And the temperature at which this occurs is the glass transition temperature, designated T_g. This temperature is determined by the interactions among the molecular chains (hydrogen bonding, ionic or polar interactions, steric factors, etc.). Strong interactions lead to high values of T_g, as do bulky side groups and rigid main chains. Some glass transition temperatures and melting temperatures of various plastics are given in Table 2.6.

TABLE 2.6 Glass Transition and Crystalline Melting Temperatures of Some Common Polymers

Polymer	T_g, °C	T_m, °C
Poly(α-methylstyrene)	175	—
Polycarbonate	150	220
Polymethacrylonitrile	120	—
Poly(acrylic acid)	106	—
Poly(methyl methacrylate), syndiotactic	105	>200
Polyacrylonitrile	104	317
Polystyrene	100	—
Poly(vinyl chloride)	83	220–240
Poly(ethylene terephthalate)	69	267
Poly(ethyl methacrylate)	65	—
Polycaproamide (nylon 6)	50	225
Poly(hexamethylene adipamide) (nylon 66)	50	265
Poly(ω-aminoundecanoic acid) (nylon 11)	46	194
Poly(hexamethylene sebacamide) (nylon 610)	40	227
Poly(methyl methacrylate), isotactic	45	160
Poly(vinyl acetate)	29	—
Poly(n-butyl methacrylate)	22	—
Poly(methyl acrylate)	9	—
Poly(vinylidene chloride)	−17	190–198
Polypropylene	−19	176
Poly(ethyl acrylate)	−22	—
Poly(butyl acrylate)	−55	—
Polyoxymethylene	−68	180–200
Poly(cis-isoprene)	−73	28
Polyethylene, branched	−80	105–125
Polyethylene, linear	−80	137
Polybutadiene, syndiotactic	−85	154
Poly(dimethyl siloxane)	−123	—

SOURCE: Ref. 6.

Synthesis

There are four basic methods of producing a polymer. Many factors influence the choice of a particular method of producing a polymer. In many instances, the nature of the reaction chemistry dictates the specific method to be used. In other instances, the resultant polymer (low or semiviscous liquid, friable or rigid solid) may limit the choice. The interested reader is referred to any basic organic polymer chemistry text for more detailed descriptions.

Bulk Polymerization

From the point of view of equipment, complexity, and economics, the simplest method is mass or bulk polymerization. This procedure merely allows the monomer to react at a predetermined reaction temperature, with or without catalysts, to form the polymer.

Theoretically, the monomer can be a gas, liquid, or solid, but, practically, almost all mass polymerizations take place in a liquid phase.

The polymer may be either soluble or insoluble in the monomer. If the former, then the mass viscosity continually increases until the final degree of polymerization is obtained. If the latter, the polymer will precipitate from the remaining unreacted monomer and can be subsequently separated.

A serious drawback to bulk polymerization is the control of the heat of reaction. The generated exothermic heat tends to stay within the mass and is not easily withdrawn. Stirring the mass helps, but as the viscosity continues to increase, stirring becomes more difficult with a less efficient heat dissipation mechanism. This lack of control causes difficulty in the control of the molecular weight and the molecular weight distribution (MWD) of the final polymer.

The method does, however, lend itself to use in small casting or batch production. Gaseous-phase bulk polymerization takes place under pressure, often requiring specific catalysts for conversion.

In summary, mass or bulk polymerization uses simple equipment, is highly exothermic with difficult heat control, and yields a polymer with a broad MWD.

Solution Polymerization

If we now carry out the polymerization in a suitable solvent, we can simplify the reaction heat removal, since the solution of solvent, monomer, and polymer is less viscous than molten polymer. This technique is called *solution polymerization*.

If a solvent can be found in which the monomer is soluble but the polymer insoluble, the resultant polymer precipitation facilitates the separation steps.

In summary, one can control heat more readily in solution polymerization, although higher-molecular-weight polymers are difficult to produce. A solution of the polymer itself may be marketable, but the purification of solid polymer may involve complex procedures.

Emulsion Polymerization

If the monomer can be polymerized in a water emulsion, then we can retain the low viscosity needed for good heat control without the hazards associated with the handling of solvents. Such a procedure is called *emulsion polymerization*.

Reaction rates and molecular weights are usually higher with this method than with mass or solution polymerization. The MWD is often quite narrow; water is cheaper and less hazardous than solvent, and recovery steps are not as complex. However, ingredients must be added to aid emulsification (emulsifying and stabilizing agents). This added contamination and the requirement of a drying step for the polymer constitute significant disadvantages of the process.

Suspension Polymerization

Finally, there is *suspension polymerization* in which the monomer and the forming polymer globules are maintained in suspension by agitation without the use of an emulsifying agent. The polymer beads are formed by coalescence, and their size is regulated by suspension stabilizers and the amount and intensity of agitation. The final beads must be screened out of the liquid phase, washed, and dried before they can be used, although suspensions can be, and are, marketable. Control of exothermic heat is good, and high-molecular-weight polymers with relatively narrow MWDs are possible.

Terminology

To acquaint those unfamiliar with the language of polymers, Table 2.7 and the following glossary[7] present terms associated with polymers and their use in the electrical industry. This material is not all-inclusive and serves only to acquaint the reader with some basic information in the area. An excellent source for more in-depth descriptions of the terms, concepts, definitions, materials test instruments, etc., is found in Ref. 8.

TABLE 2.7 **Significance of Important Electrical Insulation Properties**

Property and definition	Significance of values
Dielectric strength All insulating materials fail at some level of applied voltage for a given set of operating conditions. The dielectric strength is the voltage that an insulating material can withstand before dielectric breakdown occurs. Dielectric strength is normally expressed in voltage gradient terms, such as volts per mil. In testing for dielectric strength, two methods of applying the voltage (gradual or by steps) are used. Type of voltage, temperature, and any preconditioning of the test part must be noted. Also, thickness of the piece tested must be recorded because the voltage per mil at which breakdown occurs varies with thickness of test piece. Normally, breakdown occurs at a much higher volt-per-mil value in very thin test pieces (a few mils thick) than in thicker sections (⅛ in thick, for example).	The higher the value, the better the insulator. Dielectric strength of a material (per mil of thickness) usually increases considerably with decrease in insulation thickness. Materials suppliers can provide curves of dielectric strength versus thickness for their insulating materials.
Resistance and resistivity Resistance of insulating material, like that of a conductor, is the resistance offered by the conducting path to passage of electric current. Resistance is expressed in ohms. Insulating materials are very poor conductors, offering high resistance. For insulating materials, the term "volume resistivity" is more commonly applied. Volume resistivity is the electrical resistance between opposite faces of a unit cube for a given material and at a given temperature. The relationship between resistance and resistivity is expressed by the equation $\rho = RA/l$, where ρ = volume resistivity in ohm-centimeters, A = area of the faces, and l = distance between faces of the piece on which measurement is made. This is not resistance per unit volume, which would be ohms per cubic centimeter although this term is sometimes erroneously used. Other terms are sometimes used to describe a specific application or condition. One such term is *surface resistivity,* which is the resistance between two opposite edges of a surface film 1 cm square. Since the length and width of the path are the same, the centimeter terms cancel. Thus, units of surface resistivity are actually ohms. However, to avoid confusion with usual resistance values, surface resistivity is normally given in ohms per square. Another broadly used term is *insulation resistance* which, again, is a measurement of ohmic resistance for a given condition, rather than a standardized resistivity test. For both surface resistivity and insulation resistance, standardized comparative tests are normally used. Such tests can provide data such as the effects of humidity on a given insulating material configuration.	The higher the value, the better or a good insulating material. The resistance value for a given material depends on a number of factors. It varies inversely with temperature and is affected by humidity, moisture content of the test part, level of the applied voltage, and time during which the voltage is applied. When tests are made on a piece that has been subjected to moist or humid conditions, it is important that measurements be made at controlled time intervals during or after the test condition has been applied, since dry-out and resistance increase occur rapidly. Comparing or interpreting data is difficult unless the test period is controlled and defined.

TABLE 2.7 Significance of Important Electrical Insulation Properties (*Continued*)

Property and definition	Significance of values

Dielectric constant

The dielectric constant of an insulating material is the ratio of the capacitance of a capacitor containing that particular material to the capacitance of the same electrode system with air replacing the insulation as the dielectric medium. The dielectric constant is also sometimes defined as the property of an insulation which determines the electrostatic energy stored within the solid material. The dielectric constant of most commercial insulating materials varies from about 2 to 10, air having the value 1.

Low values are best for high-frequency or power applications, to minimize electric power losses. Higher values are best for capacitance applications. For most insulating materials, dielectric constant increases with temperature, especially above a critical temperature region which is unique for each material. Dielectric constant values are also affected (usually to a lesser degree) by frequency. This variation is also unique for each material.

Power factor and dissipation factor

Power factor is the ratio of the power dissipated (watts) in an insulating material to the product of the effective voltage and current (voltampere input) and is a measure of the relative dielectric loss in the insulation when the system acts as a capacitor. Power factor is nondimensional and is a commonly used measure of insulation quality. It is of particular interest at high levels of frequency and power in such applications as microwave equipment, transformers, and other inductive devices.

Dissipation factor is the tangent of the dielectric loss angle. Hence, the term "tan delta" (tangent of the angle) is also sometimes used. For the low values ordinarily encountered in insulation, dissipation factor is practically the equivalent of power factor, and the terms are used interchangeably.

Low values are favorable, indicating a more efficient system, with lower losses.

Arc resistance

Arc resistance is a measure of an electrical breakdown condition along an insulating surface, caused by the formation of a conductive path on the surface. It is a common ASTM measurement, especially used with plastic materials because of the variations among plastics in the extent to which a surface breakdown occurs. Arc resistance is measured as the time, in seconds, required for breakdown along the surface of the material being measured. Surface breakdown (arcing or electrical tracking along the surface) is also affected by surface cleanliness and dryness.

The higher the value, the better. Higher values indicate greater resistance to breakdown along the surface due to arcing or tracking conditions.

Comparative tracking index (CTI)

This is an Underwriters Laboratories test which is run similarly to arc resistance except that an electrolyte solution (ammonium chloride) is put on the surface. The CTI is the value of the voltage required to cause a conductive path to form between electrodes.

The test is useful because it measures the arc resistance on a contaminated surface, which is often the case with actual electrical and electronics equipment.

SOURCE: Ref. 7.

Glossary

Accelerator A chemical used to speed up a reaction or cure. The term *accelerator* is often used interchangeably with the term *promoter*. An accelerator is often used along with a catalyst, hardener, or curing agent.

Adhesive Broadly, any substance used in promoting and maintaining a bond between two materials.

Aging The change in properties of a material with time under specific conditions.

Arc resistance The time required for an arc to establish a conductive path in a material.

Blowing agent Chemical that can be added to plastics and that generates inert gases upon heating. This blowing or expansion causes the plastic to expand, thus forming a foam. Also known as a *foaming agent*.

Bond strength The amount of adhesion between bonded surfaces.

B stage An intermediate stage in the curing of a thermosetting resin. In this state, a resin can be heated and caused to flow, thereby allowing final curing in the desired shape. The term *A stage* is used to describe an earlier stage in the curing resin. Most molding materials are in the B stage when supplied for compression or transfer molding.

Capacitance (capacity) That property of a system of conductors and dielectrics which permits the storage of electricity when a potential difference exists between the conductors. Its value is expressed as the ratio of quantity of electricity to a potential difference. A capacitance value is always positive.

Cast To embed a component or an assembly in a liquid resin, by using molds that separate from the part for reuse after the resin is cured. See *Embed* and *Pot*.

Catalyst A chemical that causes or speeds up the cure of a resin, but that does not become a chemical part of the final product. Catalysts are normally added in small quantities. The peroxides used with polyester resins are typical catalysts.

Coat To cover with a finishing, protecting, or enclosing layer of any compound (such as varnish).

Coefficient of expansion The fractional change in dimension of a material for a unit change in temperature.

Cold flow (creep) The continuing dimensional change that follows initial instantaneous deformation in a nonrigid material under static load.

Compound Some combination of elements in a stable molecular arrangement.

Contact bonding A type of adhesive (particularly nonvulcanizing natural rubber adhesives) that bonds to itself on contact even though solvent evaporation has left it dry to the touch.

Cross-linking The forming of chemical links between reactive atoms in the molecular chain of a plastic. It is this cross-linking in thermosetting resins that makes them infusible.

Crystalline melting point The temperature at which the crystalline structure in a material is broken down.

Cure To change the physical properties of a material (usually from a liquid to a solid) by chemical reaction, by the action of heat and catalysts, alone or in combination, with or without pressure.

Curing agent See *Hardener*.

Curing temperature The temperature at which a material is subjected to curing.

Curing time In the molding of thermosetting plastics, the time it takes for the material to be properly cured.

Dielectric constant (permittivity or specific inductive capacity) That property of a dielectric which determines the electrostatic energy stored per unit volume for a unit potential gradient.

Dielectric loss The time rate at which electric energy is transformed to heat in a dielectric when it is subjected to a changing electric field.

Dielectric loss angle (dielectric phase difference) The difference between 90° and the dielectric phase angle.

Dielectric loss factor (dielectric loss index) The product of the dielectric constant and the tangent of the dielectric loss angle for a material.

Dielectric phase angle The angular difference in phase between the sinusoidal alternating potential difference applied to a dielectric and the component of the resulting alternating current having the same period as the potential difference.

Dielectric power factor The cosine of the dielectric phase angle (or sine of the dielectric loss angle).

Dielectric strength The voltage that an insulating material can withstand before breakdown occurs, usually expressed as a voltage gradient (such as volts per mil).

Dissipation factor (loss tangent, tan δ, approximate power factor) The tangent of the loss angle of the insulating material.

Elastomer A material which at room temperature stretches under low stress to at least twice its length and snaps back to its original length upon the release of stress. See *Rubber*.

Electric strength (dielectric strength or disruptive gradient) The maximum potential gradient that a material can withstand without rupture. The value obtained for the electric strength will depend on the thickness of the material and on the method and conditions of test.

Embed To encase completely a component or assembly in some material—a plastic, for current purposes. See *Cast* and *Pot*.

Encapsulate To coat a component or assembly in a conformal or thixotropic coating by dipping, brushing, or spraying.

Exotherm The characteristic curve of a resin during its cure, which shows heat of reaction (temperature) versus time. Peak exotherm is the maximum temperature on this curve.

Exothermic Heat-producing.

Filler A material, usually inert, that is added to plastics to reduce cost or modify physical properties.

Film adhesive Thin layer of dried adhesive. Also describes a class of adhesives provided in dry-film form with or without reinforcing fabric, which are cured by heat and pressure.

Flexibilizer A material that is added to rigid plastics to make them resilient or flexible. Flexibilizers can be either inert or a reactive part of the chemical reaction. It is also called a *plasticizer* in some cases.

Flexural modulus The ratio, within the elastic limit, of stress to corresponding strain.

Flexural strength The strength of a material in bending, expressed as the tensile stress of the outermost fibers of a bent-test sample at the instant of failure.

Fluorocarbon An organic compound having fluorine atoms in its chemical structure. This property usually lends stability to plastics. Teflon* is a fluorocarbon.

Gel The soft, rubbery mass that is formed as a thermosetting resin goes from a fluid to an infusible solid. This is an intermediate state in a curing reaction, and a stage in which the resin is mechanically very weak. The *gel point* is defined as the point at which gelation begins.

Glass transition point Temperature at which a material loses its glasslike properties and becomes a semiliquid.

Hardener A chemical added to a thermosetting resin for the purpose of causing curing or hardening. Amines and acid anhydrides are hardeners for epoxy resins. Such hardeners are a part of the chemical reaction and a part of the chemical composition of the cured resin. The terms *hardener* and *curing agent* are used interchangeably. Note that these can differ from catalysts, promoters, and accelerators.

Heat-distortion point The temperature at which a standard test bar (ASTM D 648) deflects 0.010 in under a stated load of either 66 or 264 lb/in^2.

Heat sealing A method of joining plastic films by simultaneous application of heat and pressure to areas in contact. Heat may be supplied conductively or dielectrically.

Hot-melt adhesive A thermoplastic adhesive compound, usually solid at room temperature, which is heated to a fluid state for application.

Hydrocarbon An organic compound having hydrogen atoms in its chemical structure. Most organic compounds are hydrocarbons. Aliphatic hydrocarbons

---

*Teflon is a registered trademark of the du Pont Co.

are straight-chained hydrocarbons, and aromatic hydrocarbons are ringed structures based on the benzene ring. Methyl alcohol and trichloroethylene are aliphatic; benzene, xylene, and toluene are aromatic.

Hydrolysis Chemical decomposition of a substance involving the addition of water.

Hygroscopic Tending to absorb moisture.

Impregnate To force resin into every interstice of a part. Cloths are impregnated for laminating, and tightly wound coils are impregnated in liquid resin by using air pressure or vacuum as the impregnating force.

Inhibitor A chemical added to resins to slow down the curing reaction. Inhibitors are normally added to prolong the storage life of thermosetting resins.

Insulation resistance The ratio of the applied voltage of the total current between two electrodes in contact with a specific insulator.

Modulus of elasticity The ratio of stress to strain in a material that is elastically deformed.

Moisture resistance The ability of a material to resist absorbing moisture, either from the air or when immersed in water.

Mold To form a plastic part by compression, transfer, injection molding, or some other pressure process.

NEMA standards Property values adopted as standard by the National Electrical Manufacturers Association.

Organic Composed of matter originating in plant or animal life, or composed of chemicals of hydrocarbon origin, either natural or synthetic. Used in referring to chemical structures based on the carbon atom.

Permittivity Preferred to *dielectric constant*.

pH A measure of the acid or alkaline condition of a solution. A pH of 7 is neutral (distilled water), pH values below 7 are increasingly acid as pH values go toward 0, and pH values above 7 are increasingly alkaline as pH values go toward the maximum value of 14.

Plastic An organic resin or polymer.

Plasticizer Material added to resins to make them softer and more flexible when cured.

Polymer A high-molecular-weight compound (usually organic) made up of repeated small chemical units. Polymers can be thermosetting or thermoplastic.

Polymerize To unite chemically two or more monomers or polymers of the same kind to form a molecule with higher molecular weight.

Pot To embed a component or assembly in a liquid resin, by using a shell, can, or case which remains as an integral part of the product after the resin is cured. See *Embed* and *Cast*.

Pot life The time during which a liquid resin remains workable as a liquid after catalysts, curing agents, promoters, and so on are added; roughly equivalent to gel time. Sometimes also called *working life*.

Power factor The cosine of the angle between the voltage applied and the current resulting.

Promoter A chemical, itself a feeble catalyst, that greatly increases the activity of a given catalyst.

Resin High-molecular-weight organic material with no sharp melting point. For current purposes, the terms *resin, polymer,* and *plastic* can be used interchangeably.

Resistivity The ability of a material to resist passage of electric current either through its bulk or on a surface. The unit of volume resistivity is the ohm-centimeter ($\Omega \cdot$ cm), and the unit of surface resistivity is the ohm (Ω).

Rockwell hardness number A number derived from the net increase in depth of impression as the load on a penetrator is increased from a fixed minimum load to a higher load and then returned to minimum load. Penetrators include steel balls of several specified diameters and a diamond cone penetrator.

Rubber An elastomer capable of rapid elastic recovery.

Shore hardness A procedure for determining the indentation hardness of a material by means of a durometer. Shore designation is given to tests made with a specified durometer.

Solvent A liquid substance that dissolves other substances.

Storage life The period of time during which a liquid resin or adhesive can be stored and still remain suitable for use. Also called *shelf life*.

Strain The deformation resulting from a stress, measured by the ratio of the change to the total value of the dimension in which the change occurred.

Stress The force producing or tending to produce deformation in a body, measured by the force applied per unit area.

Surface resistivity The resistance of a material between two opposite sides of a unit square of its surface. The surface resistivity may vary widely with the conditions of measurement.

Thermal conductivity The ability of material to conduct heat; the physical constant for the quantity of heat that passes through a unit cube of a material in a unit of time when the difference in temperature of the two faces is 1°C.

Thermoplastic A classification of resin that can be readily softened and resoftened by repeated heating. Hardening is achieved by cooling.

Thermosetting A classification of resin that cures by chemical reaction when heated and, when cured, cannot be resoftened by heating.

Thixotropic Describing materials that are gellike at rest but fluid when agitated.

Vicat softening temperature A temperature at which a specified needle point will penetrate a material under specified test conditions.

Viscosity A measure of the resistance of a fluid to flow (usually through a specific orifice).

Volume resistivity (specific insulation resistance) The electrical resistance between opposite faces of a 1-cm cube of insulating material, commonly expressed in ohm-centimeters ($\Omega \cdot$ cm). The recommended test is ASTM D 257-54T.

Vulcanization A chemical reaction in which the physical properties of an elastomer are changed by causing it to react with sulfur or other cross-linking agents.

Water absorption The ratio of the weight of water absorbed by a material to the weight of the dry material.

Wet To adhere to a surface immediately upon contact.

Working life The period of time during which a liquid resin or adhesive, after mixing with catalyst, solvent, or other compounding ingredients, remains usable. See *Pot life*.

References

1. G. Odian, *Principles of Polymerization*, McGraw-Hill, New York, 1970.
2. B. Golding, *Polymers and Resins*, Van Nostrand, Princeton, NJ, 1959.
3. E. Baer, ed., *Engineering Design for Plastics*, Reinhold, New York, 1964.
4. *Designing with Plastic (The Fundamentals)*, Design Manual TDM-1, Hoechst Celanese Corporation, Chatham, NJ, 1989.
5. "Alathon," du Pont bulletin, Wilmington, DE, 1957.
6. D. S. Soane and Z. Martynenko, eds., *Polymers in Microelectronics*, Elsevier, New York, 1989.
7. C. A. Harper, ed., *Handbook of Materials and Processes for Electronics*, 2d ed., McGraw-Hill, New York, 1993.
8. *Insulation Circuits Desk Manual*, Lake Publishing Corp., Libertyville, IL, 1982.

Chapter 3

Thermoplastics

Thermoplastics are polymers that can be repeatedly softened when heated and hardened when cooled. In the melt, thermoplastics can be forced into a mold and cooled to preserve the desired shape. Scrap material can be reground and reused by mixing it with virgin material. Softening temperatures for thermoplastics vary widely (<100 to >200°C) and depend on the polymer type. Because the high temperatures required for melting can cause degradation of thermoplastics, there is a limit to the number of reheat cycles. Thermoplastics have higher creep, are more sensitive to chemicals, and have less resistance to high temperatures compared to thermoset materials. These plastics are almost always mixed with other additives such as fillers, thermal stabilizers, ultraviolet radiation stabilizers, and pigments to impart special properties. Thermoplastics are rarely used in their pristine form.

Fillers are relatively inert materials added to plastics to reduce cost and/or improve physical properties such as hardness, stiffness, and impact strength. Note that a filler generally has a much smaller particle size than a reinforcement and as such does not significantly improve the tensile strength of the plastic. However, fibrous reinforcements do markedly improve the tensile strength. A list of fillers and their impacts on certain plastic properties is shown in Table 3.1.[1] Nonmetallic minerals such as kaolin, talc, calcium carbonate, and barytes are the inexpensive fillers. Metallic fillers such as aluminum, bronze, zinc, and nickel are more expensive and are generally used to increase thermal and electrical conductivity. Glass fillers such as solid and hollow microspheres are used to improve strength or reduce density. The types of fillers and reinforcements and the resin types most often used are listed in Figs. 3.1 and 3.2.

There are approximately 27 major types of thermoplastic polymers including the categories of alloys and blends and thermoplastic elas-

TABLE 3.1 Filler Contributions to Product

Typical contribution	Talc	Mica	Wollastonites	Kaolin	Calcined kaolin	Diatomite	Crystalline silica	Precipitated silica	Alumina trihydrate	Calcium sulfate dihydrate	Calcium carbonate	Barytes	Glass spheres	Hollow glass spheres	Metallics	Organic fillers	Molybdenum disulfide	Coal	Nepheline	Feldspar	Chlorite	Antimony oxide
Low cost				✓		✓				✓	✓					✓		✓		✓		
Modulus	✓	✓	✓	✓																	✓	
High heat deflection temperature	✓	✓	✓	✓																	✓	
Tensile strength	✓	✓	✓	✓																		
Dimensional stability	✓	✓	✓	✓	✓		✓		✓		✓	✓	✓	✓					✓	✓	✓	
Chemical/moisture resistance	✓	✓		✓			✓	✓	✓		✓	✓							✓	✓	✓	
High density												✓			✓							
Low density														✓		✓		✓				
Fire retardancy									✓	✓												✓
Impact strength									✓													
Low liquid/gas permeability	✓	✓																				
Wear resistance	✓	✓														✓	✓					
Surface uniformity							✓		✓		✓	✓							✓	✓		
High melt viscosity					✓			✓														
Low melt viscosity							✓		✓		✓		✓						✓	✓		
Hardness								✓											✓	✓		
Brightness				✓							✓											
Film antiblock	✓					✓					✓											
Electrical insulation		✓					✓	✓														
Electrical conduction																✓						
Thermal conduction															✓		✓					
Translucency								✓	✓	✓				✓					✓	✓		
Thermal stability					✓		✓									✓			✓	✓		
X-ray opacity												✓			✓							
Uniformly nucleated pores in foams	✓						✓															

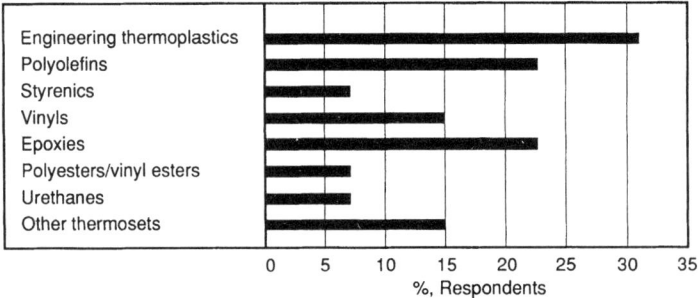

Figure 3.1 Breakdown of resins used by the survey respondents. (Source: Ref. 2. Reprinted with permission from Advanstar Communications.)

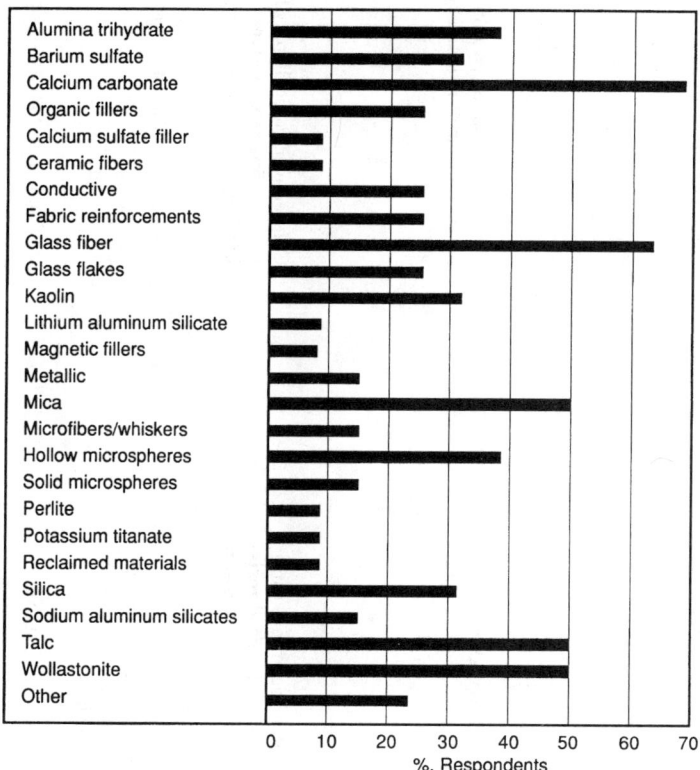

Figure 3.2 Types of fillers and reinforcements used by survey respondents. (Source: Ref. 2. Reprinted with permission from Advanstar Communications.)

tomers. Within each class there are many variations, leading to well over several hundred resin formulations. Not all these resins are used in the electronics industry. Thermoplastics are fabricated by blow molding, compression molding, extrusion, foaming, injection and rotational molding, stamping, and vacuum forming. Table 3.2 lists the general characteristics and applications of all thermoplastic polymers. Specific properties of these polymers are given in Tables 3.2 to 3.5.[3,4] Although the tables apply to all thermoplastic polymers, this chapter will describe in detail only those plastics used in electrical and electronic applications. Detailed descriptions of all thermoplastics can be found in Refs. 4 through 7.

(Text continues on p. 52.)

TABLE 3.2 General Characteristics of Thermoplastics

Material	Key advantages	Limitations	Processing*	Applications
ABS	Balanced combinations of good mechanical properties and heat and chemical resistance. Good durability; high impact strength and mar resistance. Dimensionally stable, including creep resistance. Relatively low in price. Versatility and ease of processing, fabrication, and finishing	Low weather resistance. Some self-extinguishing grades display low heat resistance	1, 2, 14, 17	Appliances, automotive parts, pipe, business machine, and telephone components
Acetals	Good creep resistance, fatigue resistance, and dimensional stability. Heat resistance to over 200°F for extended use. Very good hydrolytic stability. Low coefficient of friction. Excellent resiliency	Not self-extinguishing. Limited resistance to acids and alkalies	1, 2, 5, 8	Consumer products, automotive parts, industrial machinery parts
Acrylics	Excellent optical properties, including high transparency. Excellent long-term resistance to sunlight and weathering	Low scratch resistance compared to glass. Limited resistance to alkalies and solvents	1–6	Colored electronic display filters, glazing automotive parts, lighting-fixture coatings
Cellulose acetates	Good optical properties, including high transparency. Good toughness. Low cost	Low heat resistance. High moisture absorption. Not resistant to alkalies and strong acids. Not flame-retardant	1, 4, 5, 7, 17	Packaging, automotive parts
Cellulose acetate butyrates	Very good optical properties, including high transparency. Good toughness. Good weathering and aging resistance	Low tensile strength. Not resistant to strong acids and solvents	1, 2, 4, 5, 7	Packaging, automotive parts

Ethyl cellulosics	Good toughness (even at low temperature). Good sunlight and weathering resistance. High transparency	Low strength. Readily oxidized. Not flame-retardant	1, 2, 5, 7	Packaging, automotive parts
Cellulose propionates	Good toughness; slightly better than cellulose acetate. High transparency. Excellent molding characteristics	Low tensile strength. Not flame-retardant	1, 2, 5, 7	Packaging, automotive parts
Ethylene vinyl acetate	Flexible, high impact strength. Useful to −150°F		1, 2, 4	Packaging, film, adhesives, wire and cable insulation
PTFE fluoropolymer	Outstanding chemical resistance, even at high temperature. Excellent sunlight and weathering resistance. Temperature-resistant up to 500°F in continuous use. Coefficient of friction lowest of any solid material. Very good dielectric properties	Very expensive. Low strength. Cannot be processed by standard plastic methods	2, 5, 9, 10, 15, 16	Wire and cable insulation, electrical components, liners, filters
FEP fluoropolymer	Excellent chemical resistance, even at high temperature. Excellent sunlight and weathering resistance. Temperature-resistant up to 400°F in continuous use. Very good dielectric properties	Very expensive. Low strength	2, 7, 11–13	Wire and cable insulation, electrical components, liners, filters

*1 = injection molding, 2 = extrusion, 3 = thermoforming, 4 = blow molding, 5 = machining, 6 = casting, 7 = compression molding, 8 = rotational molding, 9 = powder metallurgy, 10 = sintering, 11 = dispersion coating, 12 = fluidized-bed coating, 13 = electrostatic coating, 14 = calendering, 15 = hot forming, 16 = cold forming, 17 = vacuum forming, 18 = vapor deposition.

TABLE 3.2 General Characteristics of Thermoplastics (*Continued*)

Material	Key advantages	Limitations	Processing*	Applications
CTFE fluoropolymer	Very good chemical resistance. Excellent sunlight and weathering resistance. Temperature-resistant up to 390°F in continuous use. Stronger and stiffer than PTFE or FEP fluoropolymers	Very expensive. Dielectric properties inferior to PTFE or FEP	2, 7	Wire and cable insulation, electrical components, liners, filters
ETFE and ECTFE fluoropolymers	Continuous use up to 300°F. Excellent weathering and chemical resistance. Good injection molding and extrusion characteristics	High cost	1, 2, 7, 8, 12, 13	Wire and cable insulation, electrical components, liners, filters
PVF_2 fluoropolymer	Very good chemical resistance. Excellent sunlight and weathering resistance. Strongest and most rigid fluorocarbon-type plastic	High cost. Lower heat resistance than other fluoropolymers	2	Wire and cable insulation, electrical components, liners, filters
Ionomer	Good toughness and barrier resistance	Not self-extinguishing. Not weather-resistant. Attacked by acids	1–4	Automotive parts, sporting goods, packaging
Nylon, type 6	Good creep resistance, high fatigue endurance. Good toughness and resiliency. Resistance to petroleum oils and greases and many chemical solvents. Good abrasion resistance	High moisture absorption. Low impact strength in dry environments	1, 2, 4, 5, 7, 16	Connectors, wire jackets, coil bobbins, packaging

Nylon, type 6/6	Very good creep resistance, high fatigue endurance. Good toughness and tensile strength. Resistant to petroleum oils and greases and many chemical solvents. Good abrasion resistance	High moisture absorption. Low impact strength in dry environments	1–5, 7	Connectors, wire jackets, coil bobbins, packaging
Nylon, type 6/10	More rigid and tougher than types 6 and 6/6	Lower tensile strength than types 6 and 6/6	1–5, 7	Connectors, wire jackets, coil bobbins, packaging
Nylon, type 8	High impact strength. Low moisture absorption compared to other nylons	Low tensile strength. High cost.	1–5, 7	Connectors, wire jackets, coil bobbins, packaging
Nylon, type 11	Low moisture absorption. Good impact strength	High cost	1–5, 7	Connectors, wire jackets, coil bobbins, packaging
Nylon, type 12	Low moisture absorption	High cost. Some grades not self-extinguishing	1–5, 7	Connectors, wire jackets, coil bobbins, packaging
Nylon copolymers	Good flexibility. Very tough	Not self-extinguishing	1–5, 7	Connectors, wire jackets, coil bobbins, packaging
Parylene	Good for insulation and protective coating	Special equipment to process. Difficult to remove for rework	18	Protective coatings
Phenoxy	Clear, tough, rigid, hard. Impact strength: 2 ft · lb/in. Adhesives	High cost	2, 7	Adhesives, coatings

*1 = injection molding, 2 = extrusion, 3 = thermoforming, 4 = blow molding, 5 = machining, 6 = casting, 7 = compression molding, 8 = rotational molding, 9 = powder metallurgy, 10 = sintering, 11 = dispersion coating, 12 = fluidized-bed coating, 13 = electrostatic coating, 14 = calendering, 15 = hot forming, 16 = cold forming, 17 = vacuum forming, 18 = vapor deposition.

TABLE 3.2 General Characteristics of Thermoplastics *(Continued)*

Material	Key advantages	Limitations	Processing*	Applications
Phenylene oxide–based materials	Good engineering properties, including creep and dimensional stability. Excellent hydrolytic stability. Temperature-resistant to 160°F in continuous use. Good self-extinguishing and nondripping characteristics. Relatively low cost	Attacked by many solvents	1–4	Connectors, fuse blocks
Polymethyl pentene	Very lightweight	Low tensile strength. Not self-extinguishing or weather-resistant	1–4	Cable, connectors, bobbins, wire coverings
Polyamide-imide	Heat-resistant to 500°F	Difficult to process	1, 2, 7, 9	Connectors, circuit boards, radomes, wire coatings, film
Polyaryl ether	Heat-resistant to 250°F. Good creep and fatigue resistance. Good chemical resistance. Excellent dimensional stability	Not self-extinguishing. Poor weathering resistance. Difficult to mold	1, 2	Electrical components, appliances, valves, housing
Polybutylene	Lightweight. Continuous use to 225°F. Very high impact strength. Good chemical resistance	Not weather-resistant. Not self-extinguishing	1, 2, 4	Pipe

Polycarbonates	Good creep and fatigue resistance and dimensional stability. Excellent toughness. Heat-resistant to 250°F in continuous use. Very good optical properties, including transparency. Very good dielectric properties	Poor resistance to solvents. Low scratch resistance	1–4	Connectors, terminal boards, bobbins
Polyethylenes, low-density	Excellent chemical resistance. Very flexible with good fatigue resistance and toughness. Very good dielectric properties. Tasteless, odorless. Low cost	Low tensile strength. Very difficult to bond or print on. Susceptible to environmental stress cracking	1–4, 7, 8	Wire and cable insulation
Polyethylene, high-density	Excellent chemical resistance. Very good dielectric properties. Highest rigidity of ethylene plastics. Good toughness. Readily molded and extruded. Low cost	Very difficult to bond or print on. Self-extinguishing grades have low properties	1–4, 7, 8	Wire and cable insulation
Polyethylene, high-molecular-weight	Excellent chemical resistance. Very good toughness. Low coefficient of friction. High dielectric strength. Good creep resistance	High cost. Difficult to process	2	High-pressure pipe
Polyimides	Retention of high mechanical strength and dielectric properties up to about 500°F. Highly resistant to damage from ionizing radiation. Very good wear resistance	High cost. Difficult to process. Low impact strength	1	Connectors

*1 = injection molding, 2 = extrusion, 3 = thermoforming, 4 = blow molding, 5 = machining, 6 = casting, 7 = compression molding, 8 = rotational molding, 9 = powder metallurgy, 10 = sintering, 11 = dispersion coating, 12 = fluidized-bed coating, 13 = electrostatic coating, 14 = calendering, 15 = hot forming, 16 = cold forming, 17 = vacuum forming, 18 = vapor deposition.

TABLE 3.2 General Characteristics of Thermoplastics (*Continued*)

Material	Key advantages	Limitations	Processing*	Applications
Polypropylenes	Strong, relatively rigid, tough, and lightweight. Retain most mechanical properties at elevated temperatures. Outstanding chemical and stress cracking resistance. Excellent dielectric properties. Low cost	Embrittles below 0°F. Not weather-resistant	1, 2, 4	Fibers, films, cable insulation
Polystyrenes	Good transparency. High hardness and rigidity. Readily molded and extruded. Low cost	Brittle. Low heat resistance to solvents	1, 2, 5, 7	Housings
Polysulfones	Good creep resistance and dimensional stability. Temperature resistant to 300°F in continuous use. Very good dielectric properties. Good chemical resistance and hydrolytic stability. Self-extinguishing	Not resistant to polar solvents	1–4	Circuit boards, connectors, TV components
Polyaryl sulfone	Continuous use at 500°F. Good creep and fatigue resistance. Good dimensional stability. Excellent chemical resistance		1	Circuit boards, connectors, TV components
Vinyls, rigid	Excellent corrosion resistance. High dielectric strength. Good toughness and abrasion resistance. Self-extinguishing. Good weatherability	Susceptible to staining	1–4, 6–8, 14	Pipe

Vinyls (chlorinated), rigid	Excellent corrosion resistance. Heat resistance 50°F higher than rigid vinyls. Good toughness and abrasion resistance. Self-extinguishing. High dielectric strength. Low cost	Difficult to process	1, 2	Pipe
Vinyls, flexible	Very flexible. Good chemical and weatherability resistance. Inherently self-extinguishing. High dielectric strength. Low cost	Stiffen at low temperatures. Susceptible to straining. Some plasticizers migrate to surface	1–4, 6–8, 14	Packaging, coatings
Polyketones	Heat and chemical resistance. High strength, resistance to burning	High melt process temperature. Affected by uv radiation	1, 2	Wire insulation, cable connectors
Liquid-crystal polymers	High temperature resistance, chemical resistance, low thermal expansion	Anisotropic properties	1	Chip carriers, socket connections, relay cases
Polyarylate	Ultraviolet radiation stability, high arc resistance	Susceptible to stress cracking	1–4	Connectors, coil bobbins, switches, relays, fuse covers
Polyester (PBT, PCT, PET)	Good electrical properties. Low moisture absorption. Temperature resistance		1, 2	Connectors, sockets, chip carriers, switches, relays, coil bobbins

*1 = injection molding, 2 = extrusion, 3 = thermoforming, 4 = blow molding, 5 = machining, 6 = casting, 7 = compression molding, 8 = rotational molding, 9 = powder metallurgy, 10 = sintering, 11 = dispersion coating, 12 = fluidized-bed coating, 13 = electrostatic coating, 14 = calendering, 15 = hot forming, 16 = cold forming, 17 = vacuum forming, 18 = vapor deposition.

TABLE 3.2 General Characteristics of Thermoplastics (*Continued*)

Material	Key advantages	Limitations	Processing*	Applications
Polyetherimide	High temperature strength. Heat resistance. Low smoke generation		1–3, 7	Connectors, low-loss radomes, printed-circuit boards
Polyphthalamide	Good combination of mechanical, chemical, electrical properties	Absorbs moisture	1	Connectors, switches

*1 = injection molding, 2 = extrusion, 3 = thermoforming, 4 = blow molding, 5 = machining, 6 = casting, 7 = compression molding, 8 = rotational molding, 9 = powder metallurgy, 10 = sintering, 11 = dispersion coating, 12 = fluidized-bed coating, 13 = electrostatic coating, 14 = calendering, 15 = hot forming, 16 = cold forming, 17 = vacuum forming, 18 = vapor deposition.

TABLE 3.3 Typical Physical Properties of Thermoplastics

Resin material	Coefficient of thermal expansion, in/in ($°C \times 10^{-3}$)	Thermal conductivity, cal/(cm$^2 \cdot$ s $\cdot$ °C $\cdot$ cm $\cdot 10^{-4}$)	Water absorption in 24 h, %	Flammability, in/min (0.125 in)	Specific gravity
ABS	6–13	4–9	0.2–0.5	1.0–2	1.01–1.07
Acrylic	−9	1.4	0.3	9–1.2	1.18–1.19
Chlorotrifluoroethylene	3.6	4–6	0	0	2.8–2.2
Fluorinated ethylene-propylene	8.3–10.5	5.9	<0.05	Nonflammable	2.16
Polytetrafluoroethylene	2–12	6	<0.01	Nonflammable	2.14–2.28
Nylon 6	1.1–8.0	5.9	1.5	Self-extinguishing	1.07–1.15
Polycarbonate	6.7–7	4.6	0.15	Self-extinguishing	1.2
Polyethylene, low-density	10–20	8	<0.01	Slow-burning	0.910–0.925
Polyethylene, medium-density	10–20	8	<0.01	Slow-burning	0.926–0.940
Polyphthalamide	0.8–3.3	1.7–2.6	0.1–0.8	Self-extinguishing	1.13–1.70
Polyamide-imide	3.0	6.2	0.33	Self-extinguishing	1.42
Polyarylate	2.7–4.0	—	0.1–0.2	Self-extinguishing	1.19–1.22
Polybutylene terephthalate	6.0–9.5	4.2–6.9	0.08	—	1.30–1.38
Polyethylene terephthalate	6.5	3.4	0.1–0.2	—	1.29–1.40
Polyethylene, high-density	5–11	11–12	<0.01	Slow-burning	0.941–0.965
Polyethylene, high-molecular-weight	7–11	8	<0.01	Slow-burning	0.93–0.94
Polyethylene, ultrahigh-molecular-weight	13–20	—	<0.01	Slow-burning	0.94
Polyimide	4.5–5.6	2.3–2.6	0.24	—	
Polypropylene	3.8–9	2.8–4	<0.01	Slow-burning to nonburning	0.90–1.4
Polystyrene	5–8	3	0.01–0.03	0.5–2.5	1.04–1.05
Polyurethane	10–20	7.4	0.60–0.80	Slow-burning to self-extinguishing	1.11–1.26
Polyvinyl chloride	7–25	3–5	0.15–0.75	Self-extinguishing	1.15–1.80
Polyvinyl chloride, rigid	5–10	3–5	0.07–0.40	Self-extinguishing	1.33–1.58
Polyvinyl dichloride, rigid	7–8	3–4	0.07–0.11	Self-extinguishing	1.50–1.54
Styrene acrylonitrile (SAN)	7	3	0.15–0.25	0.4–0.7	1.07–1.08
Polyphenylene oxide	3.8–7.0	3.8	0.06–0.1	Self-extinguishing	1.04
Polysulfone	5.6	6.2	0.3	Self-extinguishing	1.24–1.25
Polyarylsulfone	3.1–4.9	—	0.1	Self-extinguishing	1.37
Polyethersulfone	5.5	3.2–4.4	0.12–1.7	Self-extinguishing	1.37
Polyetheretherketone	4.0–4.7	—	0.1	Self-extinguishing	1.31
Polyetherketone	1.8	10.5	0.05	Self-extinguishing	1.30
Polycyclohexylene dimethylene terephthalate (PCT) (30% glass-reinforced)	2.0	6.9	0.05		1.45
Polyetherimide	4.7–5.6	1.6	0.25	Self-extinguishing	1.27
Polyphenylene sulfide	2.7–4.9	2.0–6.9	0.05	Self-extinguishing	1.35

SOURCE: Refs. 3 and 4.

TABLE 3.4 Typical Physical and Mechanical Properties of Thermoplastics

Resin material	Impact strength notched izod, ft·lb/in, ⅛ bar	Tensile strength, lb/in² × 10²	Tensile modulus, lb/in² × 10²	Elongation, %	Flexural strength, lb/in² × 10²	Compressive strength, lb/in² × 10²	Compressive modulus, lb/in² × 10²	Heat distortion temperature at 264 lb/in², °F	Heat resistance, continuous, °F
ABS	1.5–12	2.5–5.8	120–420	20–100	5–13.5	5–11	120–200	180–245	160–235
Acrylic	0.3–0.4	8.7–11.0	350–450	3–6	2–7	11–19	350–430	167–198	130–195
Chlorotrifluoroethylene	2.5–5.0	6	150–300	80–250	7.4–11	4.6–7.4	180	160–170	390
Fluorinated ethylene-propylene	No break	2–3.2	50	250–350	—	2.2	70	124	400
Polytetrafluoroethylene	No break	2–5	50–80	200–400	—	1.7	70–90	132	500
Nylon 6		0.9–4	9.5–12.4	100–380	25–300	5.8–15.7	13–16	347	150–175
Polycarbonate	12–16	8–9.5	345	110–120	13.5	10–12.5	350	265–290	250
Polyethylene, low-density	No break	1–2.4	14–38	100–965	—	—	—	—	140–175
Polyethylene, medium-density	No break	1.7–2.8	50–80	100–965	—	—	—	—	150–180
Polyethylene, high-density	0.4–4.0	2.8–5	75–200	10–1200	1–4	2.7–3.6	50–110	110–125	180–225
Polyethylene, high-molecular-weight	4.5	2.3–5.4	136	170–800	3.5	2.4	110	120	180–225
Polyimide	1.5	5–14.0	300	8–10	19–28.8	30–40	—	680	500–600
Polyethylene	0.5–1.5	4.5–6.0	150–650	100–600	6.0	5–8	—	140–205	250
Polystyrene	0.40	5.2–7.5	400–500	1.5–2.5	10–15	11.5–16	300–560	160–215	150–190
Polyurethane	No break	4.5–8	1–3.7	60–120	0–1	>20	85	—	190
Polyvinyl chloride, flexible	Varied	1–4	—	200–450	—	—	—	—	150–175
Polyvinyl chloride, rigid	0.4–22	7.5	200–600	40–80	10–15	10–11	300–400	140–175	160–165
Polyvinyl dichloride, rigid	1.0–6.0	7.5–9.0	360–450	160–240	14.2–17	13–22	—	212–235	195–210

Material									
Styrene acrylonitrile (SAN)	0.4–0.6	10	475–560	2–3	11–19	15–17.5	650	200–218	170–210
Polyphenylene oxide	3–6	6.8–7.8	380	50	8.3–12.8	15	380	375	250
Polysulfone	0.6–1.0	10.2	360	50–100	15.4	15.4	370	345	300
Polyphthalamide	1.0	>15	—	—	23.3	—	—	248	300
Polyamide-imide	2.7	22	650	7	27.4–34.9	32	—	532	430
Polybutylene terephthalate	0.7–1.0	8.2	280–430	50–300	12–16.7	8.6–14.5	—	122–185	270
Polythylene terephthalate	0.25–0.7	7–10	400–600	30–300	14–18	11–15	—	70–100	300
Polyethylene, ultrahigh-molecular-weight	No break	5.6–7.0	—	420–525	—	—	—	110–120	175–220
Polyarylsulfone	1.2	9.0	310–380	40–60	12.4–16.1	—	374	400	450
Polyethersulfone	1.4	9.8–13.8	350	6–80	17–18.7	11.8–15.6	—	395	350
Polyetherketone	1.6	13.5	520	50	24.5	20	—	368	480
Polyetheretherketone	2.0	10–15	—	30–150	16	18	—	323	470
Polyetherimide	1.1	14	430	60	22	20	420	390	330–350
Polyphenylene sulfide	<0.5	7–12	480	1–3	14–20	16	—	250	350–400
Polycyclohexylene dimethylene terephthalate (PCT), 30% glass-reinforced	1.7	18–19	—	1.9	24–28	—	—	500	300
Polyarylate	4.1–5.5	100			110–150			230–388	Up to 300

SOURCE: Refs. 3 and 4.

TABLE 3.5 Typical Electrical Properties of Thermoplastics

Resin material	Volume resistivity, $\Omega \cdot cm$	Dielectric constant, 60 cycles	Dielectric strength, $\frac{1}{8}$-in thickness, V/mil	Dissipation or power factor, 60 cycles	Arc resistance, s
ABS	10^{15}–10^{17}	2.6–3.5	300–450	0.003–0.007	45–90
Acrylic	$>10^{14}$	3.3–3.9	400	0.04–0.05	No tracking
Chlorotrifluoroethylene	10^{18}	2.65	450	0.015	>360
Fluorinated ethylene-propylene	$>10^{18}$	2.1	500	0.0002	>165
Polytetrafluoroethylene	$>10^{18}$	2.1	400	<0.0001	No tracking
Nylon 6	10^{14}–10^{15}	6.1	300–400	0.4–0.6	140
Polycarbonate	6.1×10^{15}	2.97	410	0.0001–0.0005	10–120
Polyethylene, low-density	10^{15}–10^{18}	2.98	450–1000	0.006	Melts
Polyethylene, medium-density	10^{15}–10^{18}	2.3	450–1000	0.0001–0.0005	Melts
Polyethylene, high-density	6×10^{15}–10^{18}	2.3	450–1000	0.002–0.0003	Melts
Polyethylene, high-molecular-weight	$>10^{16}$	2.3–2.6	500–710	0.0003	Melts
Polyimide	10^{16}–10^{17}	3.5	400	0.002–0.003	230
Polypropylene	10^{15}–10^{17}	2.1–2.7	450–650	0.005–0.0007	36–136
Polystyrene	10^{17}–10^{21}	2.5–2.65	500–700	0.0001–0.0005	60–100
Polyurethane	2×10^{11}	6–8	850–1100	0.276	—
Polyvinyl chloride, flexible	10^{11}–10^{15}	5–9	300–1000	0.08–0.15	60–80
Polyvinyl chloride, rigid	10^{12}–10^{16}	3.4	425–1040	0.01–0.02	—
Polyvinyl dichloride, rigid	10^{15}	3.08	1200–1550	0.018–0.0208	—
Styrene acrylonitrile (SAN)	10^{15}	2.8–3	400–500	0.006–0.008	100–500
Polyphenylene oxide	10^{17}	2.58	400–500	0.00035	75
Polysulfone	5×10^{16}	2.82	425	0.008–0.0056	122
Polyarylate	2×10^{14}	3.08	610	0.002	125
Polybutylene terephthalate (PBT)	1.4×10^{15}	3.3	420	0.002	190
Polycyclohexylene dimethylene terephthalate (PCT)	2×10^{15}	3.2	470–530	0.0018	68–136
Polyethylene terephthalate (PET)	1×10^{15}	3.8	650	0.0059	123
Polyphenylene sulfide, 40% glass	10^{16}	3.5–3.8	340–450	0.0012	124
Polyetherimide	6.7×10^{17}	3.15	750–831	0.0013	126
Polyetherketone	10^{17}	3.5	—	0.002	—

SOURCE: Refs. 3 and 4.

Acrylics

Thermoplastic acrylic resins comprise a range of polymers and copolymers that are derived from acrylic esters (CH_2=CHCOOR) or methacrylic esters [$CH_2CH(CH_3)COOR$]. The acrylic esters may be methyl, ethyl, butyl, and 2-ethylhexyl. The methacrylates are methyl, ethyl, butyl, lauryl, and stearyl. The structural unit that characterizes this family of resins is

$$\mathrm{\left[CH_2-\underset{\underset{OR}{\underset{|}{C=O}}}{\overset{\overset{R}{|}}{C}}\right]_n}$$

These polymers are made by the free-radical polymerization of the acrylic monomer using peroxide or azo (azobisisobutyronitrile) initiators at temperatures around 100°C. Initiation is also effected by using redox polymerization techniques. Bulk, suspension, solution, and emulsion polymerization processes can be used to make these polymers. Modifications are affected by incorporation of other monomers such as methyl methylacrylate, methyl and ethyl acrylate, acrylonitrile, and styrene during the polymerization process. Other modifications can be brought about by blending with other resins such as vinyls, butadiene, polyester, or other acrylics in order to alter specific properties of the resin. The higher esters yield polymers which are generally softer and have a lower softening temperature; i.e., methyl>ethyl> propyl>butyl esters with decreasing hardness going from methyl to butyl.

Acrylic resins are characterized by their exceptional transparency (92 percent light transmission) with haze measurements ranging from 1 to 3 percent. They exhibit good weatherability and are resistant to solutions of inorganic acids, alkalies, and aliphatic hydrocarbons. Acrylic resins have excellent electrical insulation properties as well as dimensional stability, low mold shrinkage, and good impact strength. Acrylics are attacked by chlorinated, aromatic hydrocarbon, ester, ketone, and freon solvents. Acrylics are prone to stress cracking, and they are flammable. The maximum service temperature is about 200°F but it can go as high as 313°F with the new acrylic-imide copolymers. The excellent arc and track resistance of the acrylics has made them a good choice in some high-voltage applications such as circuit breakers. Acrylics are one of the few plastics that exhibit an essentially linear decrease in dielectric constant and dissipation factor with an increase in frequency. The acrylics are produced in many forms including film, rod, sheet, tube, powder, solutions, and reactive syrups.

Acrylic resins are processed by injection molding, extrusion, thermoforming, blow molding, and machining applications include molding, casting, and coating formulations [both dip and sprayable for printed-circuit boards (PCBs) and other electronic components]. Acrylics can also be formulated into thermosetting resins. Some acrylic resin suppliers are Aristech, Rohm & Haas, and du Pont. Additional suppliers are listed in the Buyer's Guide section of the *Modern Plastics Encyclopedia*.[4]

Acetals

Acetals are linear, highly crystalline polymers produced by the polymerization of highly purified formaldehyde into both homopolymer and copolymer types. The structural unit that characterizes this resin is

$$\pm CH_2-O\pm_n$$

and it results in a regular highly crystalline structure which imparts to this resin its unique combination of mechanical, chemical, and electrical properties that are stable over a broad temperature range for long times. The acetal plastics have excellent creep resistance under continuous load as well as fatigue endurance. These plastics are hard, strong, and stiff with excellent toughness and a low friction coefficient. Chemical resistance is excellent; they are resistant to alcohols, aliphatics, aromatics, aldehydes, ketones, ethers, and oils.

The acetal homopolymer is resistant to mild acids and bases with pH range of 4 to 10 and to water at 140°F. The copolymer is resistant to a pH range of 4 to 14. Strong oxidizing agents (hypochlorite) in water can deteriorate the polymer.

Typical properties of acetals are shown in Table 3.6. Special grades are available with improved properties. Acetals can be injection-molded, extruded, and rotationally molded. In the electronics industry, they are used in electric switches, key tops, plungers, and cassette tape rollers. Suppliers of acetals include du Pont and Hoechst Celanese. Additional suppliers of specialty acetals can be found in Ref. 4.

TABLE 3.6 General Properties of Acetal Plastics

Property	Homopolymer	Copolymer
Heat deflection temperature, °C		
at 66 lb/in^2	170	158
at 264 lb/in^2	123	110
Specific gravity	1.42	1.41
Tensile strength, lb/in^2	10,000	8800
Continuous-use temperature, °C	84	104
Tensile modulus, lb/in^2	520,000	410,000
Dielectric strength, 90 mils thick, V/mil	500	
Dielectric constant, 10^2 to 10^7 Hz	3.7	
Volume resistivity at 73°F, $\Omega \cdot$ cm	10^{14}–10^{15}	
Coefficient of linear thermal expansion from -22 to 140°F, in/(in $\cdot$ °F)	4.2–5.3×10^5	
Flammability	UL-94 HB	

Cellulosics

This family of thermoplastic resins is produced by chemical modification of cellulose $(C_6H_{10}O_5)_n$. Included are cellophane, a regenerated cellulose made by mixing cellulose xanthate (ROCSSH) with a dilute sodium hydroxide (NaOH) solution to form a viscose, then extruding the viscose into an acid bath for regeneration; cellulose acetate, an acetic acid ester $(CH_3COOC_2H_5)$ of cellulose; cellulose acetate butyrate, a mixed ester produced by treating fibrous cellulose with butyric acid $(CH_3CH_2CH_2COOH)$, butyric anhydride $(CH_3CH_2CH_2CO_2O)$, acetic acid (CH_3COOH), and acetic anhydride $[(CH_3CO)_2O]$ in the presence of sulfuric acid (H_2SO_4); cellulose propionate, formed by treating fibrous cellulose with propionic acid $(CH_3CH_2CO_2H)$ and acetic acid and anhydrides in the presence of sulfuric acid; cellulose nitrate, made by treating fibrous cellulosic materials with a mixture of nitric (HNO_3) and sulfuric acids. Only cellulose acetate is used in electrical applications as recording tape. Eastman Chemical is the major supplier of all the cellulosics.

Fluoroplastics

These materials are analogous to polyethylene in which some of or all the hydrogen atoms attached to the carbon are replaced by fluorine, chlorine, or a fluorinated alkyl group. The most common fluoropolymers are FEP (fluorinated ethylene-propylene) from tetrafluoroethylene (C_2F_4) and hexafluoropropylene (C_3F_6); PTFE (polytetrafluoroethylene) from the polymerization of tetrafluoroethylene and ethylene (C_2H_4); PFA (perfluoroalkoxy) from tetrafluoroethylene and perfluoropropyl vinyl ether $(C_3H_7C_4OF_5)$; PCTFE (polychlorotrifluoroethylenevinylidene fluoride) from chlorotrifluoroethylene and vinylidene fluoride $(C_2H_2F_2)$; E-CTFE (polyethylene-chlorotrifluoroethylene) from chlorotrifluoroethylene and ethylene; PVDF (polyvinylidene fluoride) from vinylidene fluoride monomer; and PVF (polyvinyl fluoride) from vinyl fluoride monomer (C_2H_3F). The structural unit of each of these polymers is shown in Table 3.7, and general properties are given in Table 3.8.

The fluoroplastics are synthesized by free-radical polymerization techniques. All the fluoroplastics are characterized by their unique combination of properties. They have high heat resistance, are inert to almost all chemicals, and do not burn; they have high arc resistance, low dielectric losses, zero water absorption, and a low coefficient of friction. Their limitation is that these polymers are relatively soft, difficult to process, expensive, and subject to creep. Upper-limit service temperatures for these materials range from 110 to 260°C depending on

TABLE 3.7 Structural Composition of Fluoroplastics

Polymer	Structure
PTFE	$-(-CF_2-CF_2-)_n-$
FEP	$-(-CF_2-CF_2-CF_2-CF)_n$ $\quad\quad\quad\quad\quad\quad\quad\quad\quad\vert$ $\quad\quad\quad\quad\quad\quad\quad\quad\quad CF_3$
PFA	$-(-CF_2-CF_2-CF-CF_2-CF_2)_n$ $\quad\quad\quad\quad\quad\quad\vert$ $\quad\quad\quad\quad\quad\quad OR$
PCTFE	$-(-CF_2-CFCL-)_n-$
ECTFE	$-(-CF_2-CF_2-CF_2-CFCL-)_n-$
PVDF	$-(-CF_2-CF_2-)_n-$

TABLE 3.8 Properties of Fluoropolymers

	PTFE	FEP	PFA	ETFE	PVDF	PCTFE	ECTFE	PVF
Fluorine percentage	76.0	76.0	76.0	59.4	59.4	48.9	39.4	41.3
Melting point, °C	327	265	310	270	160	218	245	200
Upper-use temperature, °C	260	200	260	180	150	204	170	110
Density, g/cm^3	2.13	2.15	2.12	1.7	1.78	2.13	1.68	1.38
Oxygen index, %	>95	95	95	28–32	44	—	48–64	—
Arc resistance, s	>240	>300	>300	72	60	2.4	18	—
Dielectric constant	2.1	2.1	2.1	2.6	9–10	2.5	2.5	9
Dissipation factor	0.0002	0.0002	0.0002	0.0008	0.02–0.02	0.02	0.003	0.002
Tensile strength, lb/in^2	5000	3100	4300	7000	6200	6000	—	—
Specific gravity	2.2	2.17	2.17	1.7	1.78	2.2	1.68	—
Water absorption in 24 h, %	0	0	0.03	0.03	0.06	0	—	—
Electrical strength, V/mil	480	600	500	400	280	600	—	—

SOURCE: Ref. 3.

the type of fluoropolymer. The main suppliers of fluoroplastics are du Pont, ICI, Pennwalt, and Ausimont. Additional suppliers can be found in Ref. 4.

PTFE is a highly crystalline, high-melt viscosity polymer which prevents processing by conventional thermoplastic processing methods. PTFE forms are fabricated by molding and extrusion, followed by sintering at 380°C. PTFE has a useful temperature range from -200 to $+260$°C. It has high impact strength, but low tensile strength, wear, and creep resistance. Electrical properties are stable over a wide temperature range. Its degradation products are toxic. PTFE finds use as wire insulation, as a core insulator of coaxial cables, as printed-wiring board substrates for microwave applications, and as heat-shrinkable tubing sleeves for capacitor and resistor protection. FEP is also a crystalline polymer with slightly lower properties than PTFE. However, since it is a copolymer, the processability of FEP is superior to that of PTFE; FEP can be processed by compression molding and extrusion and is used as powders for fluidized-bed and electrostatic coating processes and in aqueous coating dispersions. It has low mold shrinkage. Like PTFE, it has good weatherability. FEP is used in wire and cable applications such as hookup wire, plenum cable, flat cable, and fire alarm cable.[4] FEP has toxic products upon thermal decomposition.

PCTFE is a crystalline polymer. It is dissolved by a few solvents above 100°C. This polymer has excellent barrier properties to gases and the lowest water vapor transmission. It can be molded and extruded but with difficulty because of its high melt viscosity.

E-CTFE is a crystalline copolymer that has better strength, wear resistance, and creep than PTFE, PFA, and FEP. It is processed by compression molding extrusion and rotational molding and is used in fluidized-bed and electrostatic coating processes. It is used in wire and cable applications similar to PCTFE resin.

ETFE is a crystalline copolymer with high-impact properties. It has good chemical resistance, electrical properties, and weathering resistance, but not as good as the fully fluorinated polymers. It is processed on conventional thermoplastic equipment such as in extrusion and molding (compression and rotational) and is used in electrostatic and fluidized-bed coating processes.

PVDF is a crystalline polymer with greater strength, wear, and creep resistance than PTFE, FEPA, and PFA. It has exceptional electrical properties. PVDF, if properly processed, has the highest piezoelectric and pyroelectric coefficients of any known polymer, with values comparable to those of the piezoelectric ceramics.[9] PVDF film is used in sensor and transducer applications for pressure, strain, infrared, vibration, and impact detection.[10] Typical properties of a PVDF film are shown in Table 3.9. PVDF resin can be extruded, injection-molded, transfer-molded, and applied as a coating.

PVF is a highly crystalline, tough, flexible polymer that is available only as a film. It has excellent weather resistance and maintains its properties from -70 to $+110°C$.[4] The film can be laminated to plywood, vinyl, polyesters, and metal foils.

Ionomer

Ionomer is a generic name for polymers containing interchain ionic bonding. They are based on the metal salts of ethylene methacrylic acid copolymers. The properties vary according to the type of metal cation present, molecular weight, and copolymer composition. These polymers are very strong and tough with excellent barrier properties, puncture

TABLE 3.9 Typical Properties of PVDF Piezoelectric Film

Property	Units	Value
Thickness	meters	$9\text{–}800 \times 10^{-6}$
d_{31}	$\dfrac{\text{m/m}}{\text{V/m}}$ or $\dfrac{\text{C/m}}{\text{N/m}}$	23×10^{-12}
d_{33}		-33×10^{-12}
Piezoelectric stress constant		
g_{31}	$\dfrac{\text{V/m}}{\text{N/m}^2}$ or $\dfrac{\text{m/m}}{\text{C/m}^2}$	216×10^{-3}
g_{33}		-339×10^{-3}
Electromechanical coupling factor	%	
k_{31}		12 (at 1 kHz)
k_{33}		19 (at 1 kHz)
Capacitance	pF/cm^2	380 (for 28-μm film)
Young's modulus	GPa	2
Pyroelectric coefficient	C/(m$^2 \cdot$ K)	-25×10^{-6}
Permittivity	F/m	106×10^{-12}
Relative permittivity (dielectric constant)	—	12
Volume resistivity	$\Omega \cdot$ m	10^{13}
Dissipation factor	—	1.5×10^{-2} (10 Hz)
		2.0×10^{-2} (10 kHz)
Compressive strength	MPa	60
Tensile strength at break	MPa	
Machine direction		160–300
Transverse direction		30–55
Operating temperature range	°C	-40–100
Water absorption	%	0.02
Max. operating voltage	kV/mm	30
Breakdown voltage	kV/mm	100
Specific gravity	—	1.78

SOURCE: Ref. 10.

resistance, and low-temperature brittleness. Their application in the electrical industry is as a packaging material.

Ketone Plastics

The ketone-based plastics are a family of semicrystalline polymers that contain the aromatic ketone group

$$\text{+}\!\!\bigcirc\!\!\text{–CO+}_n$$

as the element in their structure. This group imparts high thermal stability and a chemical resistance chain. There are several types of ketone polymers, as shown below. The major ketone polymers are polyetherketone (PEK), polyaryletherketone (PAEK), and polyetheretherketone (PEEK). Their structures are shown below:

Polymer	Structure
PEK	⎡⟨O⟩–O–⟨O⟩–CO⎤$_n$
PAEK	⎡⟨O⟩–O–⟨O⟩–O–⟨O⟩–CO⎤$_n$
PEEK	⎡⟨O⟩–O–⟨O⟩–CO–⟨O⟩–O–⟨O⟩–CO–⟨O⟩–CO⎤$_n$

Ether linkages (—O—) are introduced into the polymer chain to impart flexibility and improve processing. Reviews of ketone polymers can be found in Refs. 11 and 12.

Ketone polymers are synthesized by condensation polymerization. Melting temperatures are above 300°C. The thermal expansion coefficient increases substantially above their T_g values (143, 162, and 143°C for polyketone, polyetherketone, and polyetheretherketone, respectively). All the ketone polymers have low smoke emission and meet UL 94 V-O tests in their neat condition (no additives). The polymers are tough, strong, and stiff and have high impact strength and load-bearing properties. Electrical properties are excellent with only small changes in dielectric constant and dissipation factor over a wide frequency range. Chemical resistance is excellent; only concentrated anhydrous or strong oxidizing acids have an effect on these polymers. The only known solvent for PEK and PEEK is concentrated sulfuric acid. The ketone polymers are also highly resistant to hot-water hydrolysis.

The ketone polymers are affected by ultraviolet (uv) radiation, but they have high resistance to beta, gamma, and x rays over a wide temperature range. Processing methods applicable for these polymers include injection molding, extrusion, rotational molding, and powder coating. PEEK has been used for injection-molded circuit boards but has a wider range of applications and is used as high-temperature connectors, wire insulation, cable couplers, flexible film circuit substrates, adhesives in hybrid microcircuits, and semiconductor wafer carriers.

Mold shrinkage values of 0.7 to 1.2 percent must be accounted for in the neat resin. These values are reduced by one-half with the addition of glass or carbon fibers. Polyketones are molded in both natural and reinforced forms. Typical fillers include glass, carbon, graphite, and mica.

Liquid-Crystal Polymers

Liquid-crystal polymers (LCPs) belong to a class of materials that exhibit a highly ordered structure in both the melt and solid states and whose main features include high stiffness in thin sections, high temperature stability, and chemical resistance. Because of the high degree of molecular ordering, liquid-crystal polymers exhibit a high degree of anisotropy. If the liquid-crystalline phase forms on melting the polymer, it is known as a *thermotropic* liquid crystal; and if it forms in solution as the result of solvent addition, it is known as *lyotropic*. Condensation polymerization has been used to prepare these polymers. A number of polymers exhibit liquid-crystalline behavior, but the three commercially important polymers are Xydar (Amoco Performance Products, Inc.), Vectra (Hoechst Celanese Corp.), and HX series from du Pont Co. There is no one chemical structure that characterizes LCPs; however, all LCPs have these common characteristics: The molecular shape has a large aspect ratio (length or diameter to width or thickness), the molecule has a large polarizability along the rigid chain axis compared to the transverse direction, and the molecule must have good molecular parallelism of the rigid units comprising its structure. To meet these requirements, an LCP should possess a rigid molecular structure.[11] The most common types of LCPs are the wholly aromatic copolyesters and copolyester-amide. They are synthesized by melt condensation polymerization, and their unique properties result from the rigid-rod-shaped molecular orientation imparted to the polymer's chains during processing.[12] Both Xydar and Vectra meet these criteria, and their structures are as follows:

Xydar

$$\left[-CO-\underset{}{\bigcirc}-COO-\underset{}{\bigcirc}-\underset{}{\bigcirc}-OOC-\underset{}{\bigcirc}-O-\right]_n$$

Vectra

$$\left[-CO-\underset{}{\bigcirc}-COO-\underset{}{\bigcirc\bigcirc}-O-\right]_n$$

The major properties that characterize liquid-crystal polymers are low melt viscosity; exceptional tensile, compressive, and modulus values; and outstanding chemical, radiation, and thermal stability. A general comparison of flexural modulus and mold shrinkage for LCPs and other polymers is shown in Figs. 3.3 and 3.4,[13] while Table 3.10 presents a comparison of specific properties of filled liquid-crystal polymers. The Vectra polymers are made from three comonomers all based on p-hydroxylbenzoic acid and 2,6-hydroxynapthoic acid. The Xydar polymers are made from terephthalic and isophthalic acid, p,p'-dihydroxybiphenol, p-hydroxybenzoic acid, and biphenol. The general properties of liquid-crystal polymers are listed in Table 3.11. Because of the anisotropy of the unfilled LCPs, they are reinforced for use in electronic applications where 30 to 50 weight percent glass fiber is a typical loading. The high melting temperatures of LCPs (300°C range)

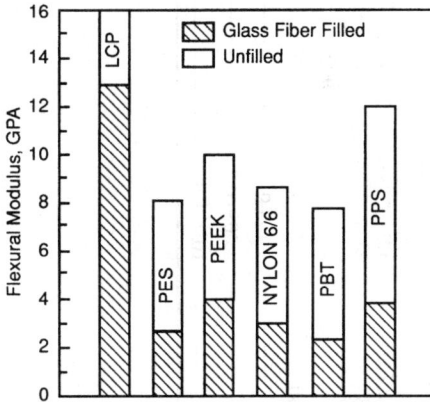

Figure 3.3 Comparison of flexural modulus of selected thermoplastics. (Source: Ref. 13. Reprinted with permission.)

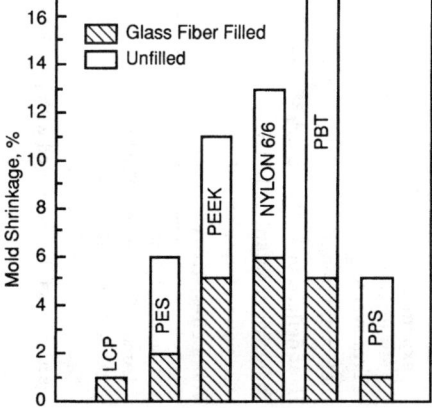

Figure 3.4 Comparison of mold shrinkage of selected thermoplastics. (Source: Ref. 13. Reprinted with permission.)

TABLE 3.10 Typical Properties of LCP Molding Compounds

Grade	Vectra* A130	Vectra B230	Vectra A420	Xydar† G-930	Xydar NG-350	Du Pont HX7180	Du Pont HX6130
Fillers	30% glass fiber	30% carbon fiber	50% glass/mineral, 5% graphite flake	30% glass fiber	50% mineral/glass	30% glass fiber	30% glass fiber
Tensile strength, klb/in²	30	35	21.5	19.6	14.2	21	21.8
@ temperature	1.9 @ 400°F	—	1.5 @ 400°F	2.8 @ 500°F	2.3 @ 500°F	8.2 @ 392°F	5.2 @ 342°F
Elongation, %	2.2	1.0	1.4	1.6	2.3	2.2	2.7
Flexural modulus, Mlb/in²	2.1	4.6	2.9	2.26	1.73	1.8	1.7
@ temperature	0.85 @ 428°F	0.41 @ 428°F	—	0.46 @ 500°F	0.29 @ 500°F	1.1 @ 300°F	0.8 @ 300°F
Impact strength, notched, ft · lb/in	2.8	1.4	1.9	1.8	1.9	4.2	2.4
Coefficient of thermal expansion, ppm/°F							
Flow direction	2.8	−1.7	3.9	2.7	5.7	8.0	7.0
Transverse	36	25	28	43	21	20.0	21
Dielectric strength, V/mil	1100‡	—	—	900	900	7104§	740
Heat deflection, at 264 lb/in², °F	446	440	437	520	493	536	493
Dielectric constant at 1 MHz	—	—	3.4–3.9	—	—	—	—
Dissipation factor at 1 MHz	—	—	0.009–0.34	—	0.39	0.029	0.026
Volume resistivity, Ω · cm	—	—	10^{14}–10^{15}	—	—	—	—

*Hoechst Celanese.
†Amoco Performance Products.
‡At 1.50-mm thickness.
§At 1.59-mm thickness.
SOURCE: Ref. 14. Reprinted with permission.

TABLE 3.11 General Properties of Liquid-Crystal Polymers

Property	Value
Flexural strength, lb/in^2	21,000–44,000
Unnotched izod, J/m	150–300
Notched izod, J/m	400–500
Rockwell hardness	M62–99
Dielectric constant, 10^3 Hz	2.9–4.5
Dissipation factor, 10^3 Hz	$4–6 \times 10^{-3}$
Volume resistivity, $\Omega \cdot$ cm	$10^{12}–10^{-13}$
Dielectric strength, V/mil	780–1000
Arc resistance, s	63–185
Water absorption, equilibrium at 23°C, %	0.02–0.04
Specific gravity	1.4–1.9
Glass transition temperature, °C	—
Melt temperature, °C	275–330
Heat distortion temperature, °C	
at 65 lb/in^2	250–280
at 264 lb/in^2	180–240
Continuous-use temperature, °C	200–240
Coefficient of thermal expansion, ppm/°C	$0–25 \times 10^{-6}$ flow direction $25–50 \times 10^{-6}$ transverse
Flammability, UL-94	V-O
Oxygen index	35–50
Tensile modulus, lb/in^2	1400–5800
Tensile strength, yield lb/in^2	20,000–35,000
Tensile elongation, %	1.2–6.9
Flexural modulus, lb/in^2	1400–5000

and high modulus impart excellent resistance to wave, vapor-phase, and infrared soldering conditions. The LCPs have a flammability rating of UL-94 V-O and have a high oxygen index. During combustion very low smoke is generated. Chemical resistance to acids, dilute bases, and organic solvents is excellent. Strong bases such as sodium hydroxide and amines will attack LCPs at elevated temperature. Most LCP material is injection-molded, and most of the LCP material is filled with glass, carbon, or minerals to minimize the material's anisotropy. The main uses for these polymers in the electrical industry are for molding of high-precision complex parts, chip carriers, sockets, connectors, pin-grid arrays, bobbins, and relay cases. Suppliers of LCPs are Amoco, du Pont, and Hoechst Celanese Corp.

Nylon

These materials, also known as *polyamides,* are characterized by having the amide group (—CONH—) as an integral part of the polymer structure. While this chemical unit is present in all nylons, the multiplicity of monomers that can be used to prepare nylons has led to a wide variety of materials with different properties. Presently there are 11 types of nylons available;[14] 9 are aliphatic and 2 are aromatic. Nylons are synthesized by both condensation polymerization (types 6/6, 6/9, 6/10, and 6/12) and addition polymerization (types 6, 11, and 12). Nylons are synthesized from intermediates such as dicarboxylic acids, diamines, amino acids, and lactams; and nylons are identified by numbers denoting the number of carbon atoms in the polymer chain derived from specific constituents, with those from the diamine being given first. The second number, if used, denotes the number of carbon atoms derived from a diacid. Commercial nylons are as follows: Nylon 4 (polypyrrolidone) is a polymer of 2-pyrrolidone [$CH_2CH_2CH_2C(O)NH$]; nylon 6 (polycaprolactam) is made by the polycondensation of caprolactam [$CH_2(CH_2)_4NHCO$]; nylon 6/6 is made by condensing hexamethylenediamine [$H_2N(CH_2)_6NH_2$] with adipic acid [$COOH(CH_2)_4COOH$]; nylon 6/10 is made by condensing hexamethylenediamine with sebacic acid [$COOH(CH_2)_8COOH$]; nylon 6/12 is made from hexamethylenediamine and a 12-carbon dibasic acid; nylon 11 is produced by polycondensation of the monomer 11-amino-undecanoic acid [$NH_2CH_2(CH_2)_9COOH$]; nylon 12 is made by the polymerization of laurolactam [$CH_2(CH_2)_{10}CO$] or cyclododecalactam, with 11 methylene units between the linking —NH—CO— groups in the polymer chain. Most nylons are partially crystalline polymers. The nylons can be modified by the addition of additives or copolymerized with other monomers to produce a wide range of materials with different properties. In addition, some blending of nylon polymers can be done with acrylonitrile-butadiene-styrene and polyphenylene ether polymers. Transparent nylon is available; and unlike the other grades of nylon polymers, this is amorphous. The nylons are strong, tough thermoplastics having good tensile, flexural, friction, and impact properties. Nylons can operate satisfactorily from 0 to 149°C. All nylons are hygroscopic, although the degree of water absorption decreases with increasing hydrocarbon chain length. This moisture absorption affects some properties; e.g., it has a plasticizing effect on the polymer and increases flexural and impact strength while decreasing tensile strength. The electrical properties of nylons are quite sensitive to moisture and deteriorate with increasing water content. Nevertheless, the electrical properties of nylons are quite adequate to allow their use in most 60-Hz power applications. Nylons have good chemical resistance to hydrocarbons and aromatic and aliphatic solvents but are attacked by strong acids, bases, and phenol. Elevated temperature and uv radiation exposure will degrade nylon depending on the duration and level of the exposure. The nylons can be processed by almost all the common thermoplastic fabrication techniques. The reader is directed to Refs. 4, 6, and 14 for additional information and specific properties. Applications for nylons include card guides, connectors, terminal blocks, antenna mounts, coil bobbins, and receptacle plugs. Typical nylon suppliers include Allied Signal Inc., BASF Corp., du Pont Co., Hoechst Celanese Corp., Miles, Inc., and Monsanto Chemical Co.

High-Temperature Nylon

In addition to the aforementioned nylons, there is another class of polyamides that is based on the presence of an aromatic ring in the chemical structure. Two materials comprise this class: Nomex [(poly)1,3-phenylene isophthalamide, structure I] and Kevlar [(poly)1,4-(poly)1,4-phenylene terephthalamide, structure II].

$$\left[\text{NH}-\bigcirc-\text{NHOC}-\bigcirc-\text{CO}\right]_n \qquad \left[\text{HN}-\bigcirc-\text{NHOC}-\bigcirc-\text{CO}\right]_n$$

I II

Kevlar is spun into fiber and is mostly used in composite (nonelectrical) applications while Nomex is processed into fiber, paper, sheet, and pressboard and is extensively used in the electrical industry as insulation for transformers, generators, motors, and other electrical equipment. In its densified form, Nomex products can withstand short-term electrical stresses of 450 to 1000 V/mil depending on product type. Nomex is strong, resilient, and flexible with good resistance to tearing and abrasion. It has excellent thermal stability and radiation resistance. Nomex is unaffected by most solvents and can be used at cryogenic temperatures. Densified Nomex papers and pressboards maintain 90 percent of their dry dielectric strength at 95 percent relative humidity under equilibrium conditions. Nomex has excellent flame resistance and meets UL-94V-O requirements. Nomex is recognized by Underwriters Laboratories as a 220°C material. Nomex is used in transformers as conductor wrap, layer, and barrier insulation; coil end filler, phase insulation, and case insulation. In rotating equipment, it is used as conductor wrap, coil wrap and interleaving, slot liners, phase insulation, and lead insulation. Nomex and Kevlar are supplied by the du Pont Co. and are registered trademarks. Table 3.12 gives some electrical properties of Nomex. Paper and pressboard are the most widely used forms of Nomex.

TABLE 3.12 Electrical Properties of Nomex*

Nomex type	Thickness, mils	Dielectric strength, V/mil, ASTM D-149	Dielectric constant, 60 Hz, ASTM D-150	Dissipation factor, 60 Hz, ASTM D-150	Volume resistivity, $\Omega \cdot$ cm, ASTM D-257
410	3	540	1.6	0.005	10^{16}
411	5	230	1.2	0.003	—
414	3.4	530	1.7	0.005	10^{16}
418	3	730	2.9	0.006	10^{16}
419	7	325	2.0	—	—
992	125	380	1.7	0.020	10^{17}
993	120	540	2.6	0.015	10^{17}
994	250		3.5	0.010	10^{16}

*Unless otherwise noted, the Nomex properties are typical values measured in air under "standard" conditions (in equilibrium at 23°C, 50% relative humidity) and should not be used as specification limits.
In this table, the dissipation factor (Nomex types 418, 419, 992, and 993) and all the volume resistivities (Nomex types 410, 411, 414, 418, 419, 992, and 993) are measured under dry conditions.
Nomex is a trademark of du Pont for its aramid products. Only du Pont makes Nomex.
SOURCE: Ref. 15. Reprinted with permission.

Polyamide-Imides

These polymers are amorphous, high-temperature engineering thermoplastics produced by the condensation reaction of trimellitic anhydride and aromatic diamines. The characteristic chemical groups in the polymer chain are the amide linkage (—CONH—) and the imide linkage (—CONCO—). The structural unit in the polymer is represented by the following structure.

This structural unit represents the fully converted product, but an intermediate product called the *amic-acid form* exists that is used for coatings on wire or can be processed into film. Heating of the amic-acid form converts the product to the fully imidized form represented by the structure shown above. These polymers possess outstanding temperature resistance (useful from cryogenic temperatures to about 230°C) and radiation stability (can withstand 10^9 rads). The polyamide-imides also possess excellent mechanical properties and low dielectric losses and wear resistance, and they have a low coefficient of thermal expansion. At liquid-nitrogen temperatures, the tensile strength of the unfilled polymer is about 31,000 lb/in^2 with 6 percent elongation, a flex modulus of 1.1×10^6 lb/in^2, and a flex strength of 40,000 lb/in^2.

Polyamide-imides are inherently flame-resistant with an oxygen index of 43 and a UL-94 rating of V-O. The polymers produce very little smoke when burned. Their chemical resistance is excellent, and they are not attacked by aliphatic or aromatic hydrocarbons, halogenated solvents, and most acids and bases at room temperature. The polyamide-imides are attacked by hot caustic, acid, and steam. The material can be used unfilled or with glass and graphite fiber reinforcement. Applications include connectors, circuit boards, radomes, wire enamels, and films for electrical insulation. The polymers are supplied by the Amoco Chemical Corp.

Polyimides

These materials are derived from the solution condensation polymerization of aromatic dianhydrides and diamines and are characterized by the presence of only the imide linkage (—CONCO—). The general structure that depicts these polymers is

The polyimides are characterized by high glass transition temperatures, excellent radiation resistance, toughness, good electrical properties, and flame resistance. The properties of polyimides can be modified by adjusting both the type and the ratio of the monomers. Fillers have also been added to polyimides to alter their properties. These modifications have produced a variety of polyimide materials. The polyimides can be processed in solution or powder form and can be converted to film, molding powders, tape, and varnishes. Polyimides can be compression- and injection-molded, but considerable expertise is required because of the high T_g values and melt viscosities of these polymers. Kapton is perhaps the most widely known of the polyimide family, but other types include Vespel, Pyralin, Pyre ML, Envex, Skybond 700, FM-34, Untratherm, Eymyd, Duramid, Avimid, Cypac, and Matrimid.* Although there are differences in the properties of the various poly-

---

*Vespel is a registered trademark of du Pont Co.; Pyralin is a registered trademark of du Pont Co.; Pyre ML is a registered trademark of du Pont Co.; Envex is a registered trademark of Rogers Corp.; Skybond 700 is a registered trademark of Monsanto Corp.; FM-34 is a product of the American Cyanamid Co.; Untratherm is a registered trademark of P.D. George Co.; Eymyd is a registered trademark of Ethyl Corp.; Duramid is a registered trademark of Rogers Corp.; Avimid is a registered trademark of du Pont Co.; Cypac is a registered trademark of American Cyanamid; Matrimid is a registered trademark of Ciba-Geigy Corp.

imides, the properties of the polyimide family are illustrated with those of Kapton film. Tables 3.13 and 3.14 show these properties.

The polyimides have temperature capability up to 350°C for short exposures; they are excellent barrier materials; they have excellent adhesion and solvent resistance. Limitations include high cost and the fact that they are attacked by alkali and are difficult to fabricate.

TABLE 3.13 Physical and Electrical Properties of Kapton 100 HN Film

Physical properties	Typical values 23°C (73°F)	Typical values 200°C (392°F)	Test method
Ultimate tensile strength, MPa (lb/in^2)	231 (33,500)	139 (20,000)	ASTM D-882-91, method A*
Yield point at 3%, MPa (lb/in^2)	69 (10,000)	41 (6000)	ASTM D-882-91
Stress to produce 5% elongation, MPa (lb/in^2)	90 (13,000)	61 (9000)	ASTM D-882-91
Ultimate elongation, %	72	83	ASTM D-882-91
Tensile modulus, GPa (machine direction) (lb/in^2)	2.5 (370,000)	2.0 (290,000)	ASTM D-882-91
Impact strength, N · cm (ft · lb)	78 (0.58)		Du Pont Pneumatic Impact Test
Folding endurance (MIT), cycles	285,000		ASTM D-2176-89
Tear strength, propagating (Elmendorf), N (lb)	0.07 (0.02)		ASTM D-1922-89
Density, g/cm^3	1.42		ASTM D-1505-90
Coefficient of friction, kinetic (film to film)	0.48		ASTM D-1894-90
Coefficient of friction, static (film to film)	0.63		ASTM D-1894-90
Refractive index (sodium D line)	1.70		ASTM D-542-90
Poisson's ratio	0.34		Average three samples elongated at 5%, 7%, 10%
Low-temperature flexural life	Pass		IPC TM 650, method 2.6.18
Dielectric strength at 60 Hz, V/mil	7700		ASTM D-149-91
Dielectric constant, 1 kHz	3.4		ASTM D-150-92
Dissipation factor, 1 kHz	0.0018		ASTM D-150-92
Volume resistivity, Ω · cm	1.5×10^{17}		ASTM D-257-91

*Specimen size: 25 × 150 mm (1 in × 6 in); jaw separation: 100 mm (4 in); jaw speed: 50 mm/min (2 in/min); ultimate refers to the tensile strength and elongation measured at break.

SOURCE: Ref. 16. Reprinted with permission.

TABLE 3.14 Thermal Properties of Kapton 100 HN Film

Thermal properties	Typical values	Test condition	Test method
Melting point	None	None	ASTM E-794-85 (1989)
Thermal coefficient of expansion	20 ppm/°C (11 ppm/°F)	−14–38°C (7–100°F)	ASTM D-696-91
Coefficient of thermal conductivity, W/(m · K)	0.12	296 K	ASTM F-433-77 (1987)
$\left(\dfrac{\text{Cal}}{\text{cm} \cdot \text{s} \cdot {}^\circ\text{C}}\right)$	(2.87×10^{-4})	(23°C)	
Specific heat	1.09 (0.261)	J/(g · K) [cal/(g · °C)]	Differential calorimetry
Flammability	UL-94 V-O		UL-94 (2/8/85)
Shrinkage, %	0.17	30 min @ 150°C	IPC TM 650, method 2.2.4A
	1.25	120 min @ 400°C	ASTM D-5214-91
Heat sealability	Not heat-sealable		
Limiting oxygen index, %	37		ASTM D-2863-87
Solder float	Pass		IPC TM 650, method 2.4.13A
Smoke generation	DM = less than 1	NBS smoke chamber	NFPA-258
Glass transition temperature T_g	A second-order transition occurs in Kapton between 360°C (680°F) and 410°C (770°F) and is assumed to be the glass transition temperature. Different measurement techniques produce different results within the above temperature range.		

SOURCE: Ref. 16. Reprinted with permission.

The fully imidized polymers are produced in a variety of forms, e.g., films, varnishes, molding powders, and fabric-supported resin. Preformed stock shapes of polyimides are also available which can be mechanically machined into various parts.

Applications include printed wiring board substrates, magnet wire enamel, wire-cable insulation, insulation spacers, flexible film and cable for interconnections, interlayer dielectrics, and multichip modules.

Polyetherimide

Although this material belongs to the polyimide family of resins and has properties similar to the all aromatic polyimides, it has lower thermal stability (UL-rated for 170°C continuous use compared to 220°C for Kapton polyimide). It is an amorphous polyimide having aromatic imide and ether repeating units in its molecular chain, as shown:

$$\left[N \underset{O}{\overset{O}{\bigotimes}} O-Ar-\underset{CH_3}{\overset{CH_3}{\underset{|}{C}}}-Ar-O \underset{O}{\overset{O}{\bigotimes}} N \right]_n$$

Polyetherimide processes much better on conventional thermoplastic equipment compared to the completely aromatic polyimides and is easily molded into complex shapes. It has a UL-94 V-O flame resistance rating and an oxygen index of 47. Polyetherimide is resistant to a wide variety of chemicals such as mineral acids, aliphatic hydrocarbons, alcohols, and completely halogenated solvents. The polymer is not resistant to partially halogenated solvents (methylene chloride, chloroform), aprotic solvents such as dimethylacetamide and N-methyl-pyrrolidone, and strong bases (sodium hydroxide). It has excellent uv and gamma radiation resistance (94 percent retention of tensile strength after exposure to 400 Mrads of cobalt irradiation). The polymer retains 85 percent of its tensile strength after 10^4 h in boiling water. Electrical properties show very good stability under various conditions of temperature, humidity, and frequency. Polyetherimide has a low dissipation factor at gigahertz frequencies. Its low dissipation factor makes it transparent to microwaves. Its resistance to wave and vapor-phase soldering as well as its excellent electrical properties, dimensional stability, and flame resistance make it especially attractive for use in electrical and electronic applications. Applications include low-loss radomes, printed-wiring boards, IC chip carriers, bobbins, and *IR* switches. General Electric supplies unfilled polyetherimide.

Polyarylates and Polyesters

The thermoplastic polyesters include polyarylate (PA), polybutylene terephthalate (PBT), polyethylene terephthalate (PET), and polycyclohexylene dimethylene terephthalate (PCT). These linear polyesters range from amorphous to crystalline materials and have the characteristic ester functional group (—COOR—) present along the polymer chain. Except for the polyarylates, the other polyesters are made by a transesterification of the appropriate alcohol and ester monomers. Polyarylate is prepared from the reaction of bisphenol A and a mixture of iso- and terephthalic acids. Because of the variety of alcohols, acids, or esters available for reaction, polyesters with a broad range of properties can be synthesized.

Polyarylate (PA) resins are aromatic, linear, amorphous polyesters with excellent toughness, uv resistance, flex strength, dimensional stability, low smoke and flame retardance, and clarity. Electrical properties are fairly constant over a broad temperature range. Chemical resistance is not a strong point for the polyarylates. These polymers are susceptible to stress cracking when exposed to ketone, aromatic hydrocarbon, aldehyde, amines, ester, and chlorinated solvents. If polyarylates are alloyed with other polymers (such as nylons and other polyesters), the stress crack resistance is improved. (Polyarylate is also added to PET as a uv stabilizer.) Polyarylates are processed by most conventional melt processes such as injection extrusion, flow molding, and thermoforming. Polyarylate has very good flexural recovery properties, making it useful for snap-fit component applications. Electrical applications include connectors, relay housings, coil bobbins, and switch and fuse covers. Suppliers include Amoco and Hoechst Celanese Corp.

Polybutylene terephthalate (PBT) is a linear, semicrystalline, aromatic/aliphatic polyester. It has excellent chemical and temperature resistance and good electrical properties which are unaffected by humidity. PBT is unaffected by water, weak acids and bases, and common organic solvents at room temperature. One cannot solvent-bond PBT because of its solvent resistance. PBT resins are processed mostly by injection molding. Suppliers include Miles, General Electric, Hoechst Celanese, Albis, and BASF.

Polyethylene terephthalate (PET) is also a linear, crystalline, aromatic/aliphatic polyester that is produced in a standard and an engineering-grade material. The latter grade has superior properties (strength, stiffness, dimensional stability, chemical and heat resistance, and electrical properties) and is therefore preferred for electrical applications. PET is attacked by chlorinated solvents and strong bases at high temperatures. It is produced by the melt condensation copolymerization of dimethylterephthalate or terephthalic acid and ethylene glycol. The material is processed by injection molding and is used for lamp sockets, coil forms (audio and video transformers), connectors, and terminal blocks. Mold shrinkage of unfilled PET is about 2 percent, but with 30 percent glass-fiber reinforcement it is 0.1 to 0.3 percent. Suppliers include Allied Signal, du Pont, Miles, Hoechst Celanese, General Electric, Albis, Thermofil, and Comalloy.

Polycyclohexylene dimethylene terephthalate (PCT) is a linear high-temperature semicrystalline material. Its high heat resistance distinguishes it from PET and PBT. PCT has a melting point of 290°C, compared to 224°C for PBT and 250°C for PET. This high-temperature resistance makes PCT useful for surface-mount electronic components. The material has an excellent balance of physical, chemical, electrical, mechanical, and thermal properties. Injection molding is the preferred processing method. Applications include sockets, chip carriers, pin-grid arrays, coil bobbins, and surface-mount components. Suppliers include General Electric, Eastern, and Chemical Co.

Polycarbonates

Polycarbonate is a linear, amorphous material having the following repeat unit in the polymer chain:

$$\left[O - \underset{\underset{CH_3}{|}}{\overset{\overset{CH_3}{|}}{C}} - \underset{}{\bigcirc} - O - \overset{\overset{O}{\|}}{C} \right]_n$$

It is synthesized by interfacial polycondensation of bisphenol A and phosgene or ester interchange between diaryl carbonate and dihydric phenols. The characteristic properties of this polymer are excellent transparency (light transmission up to 92 percent) and high impact resistance and heat resistance properties (up to 140°C). The polymer has very good electrical properties and is essentially self-extinguishing. The electrical properties are good with a stable dielectric constant over a wide temperature and frequency range. Linear as well as branched polycarbonates are produced. Specialty blends of polycarbonate with various elastomers, polyolefins, thermoplastic plasters, ABS, and sulfone polymers are also available; and these blends provide improved low-temperature toughness, notch sensitivity, and processibility. Polycarbonate is also produced in a foamable grade. Polycarbonate is notch-sensitive. Properties of selected polycarbonate materials are shown in Table 3.15. The uv and chemical resistance of polycarbonates is limited. They are attacked by alkali, amines, ketones, esters, and aromatic hydrocarbons. Stressed polycarbonate parts are sensitive to many solvents and will crack upon exposure. Polycarbonate is processed by most conventional thermoplastic processing methods. Applications include connectors, breaker boxes, and bobbins. Suppliers of polycarbonate include Dow Chemical, Miles, and General Electric.

Polyolefins

This class of materials includes the polymers and copolymers of polyethylene and polypropylene. The grades include a range of densities (0.83 g/cm^3 for polymethylpentene) to 0.96 g/cm^3 for the high-density polymers and a range of molecular weights including ultrahigh molecular weight. These polymers are characterized by the —(—CH_2—CHR) repeat unit, where R = hydrogen or an alkyl radical. The polyethylene resins are obtained by polymerizing ethylene gas (C_2H_4). Low-molecular-weight polymers of ethylene are fluids used as lubricants; medium-weight polymers are waxes miscible with paraffin; and the high-molecular-weight polymers (i.e., over 6000 up to several million) are the materials used in the plastics industry. Polymers with densities

TABLE 3.15 Properties of Selected Polycarbonate Materials

Property	ASTM test method	Calibre 300-10	Lexan GR 1110	APEC HtDP9-9340
Supplier		Dow	GE Plastics	Miles
Description		General purpose	Gamma-resistant	High heat
Specific gravity	D792	1.20	1.20	1.17
Density, lb/in^3	D792		0.043	0.042
Mold shrinkage, in/in	D955	0.005–0.007	0.005–0.007	0.008
Refractive index	D542	1.586	1.586	
Haze, %	D1003	0.7–1.5	2.0	1.0
Transmittance, %	D1003	87–91	82.9	88
Yellowness	D1925		−5.0	
Tensile modulus, lb/in^2	D638	350,000		325,000
Yield strength, lb/in^2	D638	9000	9400	9600
Ultimate tensile strength, lb/in^2	D638	10,300	9300	9300
Elongation at yield, %	D638	7	6.0	6.0
Elongation at break, %	D638	150	85.0	80.0
Flexural strength, lb/in^2	D790	350,000	330,000	330,000
Flexural strength, lb/in^2	D790	14,000	13,900	12,500
Notched Izod impact strength at 73°F, ft · lb/in	D256	17	14.0	6
Unnotched Izod impact strength at 73°F, ft · lb/in		No break	40.0	No break
Instrumented dart impact energy at 73°F, in · lb	D3763	770	40.0	
Tensile impact strength, ft · lb/in^2	D1822	118 (M73)		83
DTUL at 264 lb/in^2, unannealed, °F	D648	263	238	302
Vicat softening temperature, °F	D1525	312		341
Coefficient of linear thermal expansion, in/(in · °F)	D696	3.80		4.2

SOURCE: Ref. 17. Reprinted with permission from Advanstar Communications.

ranging from about 0.910 to 0.925 are called low-density; those with densities from 0.926 to 0.940 are called medium-density; and those from 0.941 to 0.965 and over are called high-density. The low-density types are polymerized at very high pressures and temperatures, and the high-density types at relatively low temperatures and pressures. A relatively new type, called *linear low-density polyethylene*, is manufactured through a variety of processes: gas-phase, solution, slurry, or high-pressure conversion. A high-efficiency catalyst system aids in the polymerization of ethylene and allows for lower temperatures and pressures than those required in making conventional low-density polyethylene. Copolymers of ethylene with vinyl acetate ethyl acrylate and acrylic acid are the commercially important copolymers of polyethylene.

Polypropylenes are resins made by polymerizing propylene (CH_3CHCH_2) and, in the case of copolymers with other monomers and with suitable catalysts, generally aluminum alkyl and titanium tetrachloride mixed with solvents. The monomer unit in polypropylene (PP) is asymmetric and can assume two regular geometric arrangements: isotactic, with all methyl groups aligned on the same side of the chain, or syndiotactic, with the methyl groups alternating. All other forms, where this positioning is random, are called *atactic*. Commercial polypropylene contains 90 to 97 percent crystalline or isotactic PP with the remainder being atactic. Most processes remove excess atactic PP. This by-product is used in adhesives, caulks, and cable-filling compounds. There are a multitude of materials available, but the common thread that joins all these materials is their low dielectric constant, low dissipation factor, low water absorption, low coefficient of friction, and excellent chemical resistance (resistant to acids, alkalies, and most solvents). A chart showing how certain polymer properties are affected by the polymer's molecular parameters is found in Table 3.16. Special grades of polyolefins such as the higher-molecular-weight material do exhibit increased toughness, abrasion resistance, and freedom from environmental stress cracking. The polyolefins are the lightest of all plastics. The unfilled polyolefins have high coefficients of linear thermal expansion and high mold shrinkage values. The polyolefins are difficult to bond to, and special surface treatments are required to improve adhesion. Polypropylene is notch-sensitive. The polyolefins in general are used as wire and cable insulation. Some suppliers of polyethylene are Dow, Exxon, Phillips, Union Carbide, Mobil, OxyChem, and Solvay; and suppliers of polypropylene include Amoco, Aristech, Eastman, Exxon, Phillips, Shell, and Solvay.

Polyolefin Copolymers

A number of ethylene polymers are produced by using comonomers with ethylene to yield a variety of products. These materials are known as ethylene-acrylates, ethylene acid copolymer, ethylene-acetate, ethylene vinyl alcohol, and ionomer. They are not generally used in electrical and electronic applications, but some of the ethylene-acrylates,

TABLE 3.16 Polymer Parameters and Their Influence Properties

	Density		Molecular weight		Molecular weight distribution	
	Increases	Decreases	Increases	Decreases	Broadens	Narrows
Environmental stress	↓	↑	↑	↓	↑	↓
Impact strength	↓	↑	↑	↓	↓	↑
Stiffness	↑	↓	—	—	—	—
Hardness	↑	↓	—	—	—	—
Tensile strength	↑	↓	—	—	—	—
Permeation	↓	↑	—	—	—	—
Warpage	↑	↓	—	—	↑	↓
Abrasion resistance	—	—	↑	↓	—	—
Flow processibility	—	—	↓	↑	↑	↓
Melt strength	—	—	↑	↓	↑	↓
Melt viscosity	—	—	↑	↓	↓	↑
Copolymer content	↓	↑	—	—	—	—

SOURCE: Ref. 4, Advanced Materials Group. Reprinted with permission.

when compounded with carbon black, have been used in cable insulation and microchip packaging. Additional information can be obtained in Ref. 4.

Polyphenylene Oxide

Polyphenylene oxide (PPO) is a linear amorphous polymer made by a procedure called *oxidative coupling*. The structural unit characteristic of the polymer is

$$\left[\begin{array}{c}R\\ \\ \text{—}\bigcirc\text{—O—}\\ \\ R\end{array}\right]_n$$

The glass transition temperature of pure PPO* is 210°C. This polymer is not used in its neat form, but rather is blended with a high-impact-grade polystyrene to produce the commercial material designated Noryl and Prevex. A range of properties can be obtained by varying the polystyrene content. This modification reduces both the T_g value and heat distortion temperature of the blend. The PPO alloys have good resistance to acids and alkalies but are attacked by some aromatic and

---

*Registered trademark of General Electric Company.

chlorinated solvents. This polymer also has low water absorption, and good electrical properties are maintained over a wide range of humidity and temperature conditions. A number of grades are available, all of which can be easily processed on conventional thermoplastic molding equipment. Applications include computers, connectors, fuse blocks, relays, and busbar insulation. General Electric supplies this polymer.

Polyphenylene Sulfide

Polyphenylene sulfide (PPS) is a semicrystalline material whose main features are its high temperature stability, chemical resistance, flame-retardant nature, and good electrical properties. The structure that characterizes this polymer is

$$\left[\!\!\begin{array}{c}\\ \bigcirc\!\!-\!\!S\end{array}\!\!\right]_n$$

The polymer is produced by the reaction of *p*-dichlorobenzene with sodium sulfide. In its neat form, PPS is sold as a powder which is primarily used as a coating resin. Two forms of PPS are available, a linear and a branched polymer. The former has better strength and melt viscosity properties. PPS is inert to all solvents except hot nitric acid. It is affected by chlorinated and fluorinated hydrocarbon solvents. PPS is processed like most thermoplastics and can be injection- and compression-molded. Cross-linkable grades are available with exceptional heat resistance. The PPS polymers are rated UL-94 V-O. When reinforced with glass fibers, PPS has a 200°C continuous-use temperature rating.[18] The polymers have stable electrical properties over a broad temperature range, frequency, and humidity. Both volume and insulation resistance are excellent in wet and dry environments, and the arc resistance is good. PPS exhibits low flame spread and has less smoke generation than halogenated or other aromatic polymers. It has a high NASA fire safety rating. Its glass transition temperature T_g is 88°C, and its melting temperature T_m is 285 to 290°C. PPS is quite fluid at normal melt processing temperatures and easily wets glass fibers, permitting high loading of reinforcements. Mold shrinkage values for PPS are between 0.1 and 0.2 percent. PPS is difficult to metallize. The primary uses of PPS resins are in sockets and electrical connectors where dimensional stability is of prime importance, coil forms, bobbins, yokes, terminal boards, switches, and surface-mounted component housings where it must withstand vapor-phase soldering. PPS has also been used to encapsulate integrated circuits. Suppliers include Hoechst Celanese, Phillips, Miles, and General Electric.

Styrenics

The styrene polymers include acrylonitrile-butadiene-styrene (ABS), acrylic-styrene-acrylonitrile (SA), polystyrene (PS), styrene-acrylonitrile (SAN), styrene-butadiene (SB), styrene-maleic anhydride (SMA), and high-impact polystyrene (HIPS). The properties of these polymers are dependent on the ratio of the monomeric components and lead to a broad range of material properties. The structural unit common to all these polymers is

$$\left[\begin{array}{c} -CH-CH_2- \\ | \\ C_6H_5 \end{array} \right]_n$$

However, other comonomers used will contribute their structure to the overall polymer repeat unit. All these polymers are amorphous thermoplastics. The ABS polymer has good impact resistance, rigidity, and high gloss. ABS resins have good dielectric strength and arc resistance for many electrical applications. The electrical properties of ABS are not affected by temperature or humidity. It is blended with polyvinyl chloride (PVC) to yield a flame-retardant material which is required in most electrical applications. The chemical resistance of the styrenic resins is poor, and they are attacked by halogenated solvents, aromatic hydrocarbons, esters, and ketones. These polymers are generally not used in electrical applications except as electronic housings, e.g., battery cases. PS has very low dielectric losses ($e' = 2.45$, $\tan d = 0.001$), and its chief use in electronics is in strip lines. The styrenics are amenable to all forms of thermoplastic processing. Suppliers include Dow, BASF, Monsanto, Amoco, and Thermofil.

Polysulfones

Polysulfones are high-temperature-resistant amorphous polymers. These materials contain the arylsulfone group as part of their chemical structure.

$$\left[-C_6H_4-SO_2- \right]_n$$

The family of sulfone polymers consists of polysulfone, polyarylsulfone (PAS), polyethersulfone (PES), and polyphenylsulfone (PPSO). The sulfone polymers are prepared via solution polycondensation of suitably substituted chloroaromatic and hydroxy aromatic compounds. The characteristic features of the sulfone polymers are thermal stability, stiffness, and good electrical properties. The properties of each member of the sulfone family are very similar, and the differences are due to

TABLE 3.17 Thermal Properties of Polysulfones

Material	T_g, °C	Heat distortion temperature at 264 lb/in^2, °C	Continuous-use temperature, °C
Polysulfone	190	174	145
Polyarylsulfone	220	204	190
Polyethersulfone	225	203	180
Polyphenylsulfone	288	274	>220

their glass transition temperatures. The thermal properties of the sulfone polymers are shown in Table 3.17. They have high heat deflection temperatures, excellent dimensional stability, and excellent creep resistance as well as a good combination of electrical properties. Dielectric strength and volume resistivity are high, and dielectric constant and dissipation are low, which are fairly constant over a wide temperature range. All the sulfone polymers have low flammability and can be rendered UL-94 V-O with little or no fire-retardant addition. The oxygen index values of the polymers are 32 percent for polysulfone, 34 percent for polyethersulfone, and 33 percent for polyarylsulfone. These materials are resistant to acid and alkaline hydrolysis but can exhibit stress crazing when exposed to certain chemicals (esters, ketones, chlorinated solvents, and some hydrocarbons). All the polymers have excellent radiation resistance. Processing is carried out on standard thermoplastic injection molding and extrusion. Polysulfones can be plated with copper and/or nickel. Blends with other polymers are available. Applications include printed wiring board substrates, television components, coil bobbins, connectors, and switch housings. The supplier is Amoco Chemical Corp.

Vinyls

The vinyl polymers include polyvinyl chloride (PVC), polyvinylidene chloride (PVDC), and chlorinated polyvinyl chloride (CPVC). The vinyl resins contain the structural unit —(—CH_2—CHCL—)— in the polymer chain although PVDC and CPVC contain extra chlorine atoms. PVC is produced by the free-radical addition polymerization of vinyl chloride. CPVC is produced by the postchlorination of PVC, and PVDC is produced by the free-radical addition polymerization of vinylidene chloride monomer. The vinyl polymers do not have high heat resistance and are used at temperatures of 80°C or less. All the resins show flame resistance, with PVC being the weakest member of the family with regard to this property. PVC is the most versatile of the plastics because of its wide blending capability. CPVC has excellent chemical resistance, rigidity, strength, and weatherability. PVDC has low permeability to gases and liquids, good barrier properties, and good chemical resistance. Two other products are made from PVC: organosols and plastisols. Although these products are not used in the electrical and electronics industry, they are used as coatings in a variety of applications. An *organsol* is a dispersion of PVC in an organic solvent such as an aliphatic hydrocarbon, while a *plastisol* is a dispersion formulated with plasticizers. Other ingredients are also present depending on the desired final properties of the coating. A special class of PVC resin of fine particle size (called *dispersion-grade resin*) is used to produce these coating products. All the vinyl resins are processed on conventional thermoplastic equipment. Electrical applications include wire insulation, cable jackets, sleeving, and tubing. Suppliers include B. F. Goodrich, OxyChem, Dow, and Vista Chemical.

Cross-Linked Thermoplastics[19]

Overall property enhancement is the underlying principle for the development of cross-linked thermoplastics. This enhancement manifests itself in improved resistance to thermal degradation of physical properties, improved resistance to environmental stress cracking, and improved resistance to creep. Thermoplastics are cross-linked by irradiation and/or chemical techniques.

Radiation cross-linking

For the polyolefin materials, the crystalline and amorphous content of the material is changed when it is irradiated. Low-density polyethylene, when irradiated, gives a product which is entirely amorphous from a material which has 50 to 60 percent crystalline regions before irradiation. To create this property enhancement, the irradiation process must be closely controlled. In general, when polyolefins are irradiated with x rays, gamma rays, or high-energy electrons, large changes occur in the physical properties of these polymers. For example, uncontrolled exposure to radiation at high doses can result in the liberation of hydrogen and low-molecular-weight alkanes such as methane, ethane, and propane. The polymer becomes less soluble and can harden and become brittle. The beneficial effects of irradiation of polyolefins usually occur from 0 to 150 Mrads and preferably from 0 to 60 Mrads. At these doses the radiation can be better controlled, to achieve the desired level of cross-linking necessary for property enhancement.

Chemical cross-linking

The cross-linking of adjacent macromolecular chains can also be accomplished by chemical means with the use of peroxides, which are a source of free radicals. Organic peroxides are good sources of free radicals because the peroxides are stable until heated and because their decomposition is temperature-controlled. After the peroxide is mixed with the polymer, the material is then shaped and heated to decompose the peroxide to generate the radicals, which initiates the cross-linking process in the molten state. The material is then cooled to lock in the desired shape.

The cross-linking process is described in Fig. 3.5. Typical polymers capable of being cross-linked include polyolefins, fluoroplastics, vinyls,

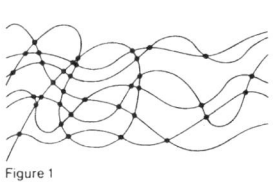

Figure 1

Thermoplastic materials are composed of extremely long, very thin molecules in a random arrangement. The strength of such materials depends upon the distance between its molecules, and the crystalline nature of its molecular structure. Figure 1 schematically illustrates the molecular structure of a thermoplastic material. The crystals formed where the molecules come close together are represented by dots. It is these crystals which provide most of the strength of the material.

As the material is heated, these crystals disappear. The molecules can then slip past each other easily and the material flows. While in this heated condition, the material may be formed into almost any desired shape. Then, when the material subsequently is allowed to cool, the crystals reform and again provide substantial strength to retain the plastic in the shape in which it has been formed.

With the advent of atomic energy, the important discovery was made that the exposure of some plastic materials to high-energy penetrating radiation can cause the permanent crosslinking, or intermolecular joining, of adjacent molecules. This crosslinking results in the chemical bonding of the plastic structure into a new three-dimensional network.

Figure 2 illustrates the molecular structure of such a system after exposure to radiation, with the crosslinks shown as heavy lines.

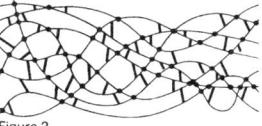

Figure 2

Once the material has been crosslinked, it will not flow at any temperature. When the material is heated, the crystals still disappear as before, but it will no longer flow or change shape because the crosslinks act as ties between the molecules. The crosslinked structure, however, is elastic. Thus, when it is heated to a temperature where the crystals have melted, the material behaves like rubber.

The unique heat-shrinkable properties of Raychem products result from the exposure of special thermoplastic formulations to radiation. Because of the resulting crosslinking, Thermofit* products have perfect elastic memory. These products are supplied in a deformed or expanded condition. When heated, they will shrink and tightly cover the object over which they have been placed. They are ideal for covering a variety of electrical and electronic components, as well as wires, lugs, terminals and connectors.

In manufacturing Thermofit products, Raychem fabricates its compounds into their final form and then subjects them to high energy radiation, thus permanently "freezing" them into the desired shape. The following illustrations demonstrate what happens to the molecular structure of Thermofit tubing during subsequent stages of manufacture and during application. Next to each illustration is an end view of a piece of heat-shrinkable tubing.

Figure 3 is an enlarged schematic view of a very small crosslinked section of extremely long molecules.

Figure 3

Once the tubing has been crosslinked, the next step in imparting elastic memory is to heat the compound above its crystalline melting point. The molecules are then tied together only by the crosslinks as shown in Figure 4.

Figure 4

While hot, the tubing is deformed by applying pressure, thus stretching the crosslinked molecule, see Figure 5.

Figure 5

While in this deformed position, the tubing is cooled; the crystals then reappear, thereby locking the structure together in this deformed condition indefinitely. This is the form in which Thermofit tubing is supplied to customers. (Figure 6).

Figure 6

The customer then heats the Thermofit tubing, melting the crystals. The crosslinks allow the material to return to its original shape as shown in Figure 7 below. This is the perfect elastic memory of Thermofit.

Figure 7

After cooling, the crystals reform and the tubing is locked in its recovered form, as shown in Figure 8.

Figure 8

Upon subsequent reheating, no further change in shape will take place, unless mechanical force is applied.

Figure 3.5 Radiation cross-linking of thermoplastics. (Reprinted with permission from the Raychem Corp.)

elastomers, and silicone resins. Cross-linked thermoplastics are used in a variety of electrical and electronic applications including wire and cable insulation, interconnection devices, identification tags, heat-shrinkable tubing and encapsulant gels, splice closures, and overall environmental protection of sensitive electronic products. A general list of cross-linked plastics and their properties and applications is shown in Table 3.18 on pp. 82–84.[20]

References

1. R. Juran, ed., *Modern Plastics Encyclopedia*, McGraw-Hill, New York, 1989.
2. "What Readers Said about Fillers and Reinforcements," *Plastics Compounding*, November/December 1992, p. 46.
3. C. A. Harper, ed., *Handbook of Materials and Processes for Electronics*, 2d ed., McGraw-Hill, New York, 1993.
4. R. Greene, ed., *Modern Plastics Encyclopedia*, McGraw-Hill, New York, 1992.
5. H. F. Mark, N. M. Bikales, C. G. Overberger, and G. Menges, eds., *Encyclopedia of Polymer Science and Engineering*, vols. 1 to 17, 2d edition, Wiley, New York, 1985–1990.
6. S. Schwartz and S. H. Goodman, *Plastics Materials and Processes*, Van Nostrand Reinhold, New York, 1982.
7. R. Greene, "A Guide to Plastics," *Modern Plastics*, January 1993.
8. C. A. Harper, ed., *Electronic Packaging and Interconnection Handbook*, McGraw-Hill, New York, 1991.
9. C. P. Wong, ed., *Polymers for Electronics and Photonic Applications*, Academic Press, San Diego, CA, 1993.
10. M. G. Broadhurst, S. Edelman, and G. T. Davis, "Piezo and Pyroelectric Applications of Plastics," *Org. Coat. and Plastics Chem.*, vol. 42, 1980.
11. Tai-Shung Chung, "Recent Developments of Thermotropic Liquid Crystalline Polymers," *Polymer Engineering and Science*, vol. 26, no. 13, July 1986, p. 901.
12. "PW Resin Profile," *Plastics World*, July 1992, p. 57.
13. A. J. Klein, "Liquid Crystal Polymers Gain Momentum," *Plastics Design Forum*, January/February 1989.
14. T. Stevens, "The Unusual World of Liquid Crystal Polymers," *Materials Engineering*, vol. 102, no. 1, January 1991.
15. *Properties of Nomex*, du Pont Co. bulletin, H-22368 to H-22375, Wilmington, DE, March 1990.
16. *Kapton Summary of Properties*, du Pont Co. bulletin, no. 231302A, Circleville, OH, August 1993.
17. L. Leonard, "Back to Basics: Polycarbonate," *Plastics Design Forum*, January/February 1993, p. 52.
18. J. Gagne, "PW Resin Profile," *Plastics World*, November 1991, p. 97.
19. S. H. Goodman, *Handbook of Thermoset Plastics*, Noyes Publications, Park Ridge, NJ, 1986.
20. Raychem Corporation, technical bulletins entitled *Inventive Solutions for Electronics*; Electronics Product Catalogue, August 1993; Heat Shrinkable Products Catalogue, August 1993.

TABLE 3.18 Cross-Linked Thermoplastics

Product	Typical applications	Basic properties	Operating temperatures, °C	Plastic type
Versafit	Strain relief; wire bundling	Low shrink temperature, quick shrink, very flexible. Highly-flame retardant	−30− +125	Polyolefin
RNF-150	Wire bundling; abrasion protection. High-temperature applications	High strength, very flexible. Good chemical resistance. High temperature. Flame-resistant	−55− +150	Polyolefin
Kynar*	Strain relief; abrasion protection. Applications requiring fuel resistance. High-temperature applications	High strength, very thin wall. Resistant to hydrocarbons, acids, and bases	−55− +175	Fluoroplastic
NT		Flexible modified elastomer	121	Elastomer
RNF-100	Strain relief; identification. Wire bundling	Abrasion-resistant; good solvent resistance; flexible. Flame-retardant	−55− +135	Polyolefin
CRN	Strain relief; component protection	Tough, semirigid	−55− +135	Polyolefin
RT-3	Terminal insulation; strain relief; semiautomated production	Tough, semirigid. Suitable for semiautomated production	−55− +135	Polyolefin
SFR		Very flexible, flame-retardant silicone elastomer	180	Silicone

MIL-LT	Strain relief, wire bundling. Covering temperature-sensitive components; hot-stamp identification	Very low shrink temperature; highly flexible. Flame retardant	−55– +135	
LSTT	Covering temperature-sensitive components. Strain relief	Extremely low shrink temperature	−45– +105	
DWTC	Abrasion and environmental protection of markers. Protection of brake lines	Adhesive-lined. Extremely low shrink temperature. Highly abrasion-resistant. Very flexible. Excellent clarity	−55– +75	Polyolefin
VTW	Strain relief; wire bundling	Low shrink temperature, very quick shrink, extremely flexible	−30– +125	Polyolefin
PD-Caps	Encapsulation of stub splices	End cap. Semirigid, meltable inner wall	−55– +110	Polyolefin
TAT-125	Sealing and protecting in-line splices and bimetallic joints. Applications requiring environmental sealing and flexibility	Adhesive-lined, flexible, thin wall. Flame-retarded jacket (colors)	−55– +110	Polyolefin
MVT	Environmental and mechanical protection of wire splices and components	Adhesive-lined, semirigid. Good chemical, abrasion resistance	−40– +121	
ATUM	High shrink ratio for connector-to-cable transitions. Environmental sealing of splices and components	Adhesive-lined, flexible, medium wall, high-shrink ratio. Flame-retarded jacket	−55– +110	

---

*Trademark of Pennwalt Corp. All other products are made by Raychem Corp.

TABLE 3.18 Cross-Linked Thermoplastics (*Continued*)

Product	Typical applications	Basic properties	Operating temperatures, °C	Plastic type
SCL	Encapsulation of components, terminations, and splices. Moisture resistance; strain relief	Semirigid, meltable encapsulant inner wall	−55– +110	Polyolefin
Viton	Tubing	High-temperature, solvent-resistant	200	Fluoroelastomer
HCTE	Tubing, conduit	Irradiated for high chemical resistance	200	Ethylene tetrafluoroethylene

Chapter 4

Thermosets

Unlike thermoplastics, thermoset materials are polymers that form a three-dimensional cross-linked network of polymer chains which cannot be softened or reheated for additional use. In general, these materials can provide higher temperature capability than the thermoplastic materials. Thermoset materials, before they are cured, are fabricated by casting, compression molding, filament winding, laminating, pultrusion, injection, and transfer molding. Most thermoset materials, before they are cross-linked, are considerably more fluid than thermoplastics during processing. A thermoset material must contain a functionality greater than 2 to facilitate cross-linking; i.e., the polymer chain must have enough reactive sites to form a three-dimensional network. Difunctional materials form linear or branched uncross-linked polymers but can be cross-linked by forming more reactive sites by the addition of either a catalyst or a curing agent. These ingredients promote the formation of active sites for further reaction. The curing and cross-linking reaction, is an exothermic reaction, and consideration must be given to control the temperature rise to prevent a runaway reaction. Thermoset materials usually shrink when they are cross-linked, but the shrinkage can be controlled with additives such as fillers and reinforcing fibers and/or fabrics. The conversion of these materials to the thermoset state can be accomplished at room or elevated temperature, with the latter giving a faster and more complete cure of the resin. It is common with some thermoset resins but not all to form a prepolymer. Prepolymerization is a method of increasing the molecular weight of the resin to some intermediate value as a means of controlling resin properties during processing. Prepolymerization should not be confused with B-staging, another process used with thermoset resins. The latter involves less control over molecular weight and produces some cross-linking. The former is usually carried out under more controlled conditions to yield specific molecular weight

TABLE 4.1 Properties of Thermosetting Plastics

	DAP (GDI-30)	Epoxy Glass-filled	Epoxy Mineral-filled	Phenolics General purpose	Phenolics Glass-filled	Phenolics Mineral-filled	Alkyds MAG	Alkyds MAI-60	Polyester GPO-3	Polyimide	Polyurethane	Mineral-filled silicone	Bismaleimides	Cynate esters	Benzocyclobutene
Dielectric constant, D-150															
60 Hz	4.2	5.0	4.0	12.0	50.0	6.0	6.3	5.6	4.5	3.5	6	3.6	—	—	—
10^6 Hz	3.5	4.6	5.0	6.0	6.0	10.0	4.7	4.6	—	3.4	3	3.7	3.5	2.66–3.10	2.65–2.70
Dissipation factor, D-150															
60 Hz	0.004	0.01	0.01	0.3	0.3	0.07	0.04	0.10	0.05	0.0025	0.1	0.005	—	—	—
10^6 Hz	0.01	0.01	0.01	0.7	0.8	0.10	0.02	0.02	—	0.01	0.04	0.003	0.007	0.01–0.005	0.0008–0.002
Dielectric strength, D-149, V/mil	400	360	400	400	350	400	400	375	300	6500	500	425	480–508	—	—
Volume resistivity, D-257, Ω · cm	10^{13}	3.8×10^{15}	9×10^{15}	10^{13}	10^{13}	10^{14}	10^{14}	10^{13}	—	10^{18}	10^{14}	10^{15}	—	10^{16}	9×10^{19}
Arc resistance, D-495, s	140	140	180	50	70	180	>180	180	>180	230	120	240	—	Excellent	—
Specific gravity, D-792	1.7	1.8	2.1	1.45	1.95	1.83	2.24	2.07	1.95	1.4	1.1	2.05	1.30	1.10–1.43	—
Water absorption, D-570, % 24 h	<0.2	0.2	0.04	0.7	0.5	0.5	0.08	0.07	0.5	2.9	0.2	0.15	4.0–4.4*	0.6–2.5*	<0.1
Heat deflection temperature, D-648, at 264 lb/in², °F	500	400	250	340	400	500	350	>400	—	680	190	>500	520	480	—
Tensile strength, D-638, lb/in²	10,000	30,000	15,000	10,000	7000	11,000	3000	6000	9000	17,000	1000	6500	12,000	13,000	—
Impact strength (Izod), D-256, ft · lb/in	5.0	10	0.4	0.3	3.5	15.0	0.3	9.5	8.0	1.5	25	0.5	0.3–0.5	0.7–0.9	—
Coefficient of thermal expansion, D-696, 10^{-5}/°F	2.6	1.7	2.2	2.5	—	0.88	3	2	2	2.8	25	2.8	40 ppm/°C	50 ppm/°C	42–70 ppm/°C
Thermal conductivity, C-177, Btu · in/h · in · ft² · °F)	—	6	—	0.3	0.34	0.2	7.2	3.6	4	6.8	0.1	3.1	—	—	—

*500-h water boil

SOURCE: Ref. 5.

materials with little to no cross-linking. The reader is referred to other texts for more detailed descriptions of all thermosetting plastics.[1-4] Properties of thermosets for electrical and electronic applications are given in Table 4.1, and general characteristics of the resins are given in Table 4.2. Fillers are also incorporated into thermoset resins, as they are in thermoplastics. See Table 3.1 for a list of fillers that are applicable to both thermoplastics and thermoset resins.

Allyl Resins

The allyl resins are thermosetting polyester materials which retain their desirable physical and electrical properties on prolonged exposure to severe environmental conditions such as high temperature and humidity. These resins have good chemical resistance and can withstand between 10^4 and 10^{12} rads of gamma radiation.[4] These polymers have the allyl radical (CH_2—CH=CH_2) as part of their chemical structure. The principal allyl resins are based on diallyl phthalate (DAP) and diallyl isophthalate (DAIP) monomers and prepolymers. There are other resins that are used alone or in combination with DAP and DAIP. They are diethylene glycol bis(allylcarbonate), allyl methacrylate, diallyl fumarate and maleate, and triallyl cyanurate. The allyl resins are converted to thermoset materials by heat and by the addition of free-radical sources such as benzoyl peroxide and t-butyl perbenzoate to the resin formulation. Curing of these resins is slow below 150°C but proceeds rapidly above this temperature. No volatiles are generated during cure of these resins. These resins are used as cross-linking agents in other polyester systems and as molding compounds, preimpregnated glass cloth, sealants, insulating coatings, and decorative laminates. Most critical electronic applications requiring high reliability under adverse conditions use allyl resins, e.g., connectors in communications, computer, and aerospace systems; insulator switches; chip carriers; and circuit boards. The allyl resins have low loss factor, high volume and surface resistivity, and high arc resistance, and these properties are retained under high-humidity conditions. The allyl resins can be compression-, transfer-, and injection-molded and can be used in prepregging operations. In general, DAP compounds are designed for continuous operation at about 176°C, while DAIP can operate at about 232°C.

Fillers are also incorporated in allyl resins with glass fibers yielding molded products with the best combination of properties. Long glass fibers provide better impact strength, and glass fibers also give the highest shock and arc resistance in DAP materials.[4] The best electrical properties under high-humidity conditions with allyl resins are obtained with acrylic fiber as the filler. Polyester fibers impart impact

TABLE 4.2 General Characteristics of Thermosetting Plastics

Material	Key advantages	Limitations	Processing*	Applications
Alkyds	Good dimensional stability. Very good dielectric properties. Temperature resistant to 300°F in continuous use	Low impact strength. Not resistant to high humidities	1, 2	Molded electrical parts
Allyl idglycol carbonates	Very high transparency. Stability of optical properties under load, heat, and many chemical environments. Good dimensional stability and radiation resistance	Available only in cast stock shapes or machined parts. High cost	6	Optical transparent products
Diallyl phthalates	Excellent dimensional stability. Retention of electrical properties at high temperature and humidity. Excellent resistance to moisture, acids, alkalies, and solvents. Self-extinguishing. Colors remain stable at high temperatures	High cost	1–5	Connectors terminals, insulators, circuit boards, switches
Epoxies	Excellent strength and toughness. Outstanding adhesion to many other materials. Low power factor and high dielectric strength. Good resistance to many acids, alkalies, and solvents. Versatility and ease of processing	Low thermal and oxidative stability	1–10	Printed wiring boards, insulator, switch gear components
Melamines	Very high hardness. Resistant to detergents, water, and staining. High arc tracking resistance. Permanency of color and molded in designs. Self-extinguishing	Fair dimensional stability. Low impact strength	1–4	Electrical parts

Resin	Properties	Limitations	Processing*	Applications
Phenolics	Temperature-resistant to 300°F (some to 600°F). Outstanding resistance to deformation under load. Dimensional stability over wide temperature range. Resistant to common solvents, weak acids, and many detergents. Low cost and ease of processing	Color limitations. Low impact strength	1–4, 7	Commitators, connectors, coil bobbins, switches
Benzocyclobutenes	High glass transition temperature (>350°C). High temperature resistance, low water absorption, good planarization and chemical resistance. Low dielectric constant and dissipation factor	Poor etchability	9–11	Interconnect dielectric
Cyanate esters	Excellent dielectric properties. Very good adhesion, dimensional and thermal stability. T_g values of 250–290°C. No volatiles during cure	Tend to be somewhat brittle	1, 2, 4	Printed wiring board substrates and radome structures
Bismaleimides	No volatiles during cure. High temperature resistance. Chemical resistance	Attached by strong base. High shrinkage during cure	1, 2, 4	Printed-wiring board
Parylene	Ultrathin coatings. Excellent electrical properties. Highly resistant to organic solvents. Excellent moisture barrier	Requires special vacuum deposition equipment	12	Conformal coatings and insulation
Polybutadienes	Low dielectric constant and dissipation factor. High dielectric strength. Good temperature stability and chemical resistance	Elevated processing temperature required		
Polyesters	Good strength and rigidity. Versatility and ease of processing. Special grades display good weather resistance, very good chemical resistance, and flame retardancy. Good dielectric properties. Low cost		1–5, 8	

*1 = Compression molding, 2 = transfer molding, 3 = extrusions, 4 = laminates, 5 = injection molding, 6 = casting, 7 = filament winding, 8 = matched die moldings, 9 = dipping, 10 = spray coating, 11 = spin coating, 12 = vapor deposition, 13 = reaction injection molding.

TABLE 4.2 General Characteristics of Thermosetting Plastics (*Continued*)

Material	Key advantages	Limitations	Processing*	Applications
Polyurethanes	Very flexible and fatigue-resistant. Excellent abrasion resistance. Very high tear strength. Good chemical and solvent resistance. Very resistant to oxygen aging. Temperature-resistant to 300°F in continuous use		13	
Silicones	Retention of mechanical and dielectrical properties at very high temperature and humidity. Self-extinguishing	Very high cost	1, 2, 4, 6, 9	

*1 = Compression molding, 2 = transfer molding, 3 = extrusions, 4 = laminates, 5 = injection molding, 6 = casting, 7 = filament winding, 8 = matched die moldings, 9 = dipping, 10 = spray coating, 11 = spin coating, 12 = vapor deposition, 13 = reaction injection molding.

resistance and strength in thin sections, and nylon-filled resins yield high durability. Cellulosics and mineral fillers (calcium carbonate, silicate, and clays) are used to reduce cost.

The stable electrical properties of allyl resins are outstanding. The insulation resistance of DAPs after 4000-h exposure at 70°C and 95 percent relative humidity is stable at about 10^{11} to 10^{12} Ω. The dielectric strength values for DAP and DAIP are good up to 190 and 205°C, respectively. The dielectric constant decreases from 3.6 to 3.2 at 25°C as a function of frequency from 60 to 10^8 Hz.[4] Suppliers include Royers Corp., OxyChem, and Cosmic Plastics.

Bismaleimides

Within the polyimide family of resins there is a class of thermosetting polymers that have a preimidized structure and form a three-dimensional network via addition polymerization without the evolution of volatile material. These materials are classified as bismaleimides (BMIs), and the monomers and prepolymers are prepared by the reaction of maleic anhydride and diamines. The chemical unit present in these polymers is

$$\left[\underset{O}{\underset{\parallel}{\overset{O}{\overset{\parallel}{\bigcirc}}}}\!\!\!N\!-\!R\!-\!N\!\!\!\underset{O}{\underset{\parallel}{\overset{O}{\overset{\parallel}{\bigcirc}}}} \right]_n$$

The material is very reactive and can be homopolymerized or copolymerized to produce a wide variety of thermosetting resins. The bismaleimide resins need cure temperatures of 175 to 232°C for several hours and then they are postcured at 232°C. The fully cured resins have T_g values above 260°C. The polymers are characterized as having the processing ease of epoxy resins but superior elevated-temperature performance properties. Epoxies operate in the 150°C temperature range while the BMIs operate from 200 to 232°C. Compression, transfer, injection molding, filament winding, and prepregging are the normal processing methods for bismaleimides. Bismaleimides are resistant to acids and common organic solvents but are attacked by strong alkali and methylene chloride. Bismaleimides are sold as powders or as solutions in polar solvents. These materials are primarily used in printed-wiring board substrates. Suppliers include Shell and Ciba-Geigy.

Epoxy Resins

Epoxy resins in the uncured state are characterized by the presence of the epoxy (oxirane) ring:

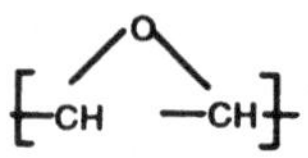

$$\text{-[-CH} \overset{\overset{\displaystyle O}{\diagup \diagdown}}{} \text{CH-]-}$$

Most commercial epoxy resins are derived from bisphenol A and epichlorohydrin, but there are many other types based on the epoxidation of multifunctional molecules that gives rise to epoxy resins with a broad range of properties. Epoxy resins can be liquids or solids. Curing of these resins is accomplished by reaction through the epoxide and hydroxyl functional groups. Curing agent type, curing agent amount, and temperature determine the condition of cure and final properties of the resin. Typical curing agents include the aliphatic amines and amides for ambient temperature cure and anhydrides, organic acids, aromatic amines, and various phenolic condensation products for elevated-temperature cure. Most common epoxy resins are solventless (100 percent solids); however, higher-molecular-weight and multifunc-

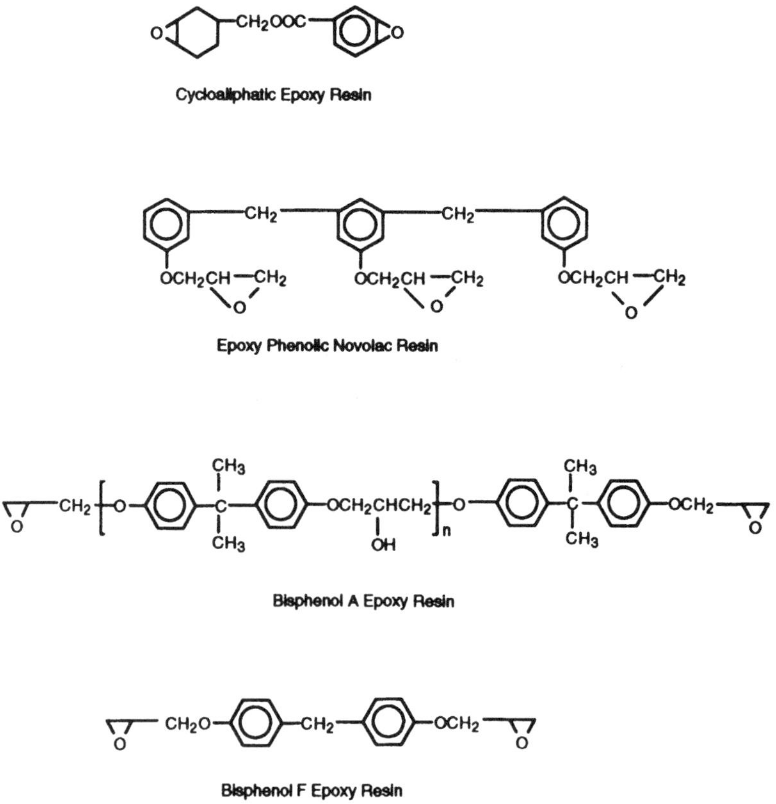

Cycloaliphatic Epoxy Resin

Epoxy Phenolic Novolac Resin

Bisphenol A Epoxy Resin

Bisphenol F Epoxy Resin

Figure 4.1 Uncured epoxy resin structures.

tional epoxies are solid and are usually processed in solution form. The curing reaction is exothermic which may be necessary to control in large batch operations. For close to 40 years epoxy resins have been used as encapsulants for transformers, coils, resistors, and capacitors, and recently some of these resins are being used for the packaging of semiconductor devices such as transistors, integrated circuits (ICs), and large-scale integration devices. The single most important characteristic of epoxy resins is their outstanding adhesion to a variety of substrates and reinforcements. The cured resins form hard, tough, cross-linked networks with an excellent combination of mechanical, electrical, chemical, and thermal properties. While all epoxy resins (uncured) contain the oxirane ring, the composition of epoxy resins varies widely and yields many different types of structures; a few are depicted in Fig. 4.1. Changes in the structure of epoxy resins are made in order to alter the final properties of both the uncured and the cured product. A list of the types of epoxy resins available is given in Table 4.3. There are many more types available for specialized applications. The operating service temperature for the standard bisphenol A type of epoxy resin is about 140 to 150°C. Specialized resins can extend that to 200°C. The chemical resistance of epoxy resins is excellent, and the resins are generally resistant to organic solvents and strong alkalies

TABLE 4.3 Epoxy Resin Types

Resin type	Comment
Diglycidyl ether of bisphenol A	Standard general-purpose resin
Novolacs	High reactivity, high cross-linking density, chemical resistance
Cycloaliphatics	Low viscosity, good weatherability
Hydantoin epoxy	High temperature resistance. High mechanical properties
Tetraglycidyl methylene dianiline	Increased cross-linking density. High temperature resistance
Tetraglycidyl bisphenol A/novolac	High reactivity, high cross-linking density, high temperature resistance
Glycerol-based epoxy	Flexibility
Triglycidyl triphenyl methane epoxy	High reactivity. Performance range of -195 to $+200°C$
Triglycidyl p-aminophenol epoxy	High-temperature properties
Halogenated epoxy	Flame-retardant properties
Silicone epoxy	High temperature. Flame-retardant properties. Flexibility

and are less resistant to strong acids and oxidizing agents. Epoxies can be compression- and transfer-molded and filament-wound. They are used in casting, prepregging, and laminating operations. Epoxies can be formulated to produce conformal coatings, adhesives, varnishes, and encapsulants and are used in the electrical industry as bobbins, connectors, chip carriers, and the matrix resin in printed-wiring board substrates.

Phenolic Resins

Phenolic resins are the reaction product of phenol and formaldehyde. The phenols that are used commercially are phenol, cresols, xylenols, p-t-butyl phenol, p-phenyl phenol, bisphenols, and resorcinol; and the aldehydes are formaldehyde and furfural.[4,6] Two kinds of phenolics are produced: the *resoles* (alkaline condensation products) and the *novolacs* (acid condensation products). The chemical structure of a resole is composed of methylene and ether bridges, as shown here:

HOH₂C—⟨◯⟩—CH₂—⟨◯⟩—CH₂—O—CH₂—⟨◯⟩—CH₂OH

And for a novolac the structure is as follows:

[structure with OH groups and CH₂ bridges between phenol rings]

The basic difference between a resole and a novolac is the presence of one or more free methylol groups on the resole, with the latter requiring additional curing agent to convert to a thermoset resin. The resins are heat-cured to form a dense cross-linked network, which gives the phenolic resins their high heat resistance and dimensional stability. Phenolic novolacs are the most widely used molding compound, and the novolacs are always combined with fillers and/or reinforcements such as mica, clay, cellulose, wood flour, mineral fibers, and chopped fabric. General-purpose phenolic molding compounds have a notched Izod impact strength of 0.3 to 0.5 ft · lb/in at room temperature. Glass-reinforced grades have values from 1 to 1.2 ft · lb/in, and high-impact grades go as high as 3 to 4 ft · lb/in. The phenolic molding resins are useful for operation at 200 to 230°C. Heat deflection temperatures range from 150 to 300°C depending on the reinforcement.[5,7] The phenolic resins have poor arc resistance but have excellent electrical insulating properties. Phenolic resins are inherently flame-resistant and resist organic solvents but are attacked by strong bases and oxidizing agents. Phenolic resins are available in solution form or as a powder, and they can be converted to molding compounds, varnishes, and laminates. They are processed by injection, compression, and transfer molding. Phenolics are used as chip carriers, connectors, bobbins, and matrix resins for printed-wiring board substrates. Suppliers include OxyChem, Plaslok, Plastics Engineering, Rogers Corp., and ICI Fiberite.

Polyesters

Thermoset polyester resins are versatile materials that are available as low-viscosity liquids to thick pastes. These polymers are formed by the reaction of a polyfunctional acid and a polyfunctional alcohol which is then reacted with an unsaturated monomer such as styrene. Unsaturation can also be present in the acid portion of the molecule. A variety of reactants are available to impart different characteristics to the final product. Some of these components are shown in Table 4.4. Fillers, pigments, fibers, and peroxide catalysts are mixed with the resin, and curing is accomplished at room temperature and/or up to 160°C. No volatile by-products are eliminated during cure. The terms *alkyl* and *polyester* are used interchangeably in connection with molding resins. Resin grades available are specified by MIL-M-14G and are defined as follows:

Type MAG	Mineral-filled, good overall dielectric properties and arc resistance
Type MAI-60	Glass-fiber-filled with high impact strength as well as the dielectric and arc resistance properties of type MAG
Type MAT-30	Glass/mineral reinforcement to impart heat and flame resistance, high impact strength and arc track resistance
Type MAI-30	Comparable in properties to type MAT-30 but with improved processability and mechanical properties

The characteristic properties of alkyds include ease of processing, low cost, good electrical properties, and high arc and track resistance (170 to 300 s). The resins are compression-, injection-, and transfer-molded, pultruded; and filament-wound. Bulk and sheet molding compounds are also produced from unsaturated polyester resins. The resins of most interest in the electrical area are the laminate grades defined by NEMA.[8]

Grade GPO-1	General-purpose
Grade GPO-2	Flame-resistant
Grade GPO-3	Flame, arc, and track resistant
Grade GPO-1P	Punch grade of GPO-1
Grade GPO-2P	Punch grade of GPO-2
Grade GPO-3P	Punch grade of GPO-3

TABLE 4.4 Unsaturated Polyester Components

Components	Ingredients	Characteristics
Unsaturated anhydrides and dibasic acids	Maleic anhydride	Lowest cost, moderately high heat deflection temperature (HDT)
	Fumaric acid	Highest reactivity (cross-linking), higher HDT, more rigidity
Saturated anhydrides and dibasic acids	Phthalic (orthophthalic) anhydride	Lowest cost, moderately high HDT, provides stiffness, high flexural, and tensile strength
	Isophthalic acid	Higher tensile and flexural strength, better chemical and water resistance
	Adipic acid, azelaic acid, sebacic acid	Flexibility (toughness, resilience, impact strength); adipic acid is lowest in cost of flexibilizing acids
	Chlorendic anhydride	Flame retardance
	Nadic methyl anhydride	Very high HDT
	Tetrachlorophthalic anhydride	Flame retardance
Glycols	Propylene glycol	Lowest cost, good water resistance and flexibility, compatibility with styrene
	Dipropylene glycol	Flexibility and toughness
	Ethylene glycol	High heat resistance and tensile strength, low cost
	Diethylene glycol	Greater toughness, impact strength and flexibility
	Bisphenol-A adduct	Corrosion resistance, high HDT, high flexural and tensile strength
	Hydrogenated bisphenol-A adduct	Corrosion resistance, high HDT, high flexural and tensile strength
Monomers	Styrene	Lowest cost, high reactivity, fairly good HDT, high flexural strength
	Diallyl phthalate	High heat resistance, long shelf life, low volatility
	Methyl methacrylate	Light stability, good weatherability, fairly high HDT
	Vinyl toluene	Low volatility, more flexibility, high reactivity
	Triallyl cyanurate	Very high HDT, high reactivity, high flexural and tensile strength
	Methyl acrylate	Light stability, good weatherability, moderate strength

The unsaturated polyester resins resist aliphatic and halogenated solvents but are attacked by strong bases, esters, ketones, and some acids. Suppliers include Reichhold, Glastic, Rostone, and Plastics Engineering.

Polyurethanes

These polymers are derived from the reaction of polyfunctional isocyanates and polyhydroxy (polyether and polyester polyols) compounds which yield linear or branched polymers. The basic chemical unit of polyurethanes is the urethane (—RNHCOOR—). These resins are produced as castable liquids (prepolymers) and are cross-linked by adjusting the stoichiometry and functionality of the isocyanate or polyol. Catalysts are added to enhance the rate of reaction. A variety of other ingredients (active hydrogen compounds) can be added to produce polyurethanes with different properties ranging from elastomeric to rigid polymers. The polyether urethanes are more hydrolytically stable than the polyester urethanes, but the latter give better strength and abrasion resistance.

The principal types of polyurethanes are classified by the American Society for Testing and Materials (ASTM) as follows:[8]

Type 1: *One-component urethane alkyds* are formed by the reaction of a diisocyanate with vegetable oils, or their fatty acids, and polyhydric alcohols. They cure rapidly by oxidation in as little as 10 min. Electrical uses include insulation and conformal coatings.

Type 2: *One-component moisture-cured urethanes* are isocyanate prepolymers made by reacting polyols with excess isocyanate. When exposed to humid air, the amino groups initially formed react with more isocyanate, generating urea linkages and releasing carbon dioxide. Films, which must be thin to prevent entrapment of gas, are extremely tough and find use in seamless floors, bowling alleys, gym floors, industrial floors, and insulating coatings.

Type 3: *One-component heat-cured urethanes* contain phenol-blocked isocyanates. When heated to 160°C, the phenol is expelled, and the liberated isocyanate then reacts with a polyol. This type is widely used in magnet wire enamels.

Type 4: *Two-component catalyst-cured urethanes* are similar to type 2 urethanes but are cured with catalysts, such as tertiary amines, forming urea linkages. This type is used for textile finishes and for floor coatings. There are no electrical applications.

Type 5: *Two-component polyol-cured urethanes* are made by reacting isocyanate prepolymers with hydroxyl-terminated polyesters or polyols such as castor oil. This is the principal type used in solvent-base and solventless insulating coatings.

Type 6: Thermoplastic urethanes are the reaction product of isocyanates with polyester or polyether diols. They have high molecular weight and are supplied as lacquers in ketone or ester solvents. They are used as textile coatings to give the "wet look." There are no electrical applications.

The principal advantages of polyurethanes are their high abrasion resistance, low-temperature properties, ambient cure, low cost, foamable prepolymers, and wide compositional latitude. Disadvantages include poor temperature stability and poor weather and solvent resistance; in addition, they are quite flammable, and the isocyanate monomers are toxic. The upper service temperature limit is about 120°C. Methylene chloride will strip polyurethane off substrates. Dielectric properties of polyurethane resins are retained better than those of epoxy resins at the upper end of their useful temperature range.[9]

Polyurethanes have poor solvent resistance. They are sensitive to chlorinated and aromatic solvents as well as to acids and bases. Urethanes are used as conformal coatings to encapsulate sensitive electronic components and as jacketing for cable and wire insulation. They are processed by reaction injection molding (RIM), compression molding, and casting.

Silicones

Silicones are polymers that consist of alternating atoms of silicone and oxygen along the backbone of the polymer chain. The backbone is modified by attaching organic side groups to the silicone atom and in so doing imparts the unique properties found in these polymers. The organic group attached to the silicone atom in the polymer chain can be aromatic, aliphatic, or vinyl. The types of groups and the amount of organic substitution determine whether the resin is a liquid or solid in its B-stage condition as well as the nature of the cured polymer. The silicone polymers are significantly different from other plastics because the polymer backbone contains no carbon. The structural unit of the silicone polymers is depicted as follows:

$$\left[-O-\underset{\underset{R}{|}}{\overset{\overset{R}{|}}{Si}}-O-\underset{\underset{R}{|}}{\overset{\overset{R}{|}}{Si}}- \right]_n$$

The silicone fluids are low-molecular-weight polymers where the organic group on the silicone is methyl, phenyl, or a mixture of both. The silicone resins are branched polymers that cure to a solid, while the elastomers are linear oils or higher-molecular-weight silicones that are reinforced with a filler and then vulcanized (cross-linked). The elastomers come in three forms: heat-cured rubber, two-component liquid injection molding compounds, and room-temperature vulcanizing (RTV) products. The conversion of silicones to cross-linked elastomers can be accomplished by free-radical condensation, addition, and ultraviolet radiation curing techniques. The silicones are characterized by their useful properties over a broad temperature range (−65 to 248°C). They exhibit excellent weatherability, arc and track resistance, and impact, abrasion, and chemical resistance. Silicones can also be copolymerized with other polymers to produce materials with a variety of interesting properties, e.g., silicone-polyimide, silicone-EPDM, and silicone-polycarbonate. Electrical applications include wire enamels, laminates, sleeving and heat-shrinkable tubing, potting of electronic components, liquid dielectrics, conformal coatings, and varnishes.

Fillers include glass fibers, mica, aluminum, and fumed silica. Carbon blacks are not used with silicone polymers at all. The fumed silica is considered to be the most effective reinforcing filler for silicones. These polymers have property profiles that remain fairly constant over a wide temperature range (−50 to +200°C). Because of the weak intermolecular attractions the silicone elastomers have low tensile strengths (400 to 1200 lb/in^2), low T_g (−120°C), and low surface energy. The silicone polymers have excellent weathering resistance. Silicone polymers are attacked by strong oxidants, chlorinated solvents and strong bases, and aromatic solvents.

Cyanate Ester Resins

Cyanate ester resins are bisphenol derivatives containing the cyanate (—O—C≡N—) functional group. Upon heating, these monomers and prepolymers cyclotrimerize to form a cross-linked network of oxygen-linked triazine rings without the evaluation of volatiles. The structural feature of these resins is the triazine ring flanked by ether linkages. These structural features impart thermal stability and some flexibility. This structural feature is illustrated below.

The cyanate ester resins range from liquids to solids and are characterized by superior dielectric properties, adhesion, low moisture absorption, flame resistance, high-temperature capability, and excellent dimensional stability. Glass transition temperatures range from 192 to 290°C. Several grades are available and can be formulated to produce laminating varnishes for impregnating inorganic and organic reinforcements. The formulations can be homopolymers, blends with other cyanate esters or with bismaleimides, and epoxy resins. Some properties of the neat resins are shown in Figs. 4.2 to 4.5 and of E-glass laminates in Tables 4.5 and 4.6.

For the electronics area, the cyanate ester resins have low dielectric loss properties, dimensional stability at molten solder temperature, and excellent adhesion. The chemical resistance of the cyanate ester resins is good, except caustic will attack the homopolymer resin. Underwriters Laboratories has rated the standard cyanate ester resin designated AROCY B at 162°C for 25,000 h.[10]

Cyanate ester resins can be processed by melt polymerization, prepregging, and lamination operations. Applications in the electrical industry include printed-wiring board substrates and radome structures. Cyanate ester resins can be toughened with thermoplastics such as polyethersulfone, polyetherimide, polyarylates, polyimides, and methylethylketone-soluble copolyesters and elastomers. Shell Chemical Co. supplies these resins.

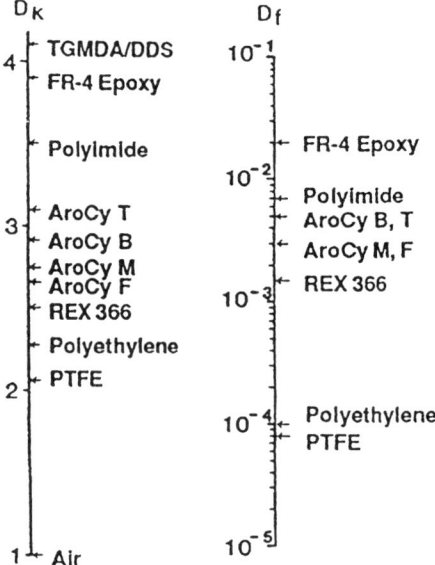

Figure 4.2 Comparison of values of dielectric constant and dissipation factor measured at 25°C and 1 MHz for representative thermoset and thermoplastic polymers. (Source: Ref. 10. Reprinted with permission.)

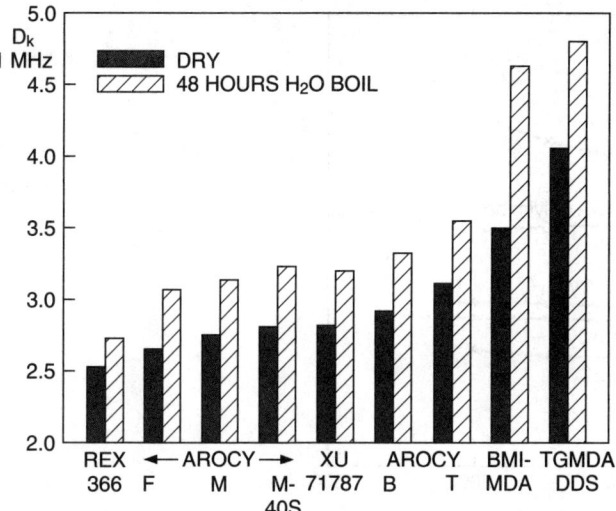

Figure 4.3 Effect of moisture conditioning on the dielectric constant of several thermoset resins. (Source: Ref. 10. Reprinted with permission from Ciba-Geigy Corp.)

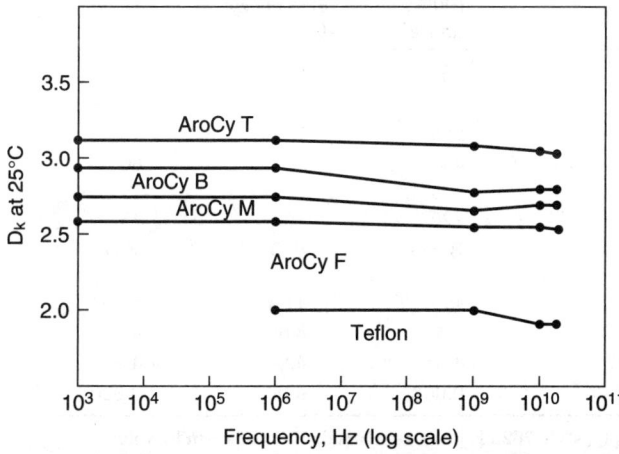

Figure 4.4 Flat dielectric-constant response of cyanate ester homopolymers to increasing test frequency. (Source: Ref. 10. Reprinted with permission.)

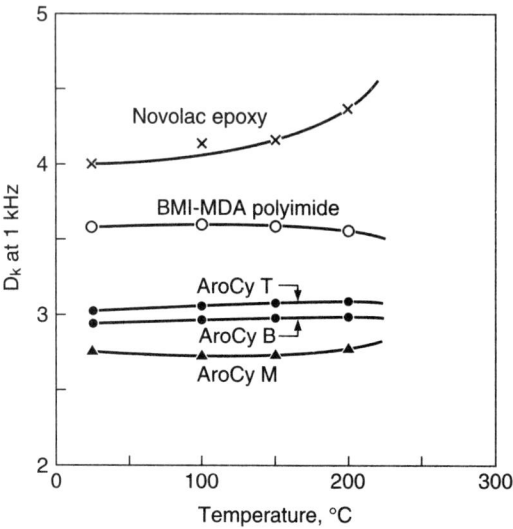

Figure 4.5 Flat dielectric-constant response of cyanate ester homopolymers over 25 to 200°C range. Epoxy novolac is the reference resin. (Source: Ref. 10. Reprinted with permission.)

TABLE 4.5 Comparison of AroCy B-40S Laminate* Properties with 60% Epoxy Modification and FR-4 Reference

Laminate property	100% cyanate†	60% epoxy‡ 40% cyanate†	100% epoxy‡
Press cure, h/°C	1/177	1/177	1/177
Postcure, h/°C	3/225	—	—
T_g (TMAS), °C	225	183	130
Coefficient of thermal expansion Z, ppm/°C	44	55	60
Steam and solder, min	120	120	45
Flammability, UL-94	Burns	V-O	V-O
Peel strength, lb/in			
25°C	12.3	11.4	12.0
200°C	9.4	8.5	4.2
D_k, 1 MHz	4.05	4.2	4.8
D_f, 1 MHz	0.003	0.008	0.020

*Laminates are eight-ply, style 7628 E-glass-reinforced, 55 ± 2% resin by volume.
†AroCy B-40S.
‡Brominated hard epoxy, WPE 500, 27% Br.
SOURCE: Ref. 10. Reprinted with permission from Ciba-Geigy Corp.

TABLE 4.6 Comparison of E-Glass Laminate Properties for Several Resin Systems at Equal Resin Volume Content*

Resin	D_k, 1 MHz, vol. % 70	55	D_f, 10^{-3}	DMA T_g, °C	TGA Onset, °C	Flammability rating UL-94	Peel strength, lb/in 25°C	200°C	Pressure cooker, min
Cyanate ester									
ArOCy F-40S	3.5	3.9	2	290	400	V-O	11	9	120
AroCyM-40S	3.6	4.0	2	290	415	VI	12	10	120
XU 71787	3.6	4.0	3	255	426	†	8	6	120
ArOCyB-40S	3.7	4.1	3	290	405	†	12	10	120
Polyimide BMI-MDA	4.1	4.5	9	312	400	VI	9	6	120
Epoxy FR-4	4.5	4.9	20	145	300	V-O	12	4	45

DMA = dynamic mechanical analysis; TGA = thermogravimetric analysis.

*Except for D_k measurements on 70 volume percent resin laminates, tests were performed on 55 volume percent, 0.060-in, eight-ply laminates prepared with 7628 E-glass and postcured 4 h at 225 to 235°C.

†Burn times exceed self-extinguishing classifications.

SOURCE: Ref. 9. Reprinted with permission from Ciba-Geigy Corp.

Benzocyclobutenes

Benzocyclobutene (BCB) polymers are derived from monomers of the generic form shown below:

The monomers are thermally polymerized to produce a cross-linked resin without the evolution of volatiles. The properties of the polymers can be varied by changing the R group. BCB is generally supplied partially polymerized in hydrocarbon solvents such as toluene or mesitylene (20 to 70 percent solids). These materials can be further processed in the neat form for coating applications or can be used in composite applications. The resins upon curing (1 h at 250°C in nitrogen) yield substantially linear polymers with very high glass transition temperatures (T_g >350°C). A general characteristic of these polymers is that they possess high thermal stability (200 to 250°C performance). The polymers exhibit excellent retention of their room-temperature mechanical properties at these temperatures.[10] These polymers all have low dielectric constant and loss properties, low water absorption, high adhesion, good planarization, and excellent chemical resistance. Thin films of BCB can withstand typical metal etching processes and can tolerate hours at elevated temperatures in acid and alkaline baths. Mixtures of hydrogen peroxide and sulfuric acid will attack these materials. The properties of some of these resins are shown in Table 4.7.

TABLE 4.7 Properties of BCB-Based Resins

Property	BCB XU13005	BCB XU130026	BCB XU130028
Glass transition temperature (TMA, DMA), °C	>350	>350	>350
Flexural modulus, klb/in^2	480	540	747
Linear coefficient of thermal expansion, 25 to 300°C, TMA, ppm/°C	65–70	42	27
Water absorption, 24-h water boil, %	0.25	0.52	0.87
Weight loss at 350°C, nitrogen, 2 h, 20-μm film on silicon wafer, %	2	1	0
Dielectric constant, 1 MHz, (+1–0.1)	2.7	2.7	2.7
Dissipation factor, 1 MHz	0.0008	0.0006	0.0004

SOURCE: Ref. 12. Reprinted with permission from Dow Chemical Co.

Polyxylylene

These polymers were developed by Union Carbide in the mid-1960s and are sold under the trade name of Parylene. The Parylene polymers are prepared by a vapor-phase deposition process of the monomer under high temperature and vacuum. The monomer is vaporized at about 175°C and 1 torr. It is then heated to about 680°C at reduced pressure (0.5 torr) where the monomer pyrolizes into a diradical, which combines and forms a high-molecular-weight polymer on the substrate to be coated, which is at room temperature and reduced pressure (0.1 torr). The Parylene polymers, of which there are three, contain the following basic structure:

$$\left[-CH_2 - \underset{R}{\overset{R}{\underset{}{\bigcirc}}} - CH_2 - \right]_n$$

This structure represents Parylene N, where R = H.

If R_1 = Cl, then the polymer is designated Parylene C; and if $R_1 + R_2$ = Cl, then the polymer is Parylene D. The molecular weight of the Parylenes is in excess of 5000. Because of the unique process by which these polymers are made extremely thin (~1000 Å), pinhole-free coatings and films are obtained. These polymers exhibit superior dielectric properties that remain stable up to 220°C in the absence of oxygen. The mono-chloro polymer has lower permeability to moisture than Parylene N, and Parylene D can withstand temperatures up to 150°C in air. The advantages of Parylene polymers are that they form ultrathin coatings with complete coverage on all substrate parts, they have superior dielectric properties (dielectric strength of 5000 V/mil in thin coatings). The Parylenes are resistant to organic solvents and provide excellent moisture and gaseous barriers. Their limitations include the need for specialized equipment to synthesize the polymers, they are subject to air oxidation above 200°C, and they are difficult to remove from electronic components when reworking is necessary. The major use of this polymer is as a protective coating for circuit boards or molecules. Other uses include the insulation of ferrites, resistors, and thermistors and for the corona and arc protection in high-voltage components. More information about the properties of the Parylenes is given in the section on conformal coatings.

Urea and Melamine Resins

These resins are formed by the condensation reaction between formaldehyde (CH_2O) and urea ($CONH_2$) to produce urea formaldehyde resins; and between CH_2O and melamine to produce melamine formaldehyde resins. The resins are supplied as aqueous or alcohol liquids, solids, and molding powders. These resins, when combined with fillers and catalyst, are converted to hard, infusible resins. The urea resins consist of the structural unit of a dimethylol urea H—(O—CH_2NH—CO—NHCH_2)—OH, and the melamine resins consist of the structural unit of hexamethylol melamine. These resins form highly cross-linked products in the presence of a catalyst. Water is a by-product of the reaction.

$$(HOCH_2)_2-N\underset{N}{\overset{N}{\bigcirc}}N-(CH_2OH)_2$$
$$N-(CH_2OH)_2$$

The melamine resins are superior to the urea resins in chemical resistance (acids, alkali, boiling water). Urea resins shrink more than melamines do and tend to crack around inserts and sharp corners. Fillers improve strength, moldability, dimensional stability, and molded-in stresses. Typical fillers include minerals, glass fibers, wood flour, alpha cellulose, and chopped cotton flock. The alpha-cellulose-filled melamines are hard, rigid, and very abrasion-resistant. Melamines are useful between −57 and +99°C while ureas are not recommended for use above 77°C. Melamines can be used outdoors; ureas cannot. These resins are attacked by strong mineral acids and alkalies. The resins are compression- and transfer-molded. Alpha-cellulose-filled urea resins are used in circuit breakers and some electrical housings, but for more critical electrical applications melamines are preferred. Wood-flour-filled melamines are used in industrial electrical parts and military specifications. MIL-M-14G lists the following melamine grades:

- *Type CMG.* A cellulose-filled, general-purpose compound with good electrical and mechanical properties, for use where good arc resistance is required.
- *Type CMI-5.* A cellulose-filled, moderate-impact compound with good all-around mechanical properties for use where resistance to arcing and moderate impact is required.
- *Type CMI-10.* A cellulose-filled, moderate impact phenol modified compound with good all-around mechanical properties suitable for tableware and similar applications. It is not intended for electrical use.
- *Type MME.* A mineral-filled compound for use where good dielectric properties and arc and flame resistance are required. Of the melamine compounds, this is the most dimensionally stable.
- *Type MMI-5.* A glass-fiber-filled compound of lower impact strength and higher dielectric constant and dissipation factor at 1 Mcycle than type MMI-30. It has better moldability than type MMI-30. Impact strength is approximately 0.5 ft · lb/in notch.
- *Type MMI-30.* A glass-fiber-filled compound of high impact strength for use where heat resistance, arc resistance, and flame resistance are required.

Other applications for these resins include switchgear, wiring devices, appliance housings and knobs, engine ignition parts, sockets, and terminal strips. Suppliers of these resins include American Cyanamid and Plastics Engineering Co.

Silicon-Carbon Thermosets

The silicon-carbon thermoset resins were introduced in 1992 by Hercules, Inc. They are characterized by the structure

$$-R_2 {\left[\begin{array}{c} Y \\ | \\ O \\ | \\ Si \\ | \\ O \\ | \\ X \end{array}\right]}_x -R_1 {\left[\begin{array}{c} Y \\ | \\ O \\ | \\ Si \\ | \\ O \\ | \\ X \end{array}\right]}_y -R_2$$

TABLE 4.8 Uncured SYCAR Resin Properties

Physical form	Viscous fluid
Viscosity, Hz	
At 25°C	3000–6000
At 110°C	60–80
Gel time, min	
At 130°C	15
At 160°C	2.5
Pot life at 25°C, months	>6

TABLE 4.9 Cured SYCAR Resin Properties

Glass transition temperature, 3 h at 180°C cure, °C	140–160
TGA 5% weight loss, °C	480
Dielectric constant at 1 MHz	2.6
Dissipation factor at 1 MHz	0.003
Moisture absorption, 24-h water boil, wt%	0.04
Coefficient of thermal expansion, -55–$0°C$, ppm/°C	90
Flexural modulus, klb/in^2	300
Flexural strength, lb/in^2	9000

and designated by the trade name *SYCAR*.* These materials are not silicone resins, although the polymer backbone resembles that of the silicones. The R groups in the structure are alkanes, and the absence of polar functional groups in the polymer imparts excellent moisture resistance and electrical properties. These polymers in the uncured condition are viscous liquids at room temperature. The properties of the uncured liquid resin as well as of the thermally cured neat resin are shown in Tables 4.8 and 4.9, respectively. These resins have applications in printed-wiring board substrate resins, encapsulants, coatings, conductive adhesives, and dielectric layers.[13] Processing techniques include automated syringe dispensing, compression molding, prepregging, and resin transfer molding. More detailed information on applications and properties can be found in Chaps. 8 and 13.

---

*SYCAR is a registered trademark of Hercules, Inc.

References

1. R. Greene, ed., *Modern Plastics Encyclopedia,* McGraw-Hill, New York, 1992.
2. H. E. Mark, N. M. Bikales, C. G. Overberger, and G. Menges, eds., *Encyclopedia of Polymer Science and Engineering,* vols. 1 to 17, 2d ed., Wiley, New York, 1985–1990.
3. S. S. Schwartz and S. H. Goodman, *Plastics Materials and Processes,* Van Nostrand Reinhold, New York, 1982.
4. S. H. Goodman, ed., *Handbook of Thermoset Plastics,* Noyes Publications, Park Ridge, NJ, 1986.
5. C. A. Harper, ed., *Electronic Packaging and Interconnection Handbook,* McGraw-Hill, New York, 1991.
6. *Facts and Figures of the U.S. Plastics Industry,* Society of Plastics Industry (SPI), Washington, August 1992.
7. "PW Resin Profile," *Plastics World,* April 1992, p. 69.
8. *Industrial Laminated Thermosetting Products,* NEMA LI-1, National Electrical Manufacturers Association, Washington, 1983, sec. II.
9. W. T. Shugg, *Handbook of Electrical and Electronic Insulating Materials,* Van Nostrand Reinhold, New York, 1986.
10. D. A. Shimp, J. R. Christenson, and S. J. Ising, "Cyanate Ester Resins—Chemistry, Properties and Applications," Hi-Tek Polymers Inc., Louisville, KY, January 1990.
11. R. A. Kirchhoff, C. Carriere, K. Bruza, N. Rondan, and R. Sammler, "Benzocyclobutenes: A New Class of High Performance Polymers," *J. Macromol. Sci., Chem.,* vol. A28, 1991, p. 1079.
12. D. Burdeaux, P. Townsend, J. Carr, and P. Garrou, "Benzocyclobutenes—Dielectrics for Fabrication of High Density, Thin Film Multichip Modules," *J. Electronic Materials,* vol. 19, no. 12, 1990.
13. J. K. Bard, and J. S. Burnier, "A New Moisture Resistant Liquid Encapsulant," NEPCON WEST Conference, Anaheim, CA, May 1992.

Chapter 5

Elastomers

Elastomers are considered apart from other polymeric materials because of their special properties. The distinguishing characteristics of elastomer materials are their ability to sustain large deformations (5 to 10 times the unstretched dimensions) and capacity to spontaneously recover nearly all that deformation without rupturing. The unique structural feature of all rubberlike substances is the presence of long polymer chains interwoven and joined together through cross-linkages (vulcanized). Generally elastomers are not as widely used as plastics for electrical and electronic applications, but elastomers are no less important in the areas that they are used. So it is both useful and instructive to review the elastomer types, properties, and applications. More detailed information on elastomers is available in Refs. 1 and 2. General property information is presented in Ref. 3, and detailed electrical property data are given in Ref. 4. Elastomers are almost always used in the compounded state. The raw rubber is blended with a variety of ingredients such as fillers, plasticizers, accelerators, and vulcanizing agents to cure and enhance the properties of the elastomer. The electrical properties of elastomers depend upon the compound ingredients and can range from strong insulating properties to fair conductors.

An elastomer is rated by its oil, chemical, and temperature resistance. Tensile properties vary a great deal with compound ingredients and vulcanization. The tensile strength of some elastomers is enhanced by the addition of fillers. Elevated temperature lowers strength while low temperatures can cause brittleness in elastomers. Resilience is a fundamental property of elastomers. Compression set is a measure of the material's ability to resist permanent deformation. Elastomers are affected by oils and solvents, which can cause them to swell. Elastomers are also subject to oxidation. Ozone, radiation, temperature, and some metals (copper, manganese, and iron) can accelerate the oxidation process in elastomers. Hardness and surface cracking are the manifestations of elastomer oxidation.

Elastomer Types

There are a large number of elastomers, and ASTM D-1418 describes many of these materials.[5] A summary of each elastomer type and its significant property is given in Table 5.1. A more detailed description of each elastomer is given in the following section. Elastomers are included regardless of their applications in nonelectrical areas so as to give the reader a broader perspective on those materials that are distinctly separated from plastics, and perhaps the reader may generate some new ideas for their use in the electronics industry.

Butyl rubber (IIR)

Butyl rubber is a polymer of isobutylene with a small amount of isoprene. It is highly impermeable to gases, has excellent dielectric properties due to its nonpolar nature, and has good tear, weather, corona, and ozone resistance. It has a low degree of unsaturation and requires vigorous but careful vulcanization. Reinforcing agents do not significantly affect tensile strength, but tear resistance is improved. Because of its excellent electrical properties and its resistance to water, weathering bacteria, fungi, and ozone, it is an important material for wire and cable insulation. When compounded with aluminum oxide trihydrate, butyl has excellent arc and track resistance. Butyl rubbers have low resilience at room temperature, but at 93°C the resilience approaches that of natural rubber. Extended aging at 121°C shows very little degradation of tear resistance. Butyl rubber is resistant to concentrated sulfuric and nitric acid but swells in petroleum solvents. Suppliers include Exxon Chemical Co., Miles, Inc., and Goldsmith & Eggleton, Inc.

Chlorosulfonated polyethylene (CSM)

The elastomer is prepared by reacting polyethylene with chlorine and sulfur dioxide. The resulting vulcanized material has very good resistance to photochemical oxidation and high concentrations of ozone. Overall CSM has good temperature, weathering, and chemical resistance (strong oxidizing chemicals). Compounding produces a wide variety of products with varied properties. Electrical properties are good with dielectric strengths ranging from 400 to 700 V/mil and direct-current (dc) resistivity about 10^{14} $\Omega \cdot$ cm. It does not support combustion and is not resistant to aromatic and chlorinated hydrocarbons. CSMs

TABLE 5.1 Elastomer Type and Significant Property

ASTM D-1418	Chemical type	Properties
NR	Natural rubber polyisoprene	Excellent physical properties
IR	Polyisoprene synthetic	Same as NR, but more consistency and better water resistance
ABR	Arylate butadiene	Mechanical elastomer; excellent heat and ozone resistance
BR	Polybutadiene	Copolymerizes with NR and SBR; abrasion resistance
CO	Epichlorohydrin	Chemical resistance
COX	Butadiene-acrylonitrile	Used with NBR to improve low-temperature performance
CR	Chloroprene, neoprene	Withstands weathering; flame-retardant; chemical resistance
CSM	Chlorosulfonated polyethylene	Colors available; weathering and chemical resistance; poor electrical properties
EPDM	Ethylenepropylene terpolymer	Similar to EPM; good electrical properties; resists water and steam
EPM	Ethylene-propylene copolymer	Similar to EPDM: good heat resistance; wire insulation
FPM	Fluorinated copolymers	Outstanding heat and chemical resistance
IIR	Isobutyleneisoprene, butyl	Outstanding weather resistance; low physical properties, track resistance
NBR	Butadiene-acrylonitrite, nitrile, Buna N	General-purpose elastomer; poor electrical properties
PVC/NBR	Polyvinyl chloride and NBR	Colors available; weather, chemical, and ozone resistance
SBR	Styrene: butadiene, GRS, Buna S	General-purpose elastomer; good physical properties; poor oil and weather resistance
SI (FS1, PS1, VS1, PVS1)	Silicone copolymers	Outstanding at high and low temperatures; arc- and track-resistant; resist weather and ozone; excellent electrical properties, poor physical properties
T	Polysulfide	Excellent weather resistance and solvent resistance
U	Polyurethane	High physical and electrical properties
TPE	Thermoplastic elastomers	Thermoplastic, injection-moldable, extrudable; no compounding required

SOURCE: Ref. 5.

are available in a variety of colors and are often used in high-voltage applications. Suppliers include du Pont and Goldsmith & Eggleton, Inc.

Epichlorohydrin (CO, ECO)

These elastomers are homopolymers of epichlorohydrin, copolymers of epichlorohydrin with ethylene oxide, or terpolymers with epichlorohydrin, ethylene oxide, and a third monomer; the physical properties are good over a wide temperature range, and they have very low permeability to gases and excellent oil resistance. Epichlorohydrin's low-temperature properties are excellent, and it can withstand temperatures up to 121°C for extended periods. These elastomers are attacked by ketones, esters, aldehydes, nitroaromatics, and chlorinated solvents. Compression set and resilience are good, and torsional stiffness at subzero temperatures is much better than that of nitrile or neoprene elastomers.

Ethylene propylene elastomers (EPM, EPDM)

These two materials represent very distinct types of elastomers within the same class. EPM, which is a copolymer of ethylene and propylene, is a fully saturated polymer that requires organic peroxides or radiation to cross-link the material. EPDM is a terpolymer of ethylene, propylene, and a diene. This polymer has some unsaturation and can be vulcanized with sulfur as well as with peroxides. EPDM is preferred because of its better processing. The elastomers have excellent resistance to ozone and weathering, good heat resistance (120 to 150°C), good resilience and low-temperature properties, and fair to good tear strength. Flame and oil resistance are poor. Electrical properties are excellent, e.g., with a dielectric strength of 900 V/mil, dielectric constant of 3.17 to 3.34, and a volume resistivity of 10^{15} to 10^{17} $\Omega \cdot$ cm. They are used in wire and cable insulation and jacketing as well as molded terminal covers, plugs, and connectors. Suppliers include du Pont, Exxon, Uniroyal, and Copolymer Rubber and Chemical Corp.

Chloroprene (CR)

Neoprene is the generic name for this elastomer, which is synthesized by the polymerization of chloroprene. It has superior properties compared to natural rubber, especially its resistance to oil, grease, weathering, ozone, and temperature. Neoprene has excellent abrasion resistance and flame resistance. It will not support combustion. Properly compounded stocks are useful over the temperature range from −50 to +121°C. The electrical properties of neoprene rank below those of natural rubber because of the polar chlorine group in the rubber molecule. The electrical properties are good enough for low-voltage wire and cable insulation. Suppliers include Miles, Inc., and Polysar Rubber Division.

Fluorinated elastomers (FKM)

These elastomers are based on the monomers of vinylidene fluoride and hexafluoropropylene to produce the Fluorel (3M Co.) or Viton brand (du Pont) with increased fluorine content via use of tetrafluoroethylene. This gives improved fluid resistance or the perfluoromethylvinyl ether group is used for improved low-temperature properties. These fluorinated elastomers provide excellent high-temperature resistance up to 300°C. They have excellent oil and solvent resistance. The fluoroelastomers are not recommended for use with ketones, low-molecular-weight esters, and nitro-containing compounds. They are flame-resistant (will not burn) and have excellent electrical properties. The fluoroelastomers adhere well to metals, have good abrasion and tear resistance, and have good compression set but fair resilience. They also have excellent weather resistance. Their high thermal stability is bettered only by the silicone elastomers.

Fluorosilicone elastomers (FVMQ)

These polymers have the general formula $CH_3—SiO—CH_2CH_2CF_3$ with a small amount of unsaturation provided by $CH_3SiOCH_2=CH_2$. The polymers are prepared via anionic polymerization of the corresponding cyclosiloxane. These elastomers are compounded with fillers, catalysts, and silicone oils and are vulcanized to yield cured elastomers with outstanding resistance to fuels, oils, solvents, and high temperatures. These polymers are superior to the fluorocarbon carbon elastomers in compression set below room temperature. The main characteristics of the fluorosilicone elastomers are lower hardness, higher resilience, better low-temperature flexibility, and different surface properties compared to the fluorocarbon elastomers.

Natural rubber (NR)

Natural rubber is a linear polymer known chemically as *cis*-1,4 polyisoprene. Natural rubber is obtained from the Hevea tree. It is not a uniform product but varies with the nature of the plant and the environmental conditions surrounding that plant. Variants of natural rubber include guayule, balata, and gutta percha. Natural rubber which is properly selected and compounded produces very high-quality rubber products and was the original basis of electrical insulation. Natural rubber possesses the highest tensile strength of the elastomers, highest resilience plus excellent tear and abrasion resistance, and very good electrical properties. On the other hand, its resistance to oxidation, ozone, solvents, and oil is poor to fair. Suppliers include Exxon, Goldsmith and Eggleton, Inc., and Acushnet Co.

Isoprene (IR)

Polyisoprene is the synthetic equivalent of natural rubber and is used in many of the natural-rubber applications. It processes more easily than NR. Suppliers include Goldsmith and Eggleton, Inc., and Acushnet Co.

Nitrile elastomers (NBR)

These are copolymers of acrylonitrile and butadiene and are known as *Bune N, nitrile,* or *NBR rubbers.* They have excellent resistance to fuels, oils, and solvents, i.e., better than neoprene, but have much poorer electrical properties because of the highly polar nitrile group along the polymer chain. Nitrile rubber has a very low resistivity value (10^{10} $\Omega \cdot$ cm), and by adding conductive fillers one can obtain resistances as low as 100 to 500 $\Omega \cdot$ cm^3. These compounds have been used to dissipate static charges. Nitrile rubbers are permeable to polar liquids and gases and are attacked by aromatic and chlorinated hydrocarbons, ketones, esters, and nitro compounds. Aliphatic hydrocarbons and acids do not effect nitriles. The major use of nitrile is in nonelectrical applications. Suppliers include Copolymer Rubber and Chemical Corp., Goodyear Tire and Rubber, and W. R. Grace and Co.

Acrylic elastomer (ACM, ANM)

The polyacrylate elastomers are copolymers of acrylic esters and a small amount of a reactive curative such as a reactive halogen or oxirane-containing compound. These elastomers have high heat and oil resistance. Properly compounded and cured, these elastomers operate over a temperature range from -40 to $+200°$C. Their electrical properties are fair, and the polymers are hygroscopic. As a result, they are not used for electrical insulation but rather are used in mechanical seal areas involving hot oils.

Ethylene acrylic elastomers

This elastomer is a copolymer of ethylene and methyl acrylate plus a small amount of a curative monomer. This elastomer possesses excellent resistance to oxidation, ozone, and uv radiation and offers good low-temperature properties and oil resistance. Its electrical properties are such that it can be used as insulation on low-voltage cable and as jacketing on higher-voltage applications. Dielectric strength of a mineral-filled cured stock ranges between 500 and 900 V/mil.

Styrene butadiene elastomers (SBR)

This elastomer is a copolymer of butadiene and styrene, and there are numerous copolymers available. The elastomers are not very resistant to oils and solvents and are quite susceptible to strong oxidizing agents, ozone, petroleum, and chlorinated hydrocarbons. It is not used in electrical applications.

Chlorinated polyethylene (CM)

These elastomers are produced by chlorination of high-density polyethylene. Because of the high degree of saturation of the polymer, it has some unique properties; i.e., it has outstanding ozone, weather, and heat resistance (up to 150°C). The polymer is cross-linked by the use of peroxides or radiation. It is not used in electrical applications.

Polybutadiene (BR)

This material is a controlled structure elastomer; i.e., it can be prepared in varying the cis/trans/vinyl ratio depending on the desired properties. It has outstanding resiliency and hysteresis as well as excellent abrasion resistance. It is often blended with other elastomers to improve certain properties. It is not used in electrical applications.

Polysulfides (T)

These elastomers are the result of the polymerization of organic dihalides and sodium polysulfide. An elastomer containing over 80 percent sulfur is produced. These elastomers are vulcanized with metallic oxides and sulfur. Polysulfide elastomers are produced as millable gums or as liquid (highly viscous) products which cure at room temperature. The elastomers are not significantly affected by aliphatic or aromatic hydrocarbons, although the elastomers swell in benzene. Alcohols, ketones, and esters have little effect. They have good aging characteristics and resistance to ozone. Strong oxidizing acids such as nitric acid and the alkalies do not significantly affect fully cured polysulfides. Polysulfide has low permeability to gases, water vapor, and organic liquids. Polysulfide elastomers are not used in electrical applications.

Silicones (MQ, VMQ)

The unique properties of the silicone elastomers result from the molecular structure, i.e., alternating atoms of silicon and oxygen with organic substitution on the silicon atom. Purely organic elastomers are composed of a backbone of carbon atoms rather than silicon and oxygen atoms. The representative structure characteristic of all silicones is as follows:

$$\left[O-\underset{\underset{R}{|}}{\overset{\overset{R}{|}}{Si}}-O-\underset{\underset{R}{|}}{\overset{\overset{R}{|}}{Si}} \right]_n$$

The R group is an organic group such as methyl (CH_3), vinyl (C=C—C), or phenyl:

—⟨O⟩—

By varying the nature of the R group, a variety of properties can be obtained from the resulting polymers. The replacement of some methyl groups with phenyl groups improves low-temperature flexibility and gamma-radiation resistance. Vinyl groups improve the curing properties and compression-set resistance.

The silicone elastomers have outstanding resistance to temperature (260°C) as well as ozone, corona, weathering, and ultraviolet radiation. There are two broad types of silicones: the heat-cured vulcanizable stocks and the room-temperature vulcanizing (RTV) liquid silicones. Molecular weights of the RTV silicones are lower than those of the heat-cured products (10,000 to 100,000 versus 500,000 to 1,000,000). Heat-cured silicone elastomers are cured through the methyl or vinyl group activated by peroxides. RTV silicones cure at room temperature either by a condensation mechanism with the elimination of a volatile material, such as alcohol or acetic acid, or by an addition mechanism using platinum catalysts without the elimination of volatiles.[1,6,7] Silicone elastomers are useful from −73 to 260°C. Compression-set values range from 12 to 50 percent at 150°C; tensile properties range between 600 and 1800 lb/in² with elongations from 500 to 800 percent. Tear strength is about 175 to 275 lb/in. Silicone rubber cannot be reinforced with carbon black; silica is used instead. The silicone elastomers do not burn and are arc-resistant. Silicones are attacked by strong alkali and strong oxidants. Aromatic and chlorinated hydrocarbons will swell the silicones. The silicone rubbers have good electrical properties. The dielectric constant and dissipation factor are low but vary with frequency and have limited use in coaxial cable. In the electrical industry, silicones are used as insulation for wire and cable, sealants for electrical machinery, and electronic parts exposed to adverse environmental conditions.

Polyurethanes (AU, EU)

Polyurethanes are materials prepared by the reaction of molecules containing isocyanate groups with molecules containing hydroxyl groups. This structural unit present in polyurethanes is the urethane group:

Most of the thermoplastic urethanes are block copolymers of diisocyanates and polyols, while a third type is based on polycaprolactone. The polyester types are tougher and stronger but much less resistant to hydrolysis than the polyether types. These urethanes have excellent resistance to ozone. Their main features are toughness, abrasion resistance, flexibility at low temperatures, and excellent adhesion to a variety of substrates. These urethanes cover a broad range of hardness values (shore A 70 to shore D 80). Tensile strength at yield is 3700 to 9000 lb/in^2 with elongations at break from 250 to 650 percent. Modulus in compression ranges from 4000 to 9000 lb/in^2. The operating service temperature is from -50 to $+121°C$. Electrical properties are quite good; dielectric strength ranges from 330 to 760 V/mil.

Processing is done on conventional thermoplastic equipment. The thermoplastic urethanes can be blended with PVC, ABS, SAN, and polycarbonate to improve their performance. Applications include protective coverings for electrical equipment. Suppliers include BASF, Dow, B. F. Goodrich, and Miles.

Elastomer Properties

The physical properties of elastomers are measured by using the same methods as for metals and plastics, but their response to testing is different. This section is a general discussion of terms used to characterize the properties of elastomers.

Tensile properties

The linear relationship between stress and strain that is observed in metals does not exist with plastics and elastomers. ASTM D-412 defines the special tensile tests required for elastomers. Elastomers are not usually designed for those applications requiring high tensile strength. There are properties of elastomers such as wear, tear resistance, resilience, creep, stress relaxation, and flexural fatigue that improve with tensile strength. The modulus of elastomers is expressed as the stress required to achieve a given elongation. It expresses resistance to extension, or stiffness of the elastomer.

Elongation of elastomers describes the ability of the elastomer to be stretched and is measured as the percentage of stretch at rupture. The tension set of vulcanized elastomers is another property that is measured; it indicates how well the elastomer recovered its original length after an imposed tensile force. It is expressed as a percentage.

Durometer hardness

Hardness is a measure of the elastomer's resistance to indentation and is found by pressing an instrument into the surface of the elastomer. Hardness values are measured in terms of shore durometer readings. Figure 5.1 shows the hardness of plastics and elastomers.

Tear resistance

Tear resistance in elastomers is described as the stress needed to propagate a cut or notch in an elastomer; the tear resistance of elastomers varies widely. ASTM D-624 is the test method.

Abrasion resistance

This property is the resistance to wear of the surface of the elastomer in contact with a moving abrasive wheel. It cannot be used to predict the service life of an elastomer. ASTM D-1044 is the test method.

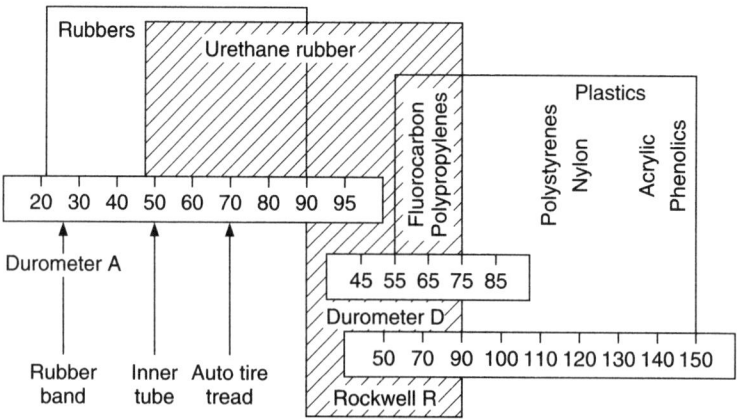

Figure 5.1 Hardness of rubber and plastics. (Source: Ref. 7.)

Flexural resistance

This property measures the resistance of the elastomer to cracking after repeated flexing. Several ASTM methods are used depending on the application. See ASTM D-430.

Hysteresis

Hysteresis is the energy loss per loading cycle resulting in a buildup of heat in elastomers. This heat emanates from the internal friction of molecular movement of the polymer chains within the elastomer. The heat buildup can be substantial and can have a significant effect on the aging of the elastomer. Heat buildup can also result at high frequencies when the dissipation factor of the elastomer is high.

Compression set

Compression set is the creep that elastomers undergo when held at constant deflection or load in a compressive mode. The most widely used method is the constant-deflection method, and the compression set is recorded as the amount (percent) by which a standard test piece of material fails to return to its original thickness after being subjected to a compressive load or deflection for a fixed time. A deflection of 25 percent is commonly used. ASTM D-395 is the test method.

Low-temperature properties

Elastomers become progressively stiffer upon cooling but are superior to plastics in low-temperature performance. The stiffening caused by thermal cooling is a first-order transition. ASTM D-746 describes a brittle-point temperature but has no relationship to stiffness. A compilation of the physical properties of elastomers used in electronic applications is given in Table 5.2, and it includes mechanical, electrical, chemical, thermal, and environmental resistance.

Electrical properties

Some elastomeric materials are designed primarily to possess reliable electrical properties to provide maximum performance for their intended use. Although physical properties of elastomers are of secondary importance in electrical applications, nevertheless they must meet certain minimum performance standards for a given application. Many elastomers find application in the electrical industry, e.g., natural rubber, SBR, butyl, EPM, EPDM, CSM, neoprene, fluoroelastomers, silicones, etc. These applications include wire and cable insulation and jacket, drop wire, and power distribution cables. Jacket or sheath

TABLE 5.2 Elastomer Properties

Property	Elastomer polybutadiene BR	Styrene-butadiene SBR	Isobutylene isoprene NR, IR	Isoprene IIR	Nitrile NBR	Hydrogenated nitrile HNBR
Specific gravity	0.91	0.94	0.92–1.037	0.92	0.98	1.1–1.3
Thermal conductivity Btu/(h · ft^2 · °F · ft)	—	0.143	0.082	0.053	0.143	—
Coefficient of thermal expansion (cubical), 10^{-5}/°F						
Hardness, durometer	45A–80A	30A–90D	30A–100A	30A–100A	30A–100A	55A–60D
Tensile strength, 10^3 lb/in^2	2.0–2.5	1.8–3.0	2.5–4.6	>2.0	1.0–3.5	1.5–6.0
Modulus, 100%, lb/in^2	300–1500	300–1500	480–850	50–500	490	300–2900
Elongation, %	450	450–500	300–750	300–800	400–600	150–550
Compression set, method B, %	10–30	5–30	10–30	25^a	5–20	10
Resilience, %						
Yerzley (ASTM 945)	50–90	20–90	80	30	—	—
Rebound (Bashore)	—	10–60	—	—	—	—
Hysteresis resistance	G	F–G	E	—	—	F–G
Flexural cracking resistance	E	G	E	—	—	G
Tear resistance	G	F	G–E	G	G	G
Abrasion resistance	E	E	E	G	E	E
Impact resistance	G	E	E	G	G	E
Volume resistivity, Ω · cm	—	5.0–8.4 × 10^{13}	—	2.0 × 10^{16}	3.5 × 10^{10}	—
Dielectric STR, V/mil	400–600	600–800	400–600	600–900	250	—
Dielectric constant						
60 Hz	—	—	—	2.31	—	—
1 MHz	3.3	—	2.9	2.25	—	—
Service temperature, °F						
Min. for continuous use	−101	−59	−59	−45	−51	−40
Max. for continuous use	93	121	71	148	101	148
Environmental resistance						
Ozone	P	P	P	E	P	E
Oxidation	G	G	G	E	F–G	E
Weathering	F	F	F	E	G	E
Water	E	E	E	E	E	E
Radiation	P	F–G	F	P	F–G	G
Alkalies	F–G	F–G	F–G	E	F–G	F–G
Aliphatic hydrocarbons	P	P	P	F	E	E
Aromatic hydrocarbons	P	P	P	F–G	G	G
Halogenated hydrocarbons	P	P	P	P	P	P
Alcohol	G	G	G	VG	VG	F
Animal, vegetable oils	P–G	P	P–G	G	E	E
Acids						
Dilute	F–G	F–G	F–G	E	E	G–E
Concentrated	F–G	F–G	F–G	G	E	F–G
Synthetic lubricants (diester)	P–F	P	P–F	F	P	F–P
Hydraulic fluids						
Silicates	P–G	P–G	P–G	F	F	F
Phosphates	P–G	P–G	P–G	G	P	P
Permeability to gases	Low	Low	Low	Very low	Very low	Extremely low
Limiting oxygen index	—	—	—	18–19	17–20	17–20

See p. 122 for abbreviations and footnotes.

Elastomer chloroprene CR	Polysulfide PTR	Ethylene propylene EPM, EPDM	Chlorinated polyethylene CM	Chlorosulfonated polyethylene CSM
1.23–1.25	1.35	0.86	1.16–1.32	1.11–1.28
0.11	1	0.15	—	0.065
34	—	32	—	27
30A–95A	20A–80A	30A–90A	50A–95-A	40A–95A
0.5–3.5	0.5–1.5	0.5–3.5	0.9–3.0	0.5–3.5
100–3000	—	100–3000	700–2200	100–3000
100–800	210–450	100–700	100–700	100–700
20–60^a	29–38	20–60^c	5–30	35–80^b
50–80	—	40–75	—	30–70
50–80	—	40–75	15–40	30–70
VG	—	G	G	F–G
VG	—	VG	E	VG
G	P–F	F–G	G	F
E	P–F	G–E	E	E
E	P–F	VG	E	VG
2.0×10^{13}	5×10^{13}	2×10^{16}–1×10^{17}	—	1×10^{14}
400–600	—	500–1000	—	650
8.0	7.3^d	2.25–3.0	—	7.0
6.7	6.8^e	2.2–2.85	—	6.0^b
−51	−45	−56	−51	−45
107	>121	148	148	135
VG	E	O	O	O
VG	E	E	O	O
VG	E	O	G	—
G	G	E	G	G
G	F	E	E	VG
E	G	G–E	E	E
G	E	P–G	E	G
F	E	P	P	F
P	F–G	F–P	—	P
G	VG	P–G	E	G
G	E	F	VG	G
E	G	E	O	E
F–G	P	F–G	E	VG
P	G	F–G	P	P
P–G	P–G	F–G	G	P
P	P–F	G–E	G	P
L–M	VL	M	L	L
38–45	—	10–20	30–35	30–36

TABLE 5.2 Elastomer Properties (*Continued*)

Property	Elastomer ethylene acrylic	Epichlorohydrin CO, ECO	Polynorborene	Polyacrylate ACM, ANM	Silicone VMQ
Specific gravity	1.08–1.02	1.27–1.49	0.96	1.09	1.1–1.6
Thermal conductivity Btu/(h · ft^2 · °F · ft)	—	—	—	—	0.13
Coefficient of thermal expansion (cubical), 10^{-5}/°F	—	—	—	—	45
Hardness, durometer	64A^f	30A–95A	15A–100A	40A–90A	20A–90A
Tensile strength, 10^3 lb/in^2	1.95	2–3	1.0–4.0	1.8–2.0	1.5
Modulus, 100%, lb/in^2	800	150–2000	100–1500	200–1500	—
Elongation, %	450	200–800	100–600	100–400	100–800
Compression set, method B, %	—	20^g	10–600	10–50	10–30
Resilience, %					
Yerzley (ASTM 945)	—	50–80	—	—	30–60
Rebound (Bashore)	20	45–75	—	—	—
Hysteresis resistance	—	G	E	—	F-G
Flexural cracking resistance	E	VG	G	F	F-E
Tear resistance	E	G	G	F-G	F-G
Abrasion resistance	E	F-G	E	G	P
Impact resistance	—	G	E	P	G
Volume resistivity, Ω · cm	1.9×10^{12}	—	—	7×10^{12}	1×10^{14}–10^{16}
Dielectric STR, V/mil	7.30	—	—	800	400–700
Dielectric constant					
60 Hz	—	—	—	—	2.95–4.0
1 MHz	—	—	—	—	2.95–4.0
Service temperature, °F					
Min. for continuous use	−34	−26−−62	−51	−40	−117
Max. for continuous use	176	163	121	176	260
Environmental resistance					
Ozone	O	E	P	E	E
Oxidation	E	E	G	E	E
Weathering	E	E	F	E	E
Water	E	G	E	F-G	E
Radiation	—	E	F-G	F	F-G
Alkalies	E	G	F-G	P	P-F
Aliphatic hydrocarbons	G	E	P	E	P-G
Aromatic hydrocarbons	G	VG	P	P	P-G
Halogenated hydrocarbons	G	G	P	P	F
Alcohol	F	G	G	P	F
Animal, vegetable oils	E	E	F-G	VG	G
Acids					
Dilute	E	G	F-G	F	VG
Concentrated	P	P	F-G	F	G
Synthetic lubricants (diester)	—	F-G	G	G	G
Hydraulic fluids					
Silicates	G	VG	F-G	G	P
Phosphates	G	P-F	F-G	P	G
Permeability to gases	—	L	VL	M	High
Limiting oxygen index	48	25–33	—	—	20–30

P = poor; F = fair; G = good; E = excellent; VG = very good; O = outstanding.

^{a} 70 h at 125°C; ^{b} 22 h at 100°C; ^{c} 70 h at 100°C; ^{d} 1 KC; ^{e} 1 MC; ^{f} at 200%; ^{g} 22 h at 70°C; ^{h} 22 h at 200°C.

SOURCE: Adapted from Ref. 3.

Elastomer fluorosilicone FVMQ	Fluoro-elastomer FKM	Perfluoro-elastomer FFKM	Poly-urethanes AU, EU	Propylene oxide PO	Recycled rubber (EPDM, SBR)
1.4	1.8–1.9	1.9–2.0	1.02–1.25	1.01	0.59–1.30
0.13	0.06–1.3	0.09	0.09–0.10	—	—
45	—	13	5–25	—	—
40A–80A	55A–95A	75A–80A	10A–80D	40A–80A	30A–60A
0.7–1.5	1.5–2.0	2.3–2.4	0.8–8.0	>2	0.44–0.22
—	200–2000	1050–1400	25–5000	—	—
200–500	150–450	135–150	250–800	500–670	70–150
10–20	15–30[h]	25–45	10–45	—	—
—	40–70	—	5–75	—	—
22–35	40–70	—	20–65	—	—
G	G	—	F–G	VG	—
G	G	—	G–E	VG	—
F–G	F–VG	—	O	E	—
P	G	—	E–O	G	—
F	G	—	E–O	E	—
10^{12}–10^{15}	2×10^{13}	—	0.3×10^{10}–4.7×10^{13}	—	—
400–700	500	>450	330–700	—	—
5–7	5.0–10.0	—	4.7–9.53	—	—
5–7	—	—	5.9–8.51	—	—
−56	−51	−40	−54	−62	—
204	260	315	121	121	—
E	O	O	E	E	—
E	O	O	E	E	—
E	E	O	G	E	—
E	VG	E	G–E	E	—
F–E	F–G	VG	G–E	P	—
VG	F–G	E	P–F	VG–E	—
E	E	E	E	F	—
E	E	E	F–G	P–F	—
E	G	E	P–F	P–F	—
G–E	VG	E	P–G	F–G	—
G	E	E	E	G	—
E	F–E	E	F	F	—
E	F–E	E	P	P	—
—	F–G	G	G	G	—
O	E	E	F	F–G	—
P	G–E	E	P	P	—
Medium	Low	Low	Medium	High	—
—	50–100	—	15–20	—	—

TABLE 5.3 Electrical Properties of Selected Elastomers

ASTM elastomer*	Dielectric strength, V/mil	Dissipation factor tan δ	Dielectric constant	Volume resistivity, Ω · cm
COX	500	0.05	10	10^{15}
CR	700	0.03	8	10^{11}
CMS	700	0.07	8	10^{14}
EPDM	800	0.007	3.5	10^{16}
FPM	700	0.04	18	10^{13}
IIR	600	0.003	3	10^{16}
NR	800	0.002	3	10^{16}
SBR	800	0.003	3.5	10^{15}
SI	700	0.001	3.6	10^{15}
T	700	0.005	9.5	10^{12}
U	500	0.03	5	10^{12}

*See Table 4.1 for an explanation of the symbols.

materials are applied over cable insulation to provide extra protection against certain environmental conditions, abrasion, and weathering and enhancement of strength properties of the overall cable insulation. Those properties of elastomers that are of electrical significance are the dielectric strength, dielectric constant, insulation resistance, and dissipation factor. A detailed description of the electrical property tests and methods applied to plastics and elastomers is given in Chap. 10. However, a comparison of selected elastomer electrical properties is shown in Table 5.3.

Thermoplastic elastomers (TPEs)

Thermoplastic elastomers (TPEs) are a class of materials that are readily processible as thermoplastics but possess the functional performance and properties of a vulcanized elastomer. Unlike rubber, the TPEs do not require vulcanization and are easy to process.

TPEs are processed almost exclusively by extrusion and injection molding but can be blow-molded and thermoformed. These processes are not available to vulcanized elastomers. Scrap from the processing operations is reusable, which is not the case with a vulcanized elastomer. TPEs have high coefficients of friction, maintain flexibility at low temperature, and have lower density and good physical properties. However, because TPEs are not vulcanized, they do melt at elevated temperatures. The usual service temperature range for those elastomers is from −50 to +121°C, although some compositions provide service from −70 to +175°C. Table 5.4 lists seven broad categories of

TABLE 5.4 Properties of Thermoplastic Elastomers

Property	Elastomer olefinics	Styrenics	Co-polymers	Amide urethanes	Polyester urethanes	Polyether ester	Copolyether polyester
Specific gravity	0.84–1.07	0.9–1.2	1.01–1.4	1.18–1.23	1.10–1.20	1.16–1.25	1.22–1.44
Thermal conductivity Btu/(h · ft^2 · °F · ft)	0.08–0.09	0.09	—	0.14–0.19	0.09–0.17	0.154	—
Coefficient of thermal expansion, cubical, 10^{-5}/°F	6–12	7.2–7.7	6.12	—	5.4–9.5	11–35	—
Hardness, durometer	35A–95A	45A–95A	60A–75D	70A–80D	80A–65D	35D–72D	30D–76D
Tensile strength, 1000 lb/in^2	0.65–4.46	0.5–3.2	3.6–5.0	5.0–8.0	3.0–8.0	1.9–8.7	1.0–4.0
Modulus, 100%, lb/in^2	200–2200	100–800	60–64,000	650–5400	650–3400	1100–95,000	—
Elongation, %	50–1000	600–800	—	390–750	350–650	350–685	380–900
Compression set, method B, %	45–92	32–59	—	27–50	25–90	41–67	—
Resilience, %							
Yerzley (ASTM 945)	—	71–75	—	—	—	—	—
Rebound (Bashore)	35–50	—	—	35	25–50	40–60	—
Hysteresis resistance	F	—	E	F	F–G	G	—
Flexural cracking resistance	F	—	E	G–E	E	E	—
Tear resistance	G–E	G	E	G–E	E–O	E	600–1340 lb/in
Abrasion resistance	F–G	—	E	E–O	E–O	E	—
Impact resistance	E–O	E	O	E	E–O	E	No break
Volume resistivity Ω · cm	1–10^{16}	2 × 10^{16}	10^3–10^{13}	10^{12} × 10^{13}	10^{12}	10^{13}	—
Dielectric strength V/mil	600	400–510	—	330–460	440–730	525–900	330–483
Dielectric constant							
60 Hz	2.41	2.5	—	5.75–6.34	6.0[a]	—	—
1 MHz	2.41	2.5	—	4.53–5.15	4.21	4.6	—
Service temperature, °C							
Min. for continuous use	−51	−54	−62	−54	−54	−51	—
Max. for continuous use	121	148	148	100	82	148	—
Environmental resistance							
Ozone	G–E	E	E	E	E	E	—
Oxidation	G	—	E	E	G–E	E	F
Weathering	E–O	—	E	G	G	G	F
Water	E	G	E	G	G–E	E	E
Radiation	G	—	VG	G	G–E	E	E
Alkalies	G–E	G	P–G	P–F	P–F	G–E	F
Aliphatic hydrocarbons	F–G	P	G	E	E	G–E	G
Aromatic hydrocarbons	P	P	G	F–G	F–G	E	G
Halogenated hydrocarbons	P	P	G	F	P–F	P	G
Alcohol	F–G	G	G–E	P–F	F–G	G	E
Animal, vegetable oils	F–G	—	—	E	E	E	G
Acid							
Dilute	G–E	G	G	F	F	G	G
Concentrated	G–E	F	G	P	P	P	P
Synthetic lubricants (diester)	G	—	G	P	P–G	E	—
Hydraulic fluids							
Silicates	F–G	—	G	P	F	E	G
Phosphates	F–G	—	G	P	P	G	—
Permeability to gases	Low	—	Low	Medium	Medium	Medium	Medium
Limiting oxygen index	17–20	—	—	—	15–20	20[b]	—

P = poor; F = fair; G = good; E = excellent; VG = very good; O = outstanding.

[a]10^3 cycles; [b]30 with additives.

SOURCE: Adapted from Ref. 3.

thermoplastic elastomers, but within some of the classes, other TPE compositions with different polymers exist. Reference 9 lists other TPEs.

These seven categories are olefinics, styrenics, elastomer alloys, thermoplastic urethanes, amides, and the copolyesters. These materials are available in a range of hardness values; and with increasing hardness the less rubberlike the properties become.

Olefinic TPEs

The olefinic thermoplastic elastomers are blends of polypropylene homopolymers or copolymers and elastomers such as EPM and EPDM. Compositions with a broad range of properties have been developed, ranging from flexible elastomers to very tough, rigid products. The flexural moduli of these materials can range from 2000 to 400,000 lb/in^2. The compositions generally have low specific gravity and good chemical and electrical properties and operate from -40 to $+130°C$. Electrical applications include wire and cable insulation and jacketing. Suppliers include D&S Plastics, Monsanto, Himont, and A. Schulman.

Styrenic TPEs

The styrene thermoplastic elastomers are block copolymers of styrene with various elastomers or olefins. The polymers produced can be linear or branched. A silicone-modified styrenic TPE is also available. The styrene TPEs are the most widely used of all the thermoplastic elastomers. These polymers are composed of hard and soft domains within their structure that affect the rheological properties of the elastomer. Mechanical properties are fairly comparable to those of vulcanized rubbers. Because of their overall excellent combination of mechanical and electrical, thermal, and flammability properties, these materials compete with cross-linked polyethylene, polyvinyl chloride, and vulcanized rubber in wire and cable insulation and vulcanized rubber in wire and cable insulation and jacketing materials for the automotive market. Suppliers include Shell Chemical, Concept Polymer Technologies, Emmont, Housmex, Inc., and Fina Oil and Chemical Co.

Polyurethane TPEs

The thermoplastic polyurethanes (TPUs) are block copolymers consisting of soft and hard segments within the molecular structure. They are synthesized by the condensation of isocyanates and polyols. Because of the many different starting materials, a wide variety of polymers with different properties are possible. For example, the thermoplastic urethanes can be either soft or hard, flexible or rigid, hydrophilic or hydropholic. As with the straight polyurethanes, there are two classes of TPUs, one based on the reaction of isocyanates with ester polyols and the other on the reaction with ether polyols. Like the polyurethanes, the polyester urethane TPEs are tougher and less hydrolytically stable than the polyether urethane TPEs. The polyurethane TPEs are noted for their flexibility even at low-temperature, abrasion resistance, and adhesive properties. The TPUs can also be aromatic or aliphatic with the latter exhibiting superior uv resistance. Compression-set values of the TPU are also very good, even up to 100°C. Most aromatics and chlorinated solvents will attack the TPUs. The electrical properties of TPUs are quite good, and because of this the TPUs are widely used as protective covering for electrical equipment, such as in wire and cable jacketing.[10] Suppliers include Miles, BASF, Dow, B. F. Goodrich, and Mortar.

Copolyester elastomers (COPEs)

The copolyester elastomers are random block copolymers containing hard and soft domains as part of their microstructure. The rigid or hard segment is composed of polyester blocks while the soft or rubbery phase is composed of polyether blocks. The polymers are semicrystalline. These elastomers have matched Izod impact strengths that range from 0.8 ft · lb/in to a no-break condition at 23°C. These materials maintain their strength at $-40°C$. Mechanical properties are excellent with flexural modulus values ranging from 4000 to 180,000 lb/in^2 and tensile yield values from 300 to 6000 lb/in^2. The upper service temperature is 120°C for continuous use. The water absorption rate is 0.2 to 1.0 percent after 24 h at room temperature. While COPEs have good dielectric strength, their low arc resistance limits their use to low-voltage applications. These elastomers are affected by strong acids and bases, chlorinated solvents, and hot water. Suppliers include Akzo/DSM, du Pont, Eastman, General Electric, and Hoechst Celanese.

References

1. R. O. Babbit, ed., *The Vanderbilt Rubber Handbook,* R. T. Vanderbilt Co., Norwalk, CT, 1978.
2. H. F. Mark, N. M. Bikales, C. G. Overberger, and G. Menges, eds., *Encyclopedia of Polymer Science and Engineering,* vols. 1–17, 2d ed., Wiley, New York, 1985–1990.
3. M. W. Hunt, ed., *Materials Engineering—Materials Selector 1993,* Penton Publishers, Cleveland, OH, 1992.
4. Chen C. Ku, and R. Liepins, *Electrical Properties of Polymers,* Hanser Publishers, New York, 1987.
5. *Practices for Rubber and Rubber Lattices Nomenclature,* ASTM-1418, American Society for Testing and Materials, Philadelphia, 1990.

6. M. LeGault, "Improved Silicone Rubber Cure Systems," *Chemtech,* January 1993, p. 5.
7. R. Tanton, G. Sullivan, and M. Marty, "Platinum Cure Technology Eases Processing of Molded Silicone Rubber," *Elastomerics,* July 1992, p. 22.
8. C. A. Harper, ed., *Electronic Packaging and Interconnection Handbook,* McGraw-Hill, New York, 1991.
9. R. Greene, ed., *Modern Plastics Encyclopedia 92,* McGraw-Hill, New York, 1991, p. 424.
10. B. S. Miller, ed., "P. W. Resin Profile," *Plastics World,* February 1993, p. 50.

Chapter

6

Alloys and Blends

This chapter is concerned with polymer mixtures. These mixtures result when two or more polymers are mixed to form a new material whose properties, processing characteristics, and morphology can be quite different from the individual components of the mixture. The ability of polymers to be mixed to form a new polymer mixture with greatly expanded properties has given rise to this group of materials. This group of materials has experienced tremendous growth because mixing of two polymers offered a way to tailor the properties of a given product that was considerably more cost-effective than the synthesis of totally new polymers; e.g., development costs are less, turnaround time to enter new and old markets is quick, and mixing is easily acceptable to changing markets. It is this flexibility that has contributed to the growth of polymeric mixtures. The polymer mixtures referred to in this chapter are either alloys or blends. In a 1990 survey,[1] that quantity was reported to be 550 million lb with a projected growth rate of 7 percent per year. Polymer blends constitute over 10 percent of all resins sold. Detailed theoretical information on polymer blending can be found in Refs. 2 and 3. In this chapter, alloys and blends refer to the combining of actual resin materials as opposed to mixing resins with reinforcing agents such as fillers and/or fiber reinforcements.

Chemistry and Properties

Two kinds of polymer mixtures can be defined, alloys and blends, with subcategories under each. The basic difference between an alloy and a blend lies in their miscibility requirements, as defined by thermodynamics. At equilibrium, a mixture of two amorphous polymers can exist as a single phase or as two distinct phases. Which of these occurs is dictated by thermodynamics and governed by the free energy of mixing:

$$\Delta G_{mix} = \Delta H_{mix} - T \Delta S_{mix}$$

where miscibility is defined by ΔG_{mix} being negative.

The ability to produce a blend having a better combination of properties than those of the individual components strongly depends on the compatibility of the components to produce a miscible system. The enthalpy ΔH_{mix} is a measure of the energy associated with intermolecular interactions and is usually not dependent on molecular weight. In high-molecular-weight polymers, enthalpy is the dominant factor determining miscibility. The entropy term ΔS_{mix} is the energy change associated with change in the polymer chain configurational arrangements.

In general, the sign of ΔG_{mix} for nonpolar polymers does not favor miscibility, although miscibility in nonpolar polymers is more favored, in blends of low-molecular-weight rather than in high-molecular-weight polymers. In polar polymers the situation is slightly different because the functional groups present in different polar polymers have a strong attraction for each other during blending. The sign of ΔG_{mix} in polar polymers generally is favorable for miscibility.[3] Miscible polymer blends are usually characterized by a single glass transition temperature T_g, and immiscible blends have separate T_g values associated with each phase in the blend. So we define an alloy as being a polymer combination that has a high level of thermodynamic compatibility, between the components. Because of thermodynamic compatibility, an alloy therefore is miscible (forms a single phase), displays strong intermolecular forces, and forms a single T_g. Polymer blends, on the other hand, have less thermodynamic compatibility than alloys and have multiple T_g values characteristic of their discrete phases (immiscibility).

Figure 6.1 illustrates the morphologies of a miscible, immiscible, and partially miscible polymer blend. In Fig. 6.1a polymer A molecules intermingle with polymer B molecules to form a single-phase system. In Fig. 6.1b, polymer A forms a separate but discrete phase from polymer B, and the result is a two-phase system. Figure 6.1c represents an intermediate level of mixing where two polymers are partially miscible. In this case some regions in the blend are truly miscible and others are not, which shows up in their morphology as discontinuous boundaries. There are very few polymers that form true one-phase systems. Two-phase systems are probably the norm, but there is an abundance of blends that fall into the category of partially miscible.

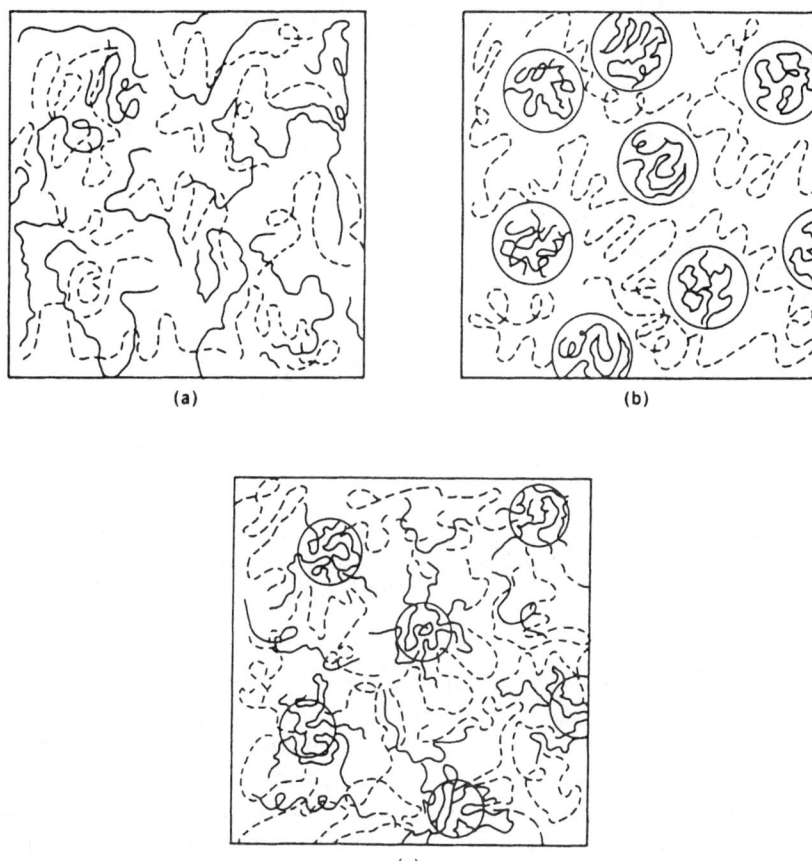

Figure 6.1 Morphologies of a blend of polymer A (solid lines) and polymer B (dashed lines): (*a*) miscible, (*b*) immiscible, (*c*) partially miscible. (Source: Ref. 3. Reprinted with permission from John Wiley & Sons.)

Preparation and Processing

Alloys and blends are prepared by mixing the individual components by melt mixing, solution blending, coprecipitation, and co-coagulation; and final processing of the blended polymers is accomplished by using standard thermoplastic techniques such as injection molding, extrusion, and blow molding. Of course, specific processing conditions are dictated by the nature of the alloy or blend. For those polymers that form a miscible blend, no other ingredients are added to enhance mixing compatibility to produce a stable one-phase system. For those blends that are formed from immiscible polymers, another ingredient

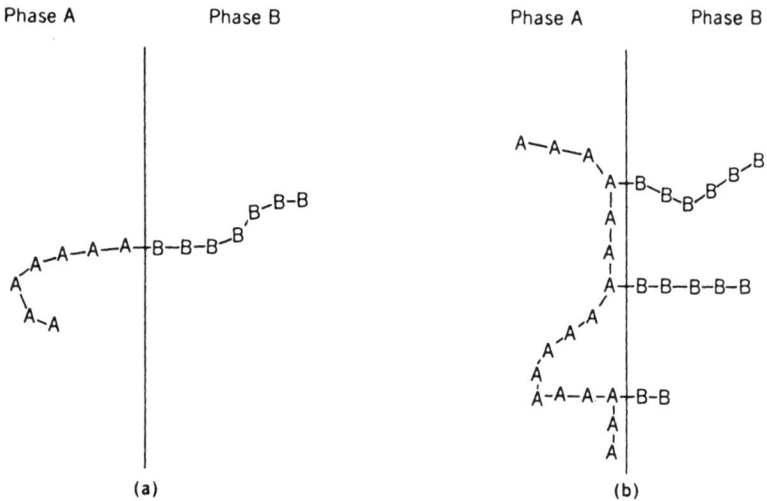

Figure 6.2 Penetration of block or graft copolymer compatibilizers into the A and B phases of a polymer blend: (a) block copolymer compatibilizer, (b) graft copolymer compatibilizer. (Source: Ref. 3. Reprinted with permission from John Wiley & Sons.)

is required to enhance the stability of the blend. Just as a surfactant is used to stabilize an oil-water mixture, a material called a *compatibilizer* is added to the polymer blend to stabilize and enhance the mixture. This compatibilizer is usually a block-and-graft copolymer[3] where one segment of the copolymer has an affinity for one part of the blend and the other segment of the compatibilizer has an affinity for the other part of the blend. This is akin to the hydrophobic and hydrophilic ends of a surfactant in an oil-water mixture. This stabilization of a blend with a compatibilizer is illustrated in Fig. 6.2.

Other Types

In addition to these physical mixtures of polymers to form various blends, there are other types that could be considered, in a broad sense, polymer blends. They are block-and-graft copolymers, interpenetrating polymer networks (IPNs), interpenetrating elastomer networks (IPENs), and liquid-crystal polymer blends (LCPBs). Block-and-graft copolymers were discussed in Chap. 2. IPNs and IPENs are a special class of materials that are combinations of two polymers in a network where at least one of the polymers is synthesized or cross-linked in the presence of the other polymer.

Most of the IPN-type polymers are thermoset materials. The two thermoset polymers are not cross-linked to each other but rather form an interpenetrating network of polymer chains. There are, however, thermoplastic IPNs and semi-IPNs where a thermoset polymer is formed within a thermoplastic polymer. LCP blends with either thermoplastic or thermoset polymers are a relatively recent development. Thermotropic liquid-crystalline polymers impart some unique properties to the blend into which they are incorporated, e.g., high strength at elevated temperature, high heat, chemical and radiation resistance, good dimensional stability, and flame resistance.[4] The LCPs form, if you will, fibrils that are encapsulated in the blend matrix, giving it a fiber-reinforcement effect. LCPs added to other polymers also have a profound effect on improving the rheology of the blend. More will be said about LCPs in Chap. 8.

Blend Properties and Characterization

The properties of a blend generally reflect the concentration of the components in the blend for simple cases only, and the property dependence on composition often varies in a complex way for most polymer blends. Properties depend on the concentration and molecular weight of components, compounding and molding procedure, and miscibility of components. Differential scanning calorimetry (DSC) is used to distinguish between compatible and noncompatible polymers. The former have a glass transition temperature T_g intermediate between those of the individual components that comprise the blend, whereas the latter exhibit two individual T_g values. The effects of varying the component fraction on the melting temperature of the blend are also determined by DSC. Furthermore, DSC is helpful in providing information so that phase diagrams can be developed that show regions of incompatibility, solution, and eutectics.

A listing of a number of blends and alloys is given in Table 6.1. Applications of some selected alloys and blends, i.e., those presently used in electrical and electronic applications, are listed in Table 6.2. Some typical properties of the blends are shown in Table 6.3. Because of the variety of formulation blends for thermoplastic as well as elastomers, it is recommended that the specific supplier of the blend or alloy be contacted for information on overall properties.

(Text continues on p. 137.)

TABLE 6.1 Alloys and Blends

Material	Type	Trade name	Supplier	Comments
TPE	Alloy	Bexloy	du Pont	High stiffness, low coefficient of thermal expansion, impact strength
PPO/PS	Alloy	Noryl	General Electric	Process ease, low moisture absorption, low cost, toughness
PPO/nylon	Alloy	Noryl GTX	General Electric	Dimensional stability, chemical resistance, low moisture absorption
Polyester/elastomer	Alloy	Valox 500	General Electric	Good flow, low coefficient of thermal expansion
PC/PBT	Blend	Xenoy	General Electric	Impact resistance, chemical resistance
ASA/PVC	Alloy	Geloy GY-1220	General Electric	Weather-resistant, process ease
ASA/PC	Alloy	Geloy XP-4001	General Electric	Impact, weather-resistant
PC/ABC	Alloy	Cycoloy	General Electric	Process ease, impact heat resistance
Modified PPO	Blend	Prevex	General Electric	Impact dimensional stability, good electrical properties
Nitrile rubber/PVC	Blend	Paracril Ozo	Uniroyal Chemical	Ozone resistance, superior processibility. Wire and cable jacketing, high abrasion and tear resistance
EPDM/silicone	Graft	Royaltherm	Uniroyal Chemical	Heat resistance, good strength, good electrical properties and compression set
PU/ABS	Blend	Prevail	Dow Chemical	Flexibility, toughness
PC/ABS	Blend	Pulse	Dow Chemical	Heat resistance, impact strength, processibility
PC/Polyester	Blend	Sabre	Dow Chemical	Heat resistance, low-temperature toughness, process ease
Halogenated olefin/ EVA/acrylate ester	Alloy	Alcryn	du Pont	Processibility
PC/ABS	Alloy	Triax 2000	Monsanto	Impact strength, processibility
Nylon/ABS	Alloy	Triax 1000	Monsanto	Impact strength, abrasion resistance, processibility
Nylon/elastomer	Blend	Zytel ST	du Pont	Toughness, fatigue resistance

System	Type	Trade name	Supplier	Features
PET/elastomer	Blend	Rynite SST	du Pont	Toughness, stiffness, solvent and high-temperature resistance
Nylon 6/PPO	Alloy	Dimension	Allied Signal	Good impact and load-bearing properties
Polyester blend	Blend	Vandar	Hoechst Celanese	Ductile, stiff, impact strength, chemical resistance
PS/ABS	Blend	Mindel A	Amoco	Hydrolytic stability, platability
PS blends	Blends	Mindel B,M,S	Amoco	Good electrical properties, hydrolytic stability, dimensional stability
Nylon/polyolefin Nylon/elastomers Nylon/polyesters	Alloy	Akuloy RM	DSM Corp.	High strength, stiffness, impact strength, heat resistance
ABS/StatRite C2300IDP	Alloy	Electrafil ABS 1200/SD	DSM Corp.	Antistatic properties
PC/ABS	Blends	Bayblend	Miles	Impact strength, dimensional stability, low moisture absorption
PC/PET	Blend	Makroblend	Miles	Impact strength, processibility, dimensional stability
PP/EPDM	Blend	Santoprene	Advanced Elastomer Systems	Good oil and heat resistance
PP/nitrile	Blend	Geolast	Advanced Elastomer Systems	High oil resistance
Polyolefin/rubber	Blend	Dytron	Advanced Elastomer Systems	Excellent electrical properties
PP/diene	Blend	Vyram	Advanced Elastomer Systems	Good combination of properties

SOURCE: Adapted from Ref. 1.

TABLE 6.2 Electrical Uses of Blends and Alloys

Material	Electrical application
Noryl	Housings
Prevex	Electronic enclosures, outlet boxes, solenoids
Makroblend	Connectors, covers, switches
Bayblend	General wiring devices, plugs, receptacles, control housings, telecommunication system connectors
Vandar	Telephone line splice cases, switches, connectors, housings
Mindel	Connectors
Dimension	Tubing, jacketing
Royaltherm	Wire and cable
Paracril	Heavy-duty cable jackets
Triax 2000	Housings
Rynite	Housings, bobbins, lamp sockets, connectors, switches
Alcryn	Wire and cable

TABLE 6.3 General Properties of Alloys and Blends

Property	Noryl[6]	Makroblend[7]	Bayblend[7]	Vandar[8]	Mindel[9]	Rynite[10]
Specific gravity	1.2	1.3	1.18	1.3	1.4	1.5
Water absorption in 24 h at 23°C, %	0.07	—	—	0.1	0.13	0.07
Tensile strength yield, lb/in^2 × 10^3	6.9–10.5	8.9	8.0	6.0	12.8	—
Tensile strength break, lb/in^2 × 10^3	6–6.9	7.4	6.5	—	—	15.5
Elongation, %	18–45	130	60	50	—	2.3
Tensile modulus, lb/in^2 × 10^3	374	—	390	—	—	—
Flexural modulus, lb/in^2 × 10^3	345	380	380	325	316	850
Izod impact, notched, ft · lb/in	3.5	13	7.5	3	1.1	1.2
Volume resistivity, Ω · cm	10^{16}	—	—	—	—	10^{15}
Dielectric constant, 10^3 H$_2$	2.6	—	—	—	3.3	3.1
Dissipation factor, 10^3 H$_2$	0.0007	—	—	—	0.009	0.004
Dielectric strength, V/mil	400	660	760	—	394	490

References

1. N. Albie, "Polymer Blending for Property Tailoring," *Plastics Compounding,* July/August 1992.
2. J. A. Manson and L. H. Sperling, *Polymer Blends and Composites,* Plenum Press, New York, 1976.
3. H. Mark, N. M. Bikales, C. G. Overberger, and G. Menges, eds., *Encyclopedia of Polymer Science and Engineering,* 2d ed., Wiley, New York, 1985–1989, vol. 3, p. 758; vol. 6, p. 125; vol. 12, p. 399; supplemental, p. 455.
4. M. McMurren, "What's Happening with ABCs, LFRTs, LCPs, IPNs?" *Plastics Compounding,* September/October 1992.
5. From *Technical Bulletin on Engineering Thermoplastics,* General Electric, Pittsfield, MA, 1993.
6. From *Technical Bulletin on Makro and Bay Blend,* Miles, Pittsburgh, PA, 1993.
7. From *Technical Bulletin on Vandar,* Hoechst Celanese, Chatham, NJ, 1993.
8. From *Technical Bulletin on Mindel,* Amoco, Alpharetta, GA, 1993.
9. From *Technical Bulletin on Rynite,* du Pont, Wilmington, DE, 1993.

Chapter 7

Processing of Plastics and Elastomers

There are approximately 14 plastic processing methods available for the conversion of compounded plastic to the finished product. The fundamental differences among the classes of polymers (i.e., thermoplastics, thermosets, and elastomers) dictate the processing method to be used. Furthermore, within each class, the differences in the thermal and melt properties of the polymers also dictate what processing methods are best suited for a given material. This chapter is designed to acquaint the reader with some basic information about polymer processing. The reader is directed to Refs. 1 to 4 for more in-depth discussions of the various processing methods. Note that the information presented here applies broadly to all classes of polymers. The plastic suppliers will provide specific details and data on their individual materials. It is strongly recommended that the plastic supplier be used as a resource to provide guidance in both design and processing of polymers for specific applications. The process sequence for processing all polymers involves heating the material to soften it, forcing the softened polymer into a mold or through a die to shape it, and cooling (thermoplastic) or curing (thermoset and elastomer) the molten polymer into its final shape.

With few exceptions, thermoplastics are supplied in the form of pellets, about 3 mm in size. Many plastics require drying before processing to eliminate absorbed water. One of the undesired consequences of excessive moisture in the plastic is bubbles in the melt which weaken the finished product.

Thermoset materials are generally liquids but can be solids. Unlike thermoplastics which solidify upon cooling to the desired shape, thermosets solidify as the result of a chemical reaction transposing the material to a cured nonremeltable product.

Elastomers can be either solid or liquid and are processed in a similar fashion to thermoplastics and thermoset plastics. Elastomers cure during processing to a cross-linked network, much as thermoset materials do, and cannot be remolded. The exception to this is the class of materials known as *thermoplastic elastomers,* which are processed like thermoplastics and have elastomeric properties but are remeltable.

Before processing, all thermoplastic, thermoset, or elastomeric materials can be compounded with a variety of additives to enhance processing and properties. These additives include fillers (organic and inorganic), vulcanizing and curing agents, accelerators, pigments, catalysts, antioxidants, antiozonants, plasticizers, tackifiers, and many more additives. These additives must be carefully chosen to be compatible with the processing method and the intended application of the finished product. Keep in mind that a plastic or elastomer is not necessarily a neat material, but can indeed be composed of many ingredients that serve many purposes, one of which is to improve the processibility of the material.

Processing Methods

Blow molding is a process whereby an extruded tube, called a *parison,* of heated thermoplastic is placed between two halves of an open split mold and is expanded against the sides of the mold by air pressure. This process produces hollow thermoplastic components.

Advantages include the low tool and die cost, fast production rate, and complex hollow-part production in one piece. Disadvantages include the limitation to hollow or tubular parts and that the wall thickness is difficult to control. Some applicable polymers are polyolefins, styrenes, vinyls, esters, amides, carbonates, and urethanes.

Variations of blow molding include blown film extrusion, injection blow molding, multilayer blow molding, and extrusion blow molding. A schematic of one type of blow molding is shown in Fig. 7.1.[5]

Autoclave molding is a process that consists of applying heat (up to 300°C) and pressure (up to 1000 lb/in^2) to a part fabricated by other methods (layup, winding, wrapping) and consolidating and curing the part in an autoclave. Other pressure-transfer media include water (hydroclaving) and a thermoclave process that employs powdered silicone rubber, which acts as a fluid under heat and pressure.

Advantages include high pressures that provide for good laminate consolidation and improved removal of volatiles for high-strength parts. Disadvantages are that capital and operating costs are high and that the part size is limited to the cavity of the autoclave. Some applic-

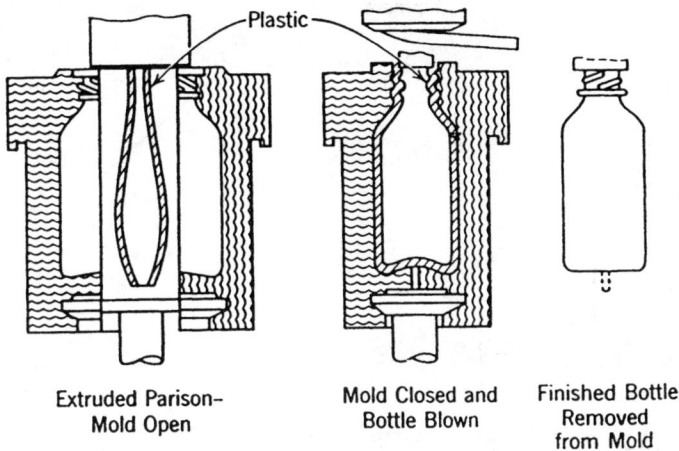

| Extruded Parison– | Mold Closed and | Finished Bottle |
| Mold Open | Bottle Blown | Removed from Mold |

Figure 7.1 Sketch of blow-molding process. (Reprinted with permission from the Society of the Plastics Industry, Inc.)

able polymers include most thermosets and some thermoplastics. Figure 7.2 is a schematic of an autoclave molding operation.

Casting is a process whereby a plastic material is heated to a fluid mass, poured into a mold, and cured without pressure. Curing is accomplished at room or elevated temperature depending on the resin used.

Advantages are the low mold cost, capacity to produce large parts with thick sections, good surface finish on parts, and the fact that very few finishing operations are required. The disadvantage is that the process is limited to simple shapes and is slow. Thermosetting polymers are mostly used in this process although some thermoplastics have been used, i.e., nylon.

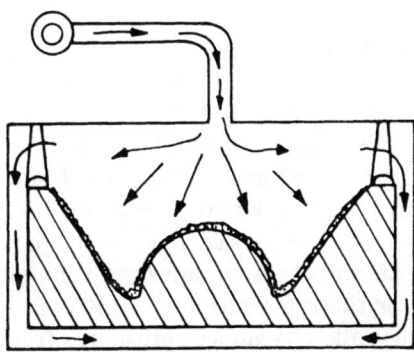

Figure 7.2 Sketch of an autoclave.

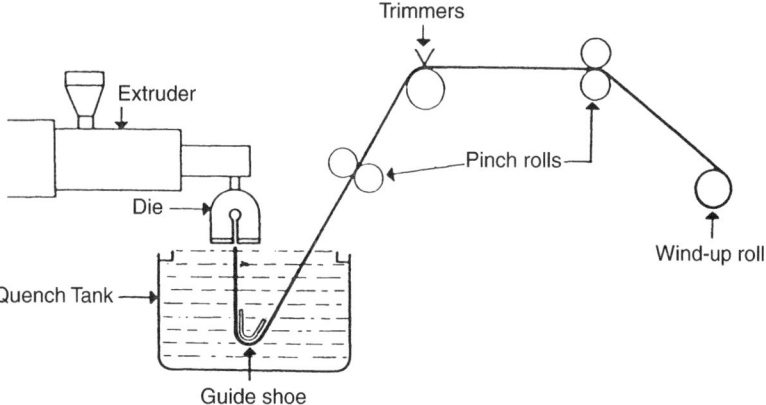

Figure 7.3 Sketch of a flat-film extrusion process. (Reprinted with permission from *Modern Plastics Encyclopedia*.)

Film casting is a particular kind of process usually limited to thermoplastics and thermosets that can be processed into thin film (10 mils or less). Casting can be done from the melt or from the liquid resin. In the former case, the melt is extruded through a slot die mold onto a chill roll where it is solidified and then removed. Films up to 100 in wide are available, and process speeds can reach 1000 ft/min for melt processible films and about 1500 ft/min for liquid cast films. Figure 7.3 is a sketch of an extrusion casting process.[6]

Centrifugal casting is a process whereby cylindrical articles such as pipe and drums can be made by rotating a mold charged with granular polymer. The granules cover the wall of the mold and are fused in place by application of heat from the outside.

Calendering is a process whereby a plastic mass is worked into a sheet of uniform thickness by passing it through and around a series of heated or cooled rolls. Precise control of temperature, pressure, and rotational speed is needed to ensure proper calendering. This process is inexpensive and can produce almost stress-free parts, i.e., the final product is isotropic. Calendering is limited to sheet materials. It is applicable to thermoplastics, mostly to PVC film and sheet products. Figure 7.4 is a sketch of a calendering process.

Compression molding is a process whereby a thermoplastic or thermosetting resin is placed in a heated mold which is then closed, and heat and pressure are applied, causing the material to flow and fill the mold. If the material is a thermoplastic, the mold is cooled and the part removed. If the material is a thermoset or an elastomer, then it is left in the mold to cure and removed after the resin hardens (fully cured) and is cooled. Compression molding usually operates at pressures from

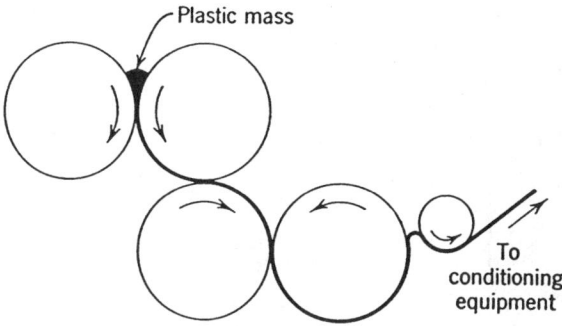

Figure 7.4 Diagram of a four-roll calender.

20 to 1000 tons, and the part size can vary from 8 in² to 5 ft². Mold temperatures usually range from 100 to 200°C, but can go higher if exotic high-temperature composite materials are used. Compression molding has a low part cost, provides fast production rates, and produces little waste. It is not suitable for intricate parts, close tolerances, undercuts, or delicate inserts. Capital equipment costs are high. Figure 7.5 illustrates several types of compression molds.

Cold forming is a process similar to compression molding, and the material is charged into an open or split mold. No heat is applied in this process—just pressure, and the molded part is placed in an oven for

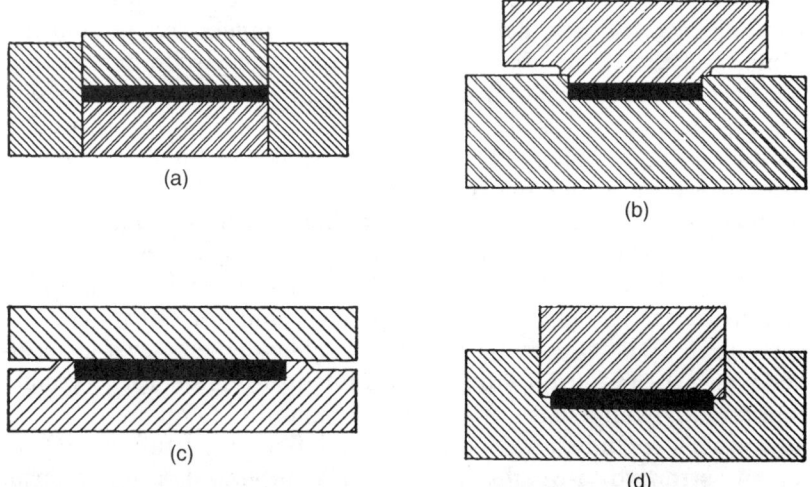

Figure 7.5 Types of compression molds: (*a*) Fully positive, (*b*) semipositive, (*c*) flash, (*d*) landed-positive. (Source: Ref. 7.)

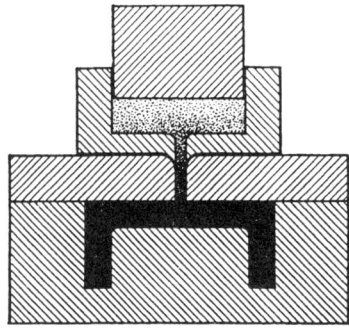

Figure 7.6 Schematic of a transfer mold. (Source: Ref. 7.)

final curing. It is low-cost and rapid. Parts have poor surface finish and do not have good dimensional accuracy.

Transfer molding is a variation of compression molding applied to thermoset materials. The material to be molded is placed in a heated cavity and forced into the mold proper when the desired viscosity is attained. A sketch is shown in Fig. 7.6. Advantages are that thin sections and delicate inserts can be easily molded, the flow control of material is superior to that in compression molding, and the production rate is rapid. Parts have good dimensional accuracy.

Extrusion is a process whereby a heated plastic is forced or driven, usually by a rotating screw, through a die having the desired dimensions of the product to be formed. The material is then cooled back to its solid state as it is held in the desired shape. Extrusion provides a rapid and efficient means of producing long, continuous shapes. Advantages are the low tool cost, variety of complex shapes that are possible, and the rapid production rate. Extrusion is limited in that close tolerances are hard to attain; openings must be in the extrusion direction, and the process is confined to shapes of uniform cross section along its length. A sketch of an extruder is shown in Fig. 7.7. Thermoplastics, thermosets, and elastomers can all be extruded.

Filament winding is a process of winding continuous fibers in the form of rovings, that have been saturated with resin, on a mandrel surface having the shape of the desired finished product in precise geometric patterns. The lay-down patterns are computer-controlled, and the mandrel is usually rotated with the fiber delivery system positioned above the mandrel. After the winding is complete, the assembly is placed in an oven to cure. After cure the mandrel is removed.

Advantages include precise control and fiber orientation and an excellent strength-to-weight ratio. Parts produced are uniform. Filament winding is limited to shapes that have a positive curvature. Thermoset materials are the applicable resins for filament winding.

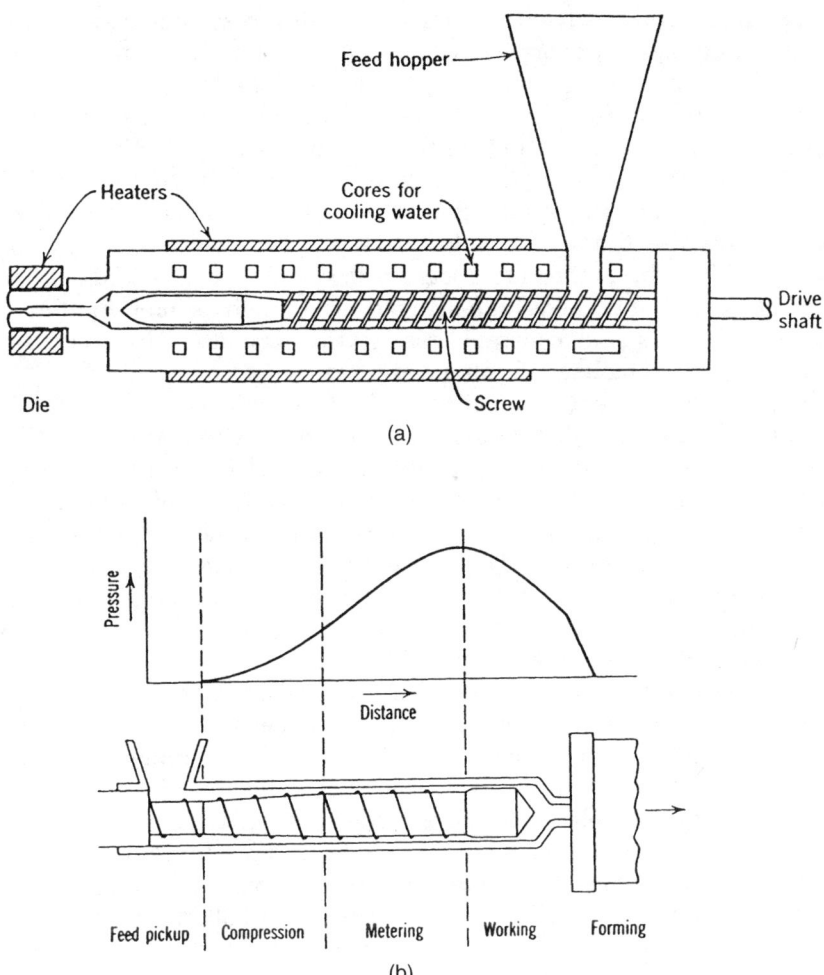

Figure 7.7 (a) Diagram of a plastics extruder and (b) details of an extruder screw. (Source: Ref. 5.)

Hand layup is a process that consists of placing a mixture of resin and reinforcement into a mold by hand. The mixture is rolled to consolidate the part. The resin-fiber mixture is allowed to cure at room temperature without pressure. Advantages include low cost and the lack of part size or shape limitation. The disadvantage is a large variability in part performance and appearance. This process is limited to room-temperature curing thermosets such as epoxies, some phenolics, and polyesters.

Spray-up is a process whereby resin and chopped fibers are sprayed simultaneously from two separate guns onto a mold surface. The sprayed mixture is then rolled flat by hand and is cured at room or elevated temperature. Advantages include the low cost, high rate of production, and capacity to produce complex parts. Limitations are the lack of reproducible parts and that the process requires skilled workers.

Pultrusion is a process whereby reinforcing filaments are saturated usually with a thermosetting resin and pulled through an orifice in a heated die. The resin is cured as it passes through the heated die. Epoxy, polyester, and some silicone resins are used in pultrusion. Advantages are that (1) complex part profiles can be produced in unlimited length, (2) the labor costs are low, and (3) the very high fiber content (75 percent) yields high-strength parts. Pultrusion is limited to noncurved parts and constant-cross-section profiles. The electrical industry is a strong user of pultruded parts.

Pulforming is an extension of the pultrusion process and was developed to produce curved shapes. Fibers saturated with resin are pulled through a rotating wagon-wheel type of die and into a recirculating mold which clamps around the saturated resin-fiber bundle and cures. As the mold moves around the table, it opens, and the part is removed and the process continued. The advantage of pulforming is the ability to form curved parts. Limitations include the fact that complex tooling arrangements are required and dies are expensive. A diagram of the pultrusion process is shown in Fig. 7.8.

Injection molding is a process whereby a polymer (thermoplastic or thermoset) is preheated in a cylindrical chamber to a temperature at which the polymer will flow, and then it is forced under pressure through a nozzle into a relatively cold, closed mold cavity. The resin solidifies, and the mold is opened and the part ejected. In the injection molding of thermosets, the resin is heated to reduce its viscosity and is forced into a hot mold and held under pressure until it cures. The cured part can be ejected while hot. Temperatures up to 300°C are common, and pressures range from 10,000 to 30,000 lb/in^2. Advantages include very high production rate, low cost per part, little finishing of part required, good part dimensions, and reproducibility. Limitations include high tool and die costs. Applicable resins include phenolic,

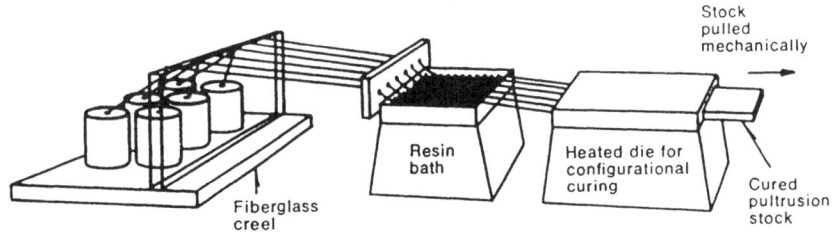

Figure 7.8 Diagram of the pultrusion process. (Source: Ref. 1.)

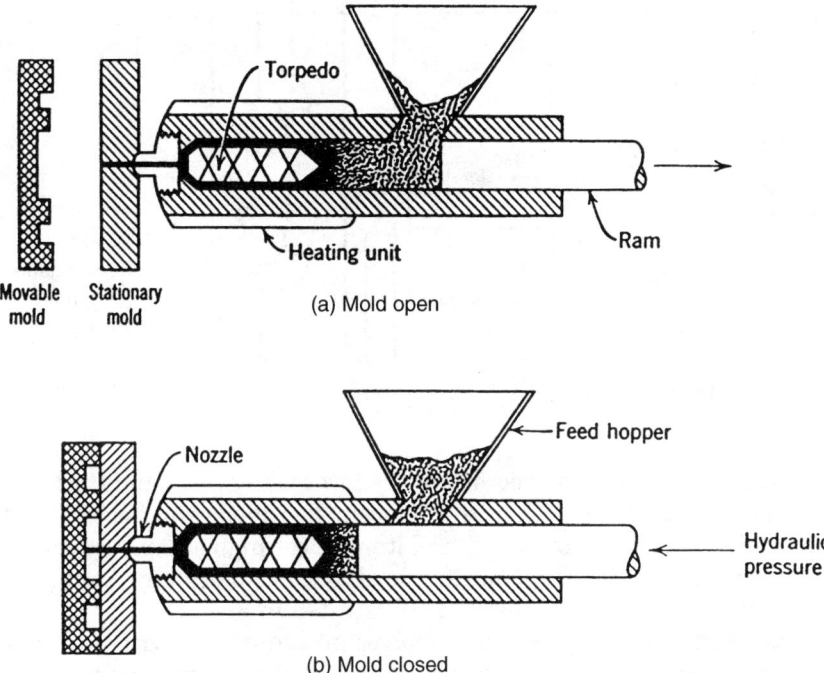

Figure 7.9 Diagram of injection molding. (*a*) Mold open; (*b*) mold closed. (Source: Ref. 5.)

epoxy, melamine, vinyl ester, and diallyl phthalates. A schematic of the injection molding process is shown in Fig. 7.9.

Reaction injection molding (RIM) consists of precisely metering and mixing two separate liquid reactant streams into a heated mold, where the part is cured under low pressure, usually about 50 lb/in^2. Advantages include low cost and the production of large parts. It is limited to polyurethane resins. A schematic of the process is shown in Fig. 7.10.

Rotational molding is a process where a powder or liquid resin is placed in a mold and heated while being rotated in the axis of two planes until the contents in the mold have fused to the inner walls of the mold. The mold is cooled, and the part is removed. The products produced by the molding method are seamless and hollow. Advantages include low cost and long-lasting molds. The process is limited to hollow parts, and production rates are slow.

Thermoforming is a process that involves the forming of a thermoplastic sheet into a desired shape by applying heat and pressure or vacuum to force the hot plastic sheet against a mold face. There are many variations of the process: plug-assisted, straight forming, mechanical drawing, drape forming, matched-mold forming, snap-back forming,

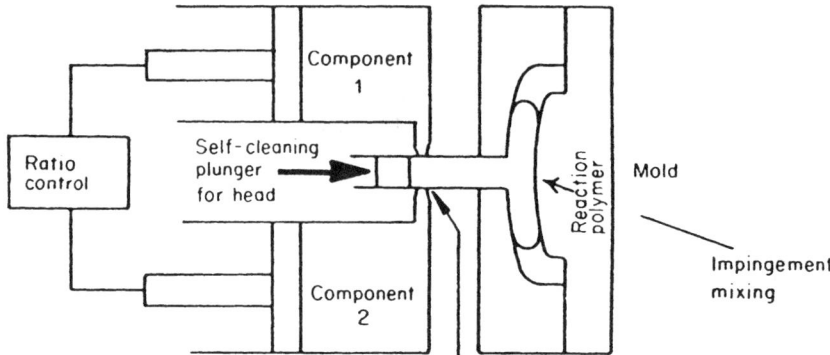

Figure 7.10 Operating principle of RIM. (Source: Ref. 2.)

etc. Advantages of this process are the low tooling costs and that large parts with thin sections can be produced. Thermoforming is limited to parts of simple configurations, and it produces high scrap. Amorphous polystyrene, PVC, and PMMA are used in thermoforming. A schematic of the thermoforming process is illustrated in Fig. 7.11.

Laminating is a process that involves pressing together two or more layers of resin-impregnated fabric, paper, or fiber under heat and pressure to cure and consolidate the stack. The resin is the binding mater-

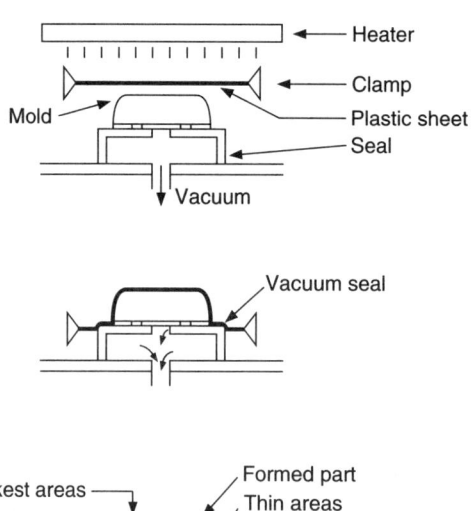

Figure 7.11 Thermoforming.

ial, and it can be thermoplastic or thermoset. It is almost always the latter for electronic applications. The reinforcement can be cotton, paper, glass, synthetic organic fibers, graphite fabric, and other inorganic fibers. The major thermoset resins used include epoxy, phenolic, melamine, polyimide, polyester, silicone, cyanate ester, and bismaleimide. Specific resins and laminates used in printed-circuit board substrates are addressed in Chap. 12. Laminating can be considered a special case of compression molding. The process starts with impregnating a reinforcement with liquid resin (solvent, melt, or 100 percent solids liquid resin) and passing the impregnated web through a drying oven to remove the solvent and to partially polymerize the resin. The dried impregnated fabric (it can range from very flexible to very rigid) is cut, stacked, and pressed. Pressures can go as high as 3000 lb/in^2 and temperatures as high as 300°C, but normally they are about 500 lb/in^2 and 175°C for 1 h. A schematic of the impregnation process is shown in Fig. 7.12. The process is inexpensive, production is rapid, and the product can be made to close tolerances. For laboratory evaluation of resins and fabrics, page-size sheets of material can be hand-dipped, dried, and pressed without difficulty.

Embedding of electronic components is another process used with plastics to protect the components from the environment. The embedding processes include casting, potting, impregnating, encapsulation, and transfer molding. *Casting* is a process involving the pouring of a liquid resin into a mold, curing it, and removing the cured product. *Embedding* is the process of surrounding a part in a liquid resin in a mold, curing the resin, and removing it.

Encapsulation is a process whereby a part or component is completely enclosed in a resin by either dipping or spraying the part with the resin. The part is usually irregular in shape.

Impregnation is a process that involves filling the interstices of an electronic component, e.g., a reinforcement, a coil, motor windings, etc., with a low-viscosity resin system in order to consolidate the component.

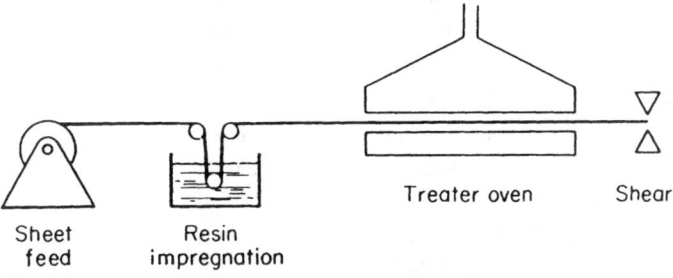

Figure 7.12 Horizontal treater tower.

Potting is very similar to embedding, except the mold in the potting operation remains a part of the entire system. Transfer molding, which was described earlier, is the process of transferring a catalyzed material under pressure into a mold which contains the part to be embedded. All the processes described above can be grouped under one term, *embedding,* and are limited only to liquid resin systems, i.e., 100 percent solids (no solvents are present in these liquid resins). More information on the resins used, properties, and applications will be given in Chap. 12.

Radiation processing is another method of processing plastics, both thermoplastics and thermosets. Processing is not limited to thermal heating methods, and radiation whether actinic or high energy can be efficiently used for plastics processing. However, just as in thermal processing methods where high temperatures can damage plastics, high radiation doses act in a similar manner. Careful control is needed in the processing of plastics using either thermal or radiation techniques. Radiation processing methods include electron beam, ultraviolet light, infrared, microwave, and high-frequency curing. The energy and wavelength associated with radiation processes are shown in Table 7.1. Bond energies of organic compounds usually lie in the range of 30 to 200 kcal/mol (1.5 to 8.5 eV per bond). The carriers of radiation, e.g., photons, electrons, or ions, can cause significant changes to take place when they collide with target materials. These changes can be ionization, excitation to higher energy states, or fragmentation of the target material; and the amount of change taking place is dependent on the energy absorbed. The motive behind the development and use of radiation processing has been the need to develop materials and processes that are more environmentally compatible. The advantages of radia-

TABLE 7.1 Classification of Radiation Spectrum

Radiation type	Energy per quantum	Wavelength
Infrared	0.01–1.6 eV	0.78–100 μm
Visible	1.6–3.3 eV	3800–7800 Å
Ultraviolet	3.3–6.2 eV	2000–3800 Å
Vacuum ultraviolet	6.2–310 eV	40–2000 Å
x rays	0.0003–1.5 MeV	0.008–40Å
γ rays	0.008–9 MeV	0.0014–1.6 Å
α rays	1.0–10 MeV	—
ß rays	0.02–13 MeV	—
Accelerated electrons	0.25–15 MeV	—

tion processing over thermal processing are that (1) the former requires less process time and less energy, is environmentally safe, does not require elevated temperatures but rather can be carried out at room temperature, (2) materials can be processed (cured) in seconds rather than hours, and (3) materials can be used with heat-sensitive parts or components which are commonly found in the electronics industry. Some limitations are that the up-front capital costs for equipment are high, the material cost is only slightly higher for radiation-curable resins than for thermally cured formulations, and operating costs are less; radiation-cured resins are thickness-limited and pigment-limited (which one is dependent on the radiation process method).

Electron beam processing involves the treatment of resins to ionizing radiation. Electrons are emitted from a heated filament in an evacuated chamber; they are accelerated and deflected by electrostatic and magnetic fields and are directed toward their target through a thin foil window.

Ultraviolet light curing of polymers involves exposing the material to radiation with wavelength from 200 to 400 nm. The radiation is generated by either medium-pressure mercury lamps or microwave-excited lamps. The opacity of the resin affects the efficiency of the cure. Thickness is usually limited to well below 10 mils. Oxygen can inhibit both uv and electron beam processes.

The electronics industry utilizes uv and electron beam processing to a great degree in the manufacture of printed-circuit boards, conformal coatings, adhesives, coil lead encapsulants, solder masks, resists, and ink processing.

Infrared, microwave, and high-frequency curing all rely on heating by electromagnetic radiation of different frequencies. The materials used must have reasonable absorption of the appropriate radiation with some degree of penetration. These processes differ from oven heating because the energy can be absorbed directly by the material at a particular frequency in depth rather than by conduction to the bulk of the material from the surface. The advantage of these three methods over convective heating is that one need not heat the intervening airspace. The frequency range used for most organic materials is from 2 to 15 μm. Infrared processing is used mostly as a heating source to drive solvents from coatings. It is not used to cure coatings of 100 percent solids.

Microwave processing covers the frequency range from 1 to 90 GHz and has been used to cure inks and coatings, cure molded parts, and fabricate foamed plastics as well as resin-impregnated fabric.

Additional information on radiation curing and equipment supplies can be found in Refs. 4 and 9 to 12.

References

1. S. S. Schwartz and S. H. Goodman, *Plastics Materials and Processes,* Van Nostrand Reinhold, New York, 1982.
2. R. Greene, ed., *Modern Plastics Encyclopedia,* McGraw-Hill, New York, 1992.
3. R. O. Babbit, ed., *The Vanderbilt Rubber Handbook,* R.T. Vanderbilt Co., Norwalk, CT, 1978.
4. D. E. Harmer and D. S. Ballentine, "Radiation Processing," *Chemical Engineering,* April 19, 1971, p. 98.
5. F. W. Billmeyer, Jr., *Textbook of Polymer Science,* Wiley, New York, 1965, p. 492.
6. B. Golding, *Polymers and Resins,* Van Nostrand Co., New York, 1959, p. 599.
7. H. R. Simonds, A. J. Weith, and M. H. Bigelow, *Handbook of Plastics,* 2d ed., D. Van Nostrand Co., Princeton, NJ, 1949.
8. C. Hepburn, *Polyurethane Elastomers,* Applied Science Publishers, New York, 1982.
9. G. A. Senrich and R. E. Florin, "Radiation Curing of Coatings," *Rev. Macromal. Chem. Phys.,* vol. C24, no. 2, p. 239, 1984.
10. *Journal of Radiation Curing,* Technology Marketing Corp., Norwalk, CT.
11. "Microwave Processing of Materials," Symposium L at the Materials Research Society Meeting, San Francisco, April 16, 1990.
12. S. H. Goodman, *Handbook of Thermoset Plastics,* Noyes Publications, Park Ridge, NJ, 1988, Chap. 10.

Chapter

8

High-Performance Polymers

High-performance polymers play an important role in both the electrical and electronics industries because each industry has extended the edge of technology which has required the development of plastics with higher-temperature capability. Examples of applications for high-performance polymers include insulation for thin-film wiring in multichip packaging and interconnect devices, passivation layers, coatings, adhesives, resists, encapsulants, and increased circuit density applications. In order for high-performance polymers to be useful, they must first be processible, which is possibly the single most important property requirement of any polymeric material. Processibility is particularly important for high-performance polymers because these materials tend to be more difficult to process than the most common engineering polymers. High-performance polymers are defined in this chapter as materials which have long-term service capabilities above 170°C and/or possess special properties that make the polymer uniquely qualified for electronic applications.

The high-temperature polymer area is a $2 billion per year business which has had a growth rate of about 15 percent per year since 1984. This area represents a small but rapidly growing segment of the plastics market and is expected to grow at about 8 percent per year in the 1990s.[1] This growth is attributed to the increasing need for high-temperature materials in the electrical, electronic, and aerospace industries, which together account for 50 percent of the high-temperature polymers market. In addition, the automotive field is creating opportunities for under-the-hood plastics to meet the needs of fuel economy standards (more lightweight materials with high-temperature properties). Furthermore, there is also a need to replace metals in corrosive and hostile environments as well as a need in niche markets such as

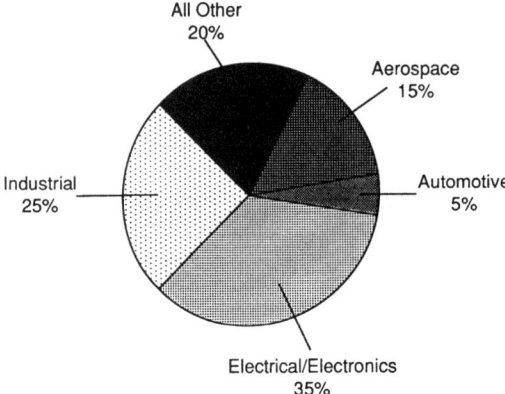

Figure 8.1 Estimated western world demand for high-temperature polymers by end use, percent of dollars, 1992. (*Source:* Ref. 2.)

medical, dental, and some unique industrial applications. The estimated distribution of high-temperature polymers in the world is shown in Fig. 8.1.[2] There are currently more than 15 different polymers that fall into the high-temperature performance category. They are the polyaramids, bismaleimides, fluoropolymers, liquid-crystal polymers, polyamide-imides, polyimides, polybenzimidazoles, polyetherimides, polyketones, polyphenylene sulfide, polysulfones, silicones, cyanate esters, benzocyclobutenes, silicone-carbon (SYCAR resins), and triazine resins. These materials are all commercially available today. Many more experimental high-temperature polymers are presently being evaluated that can easily double the above list. Some of these experimental materials are the polybenzoxazoles, polyquinoxalones, polyphenylquinoxalines, polyphenylenes, and polyspirolactams. This chapter will briefly review those classes of polymers that can provide extended service function at elevated temperatures. It includes only those materials that are used in the electrical and electronics industries.

Thermal Stability

High-performance polymers were defined previously as materials that could operate at temperatures greater than 170°C for extended periods and/or that possess a unique property such as low dielectric constant, low dissipation factor, low coefficient of thermal expansion, etc., that makes the material useful for high-performance electrical applications. Advanced design concepts especially in the electrical industry have increased the urgency of the need for more thermally stable materials. Numerous polymer systems have been studied, but only a few have achieved commercial status. The polymer chemist's goal—to make plastics as strong as steel, clear as glass, light as a feather, heat-resistant like quartz, nonflammable, processible like putty, and cheap as dirt—is admittedly difficult, exaggerated, and maybe impossible to achieve, but nevertheless it is our goal. This would be the ultimate high-performance polymer.

In examining the thermal stability of a polymer, we need a material that will have useful properties over a given temperature range. The material must maintain its properties (electrical, mechanical, etc.) over an acceptable time period at the intended-use temperature. It must not soften or melt significantly, and it must resist environmental factors such as oxidation, hydrolysis, chemical attack, and radiation. When plastics are heated to elevated temperatures, both reversible and irreversible changes occur. Melting and shaping as in thermoplastics are an example of reversible change, and curing or cross-linking of thermosetting plastics is an example of irreversible change. Undersirable decomposition processes can and will occur at sufficiently high temperatures. These processes manifest themselves as a reduction in molecular size (chain scission), yielding lower-molecular-weight products or highly cross-linked products. In either case this decomposition destroys the polymer properties. Intensive research is aimed at synthesizing better temperature-resistant plastics and has identified certain structural requirements that a more thermally stable polymer should have: (1) a high molecular weight to provide strength, flexibility, and toughness; (2) a minimum number of oxidizable groups; (3) stable aromatic or heterocyclic structural units; (4) connecting linkages to impart flexibility with high thermal stability; (5) resistance to thermal depolymerization; (6) process ease; and (7) useful properties for extended periods at elevated temperatures.

Polyimides

Among all the classes of thermally stable polymers which have emerged from the laboratory in the past 20 years, the polyimides have enjoyed the greatest commercial acceptance and the broadest utilization.[3] In the last 7 years, the imide class of polymers has undergone a renaissance with a wide range of materials that satisfy both processing and performance requirements. These materials have found their widest use in the electrical packaging industry. There are three basic types of polyimides: condensation (CPI), addition (API), and thermoplastic (TPI) with a broad subgroup under API. The polyimides as a class of materials are used in a variety of ways in the electrical indus-

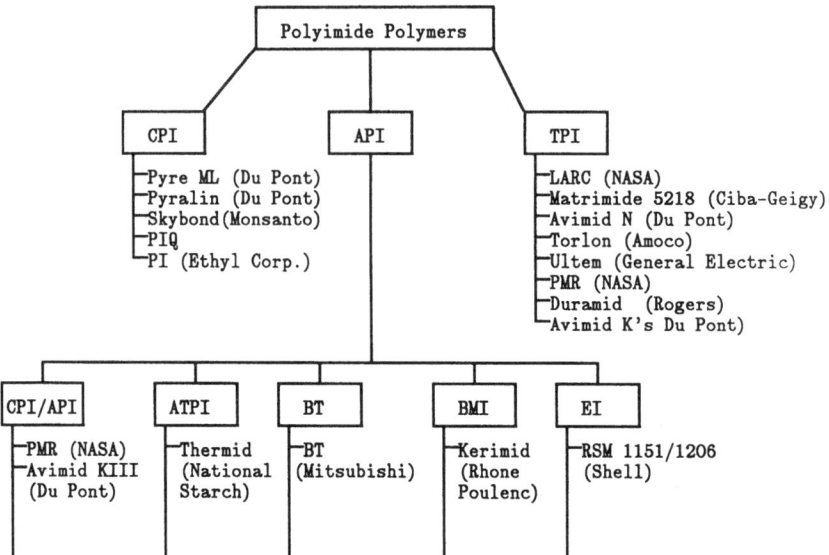

Figure 8.2 Classification of polyimides.

try, e.g., as adhesives, films, moldings, matrices for advanced composites, separation membranes, fibers, foams, coatings, resists, interlayer dielectrics, and alpha-ray shielding. Figure 8.2 shows the varied classes of polyimides. Specific products are identified also under each class. The CPI/API is a hybrid, the ATPI is a polyimide terminated with acetylene groups, BT is a bismaleimide triazine, BMI is a bismaleimide, and EI is an epoxy imide. The CPIs are mainly used in one segment of the electrical industry, the semiconductor area; however, finished forms of CPI resins such as films and moldings have broad applications in the electrical industry as a whole. The APIs are mainly used by the electrical industry as matrix resins for reinforced composites such as in printed-wiring boards. The TPIs are used broadly in the electrical industry for such applications as electronic coatings, protective coatings, insulation covering, laminates, wire enamels, adhesives, molding powders (compression and injection molding). A comparison of the various polyimide types is given in Table 8.1. The chemical structures for the three types are shown in Table 8.2. An excellent source for more detailed information on polyimides as well as supplier information is Ref. 4. The condensation polyimides with T_g values in excess of 350°C have the highest thermal stability of all the polyimides. They are also the most difficult to process, the highest in cost, and the most chemically resistant. The CPIs are available in film, powder, stock

High-Performance Polymers 157

TABLE 8.1 Comparison of Polyimide Types

Polyimide	Advantages	Disadvantages
CPI	Excellent thermal stability.	Difficult process condition. Voids. High cost
CPI/API	Wider process window. High thermal stability	Volatiles. High-temperature cures. Brittle
ATPI	High thermal stability. Process window	High-temperature cures. Volatiles. Small
BMI	Epoxylike processing. High T_g	Lower stability. Somewhat brittle. Volatiles
BT	Good thermal stability	Somewhat brittle. Volatiles
EI	Conventional solvents. Epoxylike process. High T_g	Lower thermal stability. Volatiles
TPI	Better processibility. No imidization. Volatiles	Low thermal stability. Volatiles

TABLE 8.2 Major Polyimide Types

General structure	Type
(imide ring structure)	TPI
(HOOC/COOH tetracarboxylic structure)	CPI
(bismaleimide structure)	API

shapes, or solvent solutions depending on the particular polyimide and the manufacturer. Two types of CPIs are available: fully imidized linear polymers and partially imidized polymers. In the former, no volatiles are evolved as a result of the imidization process because these polymers are already imidized. However, volatiles from solvent evaporation during the drying process are evolved. In the latter case, water of imidization as well as solvent is evolved during processing, increasing the potential for creating voids in the final product. Because of the void problem, these partially imidized CPIs cannot be used to mold thick parts. They are, however, used in the preparation of thin films and coatings. These partially imidized CPIs are processed in a form suitable for conversion to the final product. That intermediate form is called the *amic-acid* and is a solution of the polymer in a high-boiling-point solvent such as dimethylacetamide, dimethylformamide, or N-methylpyrrolidone. This solution varies in solids from 5 to about 20 percent and from 500 to 3000 cp in viscosity. A schematic representation of the amic-acid to imide conversion is shown in Fig. 8.3. As a CPI cures, i.e., from the outside in, a dense skin is formed that traps volatiles, creating voids, blisters, and delamination areas. Long, slow cure cycles are used to minimize the problems of trapped volatiles, but still up to 10 percent void volume is obtained. Because of these severe processing problems, the partially imidized CPIs have not found widespread use for thick parts in the electronics industry. Even with the preimidized polyimides there is a thickness limitation in processing these materials. A wide variety of polyimide products can be produced by varying the initial reactants, i.e., the dianhydride and the diamine; and although all these variations are not used commercially in the electronics industry, they represent an eclectic mixture of polyimide types with a broad range of processing characteristics and properties. A par-

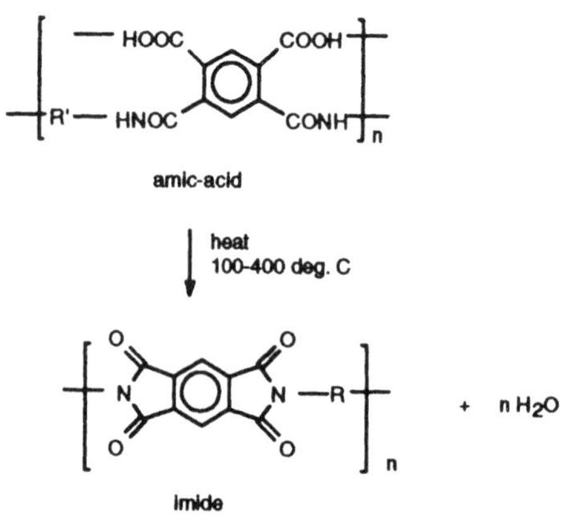

Figure 8.3 Conversion of amic-acid to imide for a CPI.

TABLE 8.3 Typical Reactants in the Synthesis of Polyimides

Diamine	Abbreviation	Dianhydride	Abbreviation
H₂N–⟨○⟩–O–⟨○⟩–NH₂	ODA	(structure)	PMDA
H₂N–⟨○⟩–C(CH₃)₂–⟨○⟩–NH₂	DABPA	(structure)	DSDA
H₂N–⟨○⟩–⟨○⟩–NH₂	DADP	(structure)	BTDA
H₂H–⟨○⟩–C(=O)–⟨○⟩–NH₂	DDBP	(structure)	6FDA
H₂N–⟨○⟩–S(=O)₂–⟨○⟩–NH₂	DDSO	(structure)	BPDA
		(structure)	ODPA
H₂N–⟨○⟩–C(CF₃)₂–⟨○⟩–NH₂	6F		

tial list of some typical dianhydrides and diamines is presented in Table 8.3. Glass transition temperatures for the CPI polyimides range from 200 to 400°C. A range of dielectric constants can also be obtained by varying the reactants. Values ranging from 2.39 for a 6FDA dianhydride and the 6F diamine to 3.22 for Kapton at 10 GHz have been obtained.[4] Furthermore, the low electrical leakage of these polyimides makes them ideal for interlayer dielectric material in multilayer integrated-circuit structures.[5]

TPIs

Most polyimides are generally classified as thermoset resins because of their processing conditions. There is a thermoplastic class of polyimides that can be processed by using conventional thermoplastic methods. These materials are preimidized so that there is no water of imidization driven off during the conversion process. This allows the production of polyimide parts with low void content. There are several ways to prepare thermoplastic polyimides, and one technique is to introduce a pendant side group on the polymer chain such as a phenyl group. Another is to introduce flexible linking units into the polymer chain, e.g., siloxane or phosphorous. Fluoroalkylene groups can also be introduced instead of siloxane and phosphorous to improve the thermoxidative stability. The polymer with the fluoroalkylene group is soluble in conventional solvents and has a T_g value around 340°C. The polymer is based on the monomeric reactants of 6FDA and 6F, shown in Table 8.3. The LARC polymer is also a thermoplastic polyimide developed by NASA Langley, and it has found use primarily as an adhesive. It is based on the monomeric reactant benzophenone tetracarboxylic dianhydride and 3,3'-diaminobenzophenone. Its T_g is about 250°C, it has a high melt viscosity, and it retains residual solvent. Rogers Corp. makes a version of this polyimide called Durimid. Other TPIs include General Electric's Ultem, which has good melt processibility and an excellent combination of properties (T_g of 220°C), and Amoco's Torlon (T_g of 227°C), a polyamide-imide. Some properties of these polyimides are listed in Tables 3.2, 3.3, 3.5, 3.13, and 3.14. Ciba-Geigy also developed a thermoplastic polyimide designated XU-218, whose structure is shown here:

It is a fully imidized polymer soluble in such solvents as methylene chloride, tetrahydrofuran, and N-methylpyrrolidone. Its T_g is 320°C. Its properties are shown in Tables 8.4 to 8.6.

APIs

Those polyimide polymers that are converted during their processing operations to the imidized structure are difficult to process because of the volatile by-products released during processing. Thus the processing difficulties encountered with the CPIs led to a search for new approaches to synthesize new materials that could be more easily processed and fabricated into a variety of product shapes. This search lead to the development of addition polyimides which, in contrast to the CPI materials, do not evolve volatile by-products from imidization. These polyimides are formed from functionalized preimidized olig-

TABLE 8.4 Matrimid 218 Polyimide Resin, 1-Mil Films

Temperature, °C	Environment	Time to embrittlement, h
200	Air	>2000
225	Air	925
250	Air	250
250	Nitrogen	>2500
300	Vacuum	>2500

TABLE 8.5 Electrical Properties of Matrimid 218 Polyimide Resin

Property	Temperature, °C	Value
Dielectric constant		
100 Hz	25	3.4
1000 Hz	25	3.3
1000 Hz	60	3.0
1000 Hz	100	3.0
1000 Hz	150	3.0
1×10^6 Hz	25	3.0
Dissipation factor		
100 Hz	25	0.0061
1000 Hz	25	0.0026
1000 Hz	60	0.0039
1000 Hz	100	0.0061
1000 Hz	150	0.0073
1×10^6 Hz	25	0.0091
Dielectric strength, kV/mil	—	5.6
Volume resistivity, $\Omega \cdot$ cm	25	4.4×10^{16}

TABLE 8.6 Tensile Properties of Matrimid 5218 Film*

Tensile strength	Temperature, °C	Tensile strength, klb/in²		Tensile modulus, klb/in²	Elongation, %	
		Yield	Break		Yield	Break
Machine direction	25	—	22.5	560	—	17
	100	14.4	18.4	370	8.8	30
	150	11.9	17.8	330	7.7	41
	204	9.1	14.4	363	6.4	47
	260	5.9	8.8	20	3.9	38
Cross-machine direction	25	—	14.6	460	0	10
	100	11.7	11.1	340	8.0	32
	150	9.0	8.3	273	6.8	49
	204	7.0	6.4	275	5.4	51
	260	4.7	4.3	260	3.4	69

*0.5-mil film uniaxially stretched from 1.0-mil film oriented in machine direction.

omers and monomers that either homopolymerize or copolymerize with other suitably substituted monomers and oligomers to form high-temperature-resistant structures without the evolution of chemical reaction by-products. This class of polymers includes the hybrid CPI/API types such as the PMR and LARC polyimides developed by NASA; the acetylene-terminated polyimides, designated Thermid and developed by National Starch; the bismaleimide developed by Rhone Poulenc, Ciba-Geigy, and now Shell Chemical; the bismaleimide triazines developed by Mitsui Toatsu; and the epoxy imides developed by Shell. A simple idealized structural representation of the various addition polyimides is shown in Fig. 8.4.

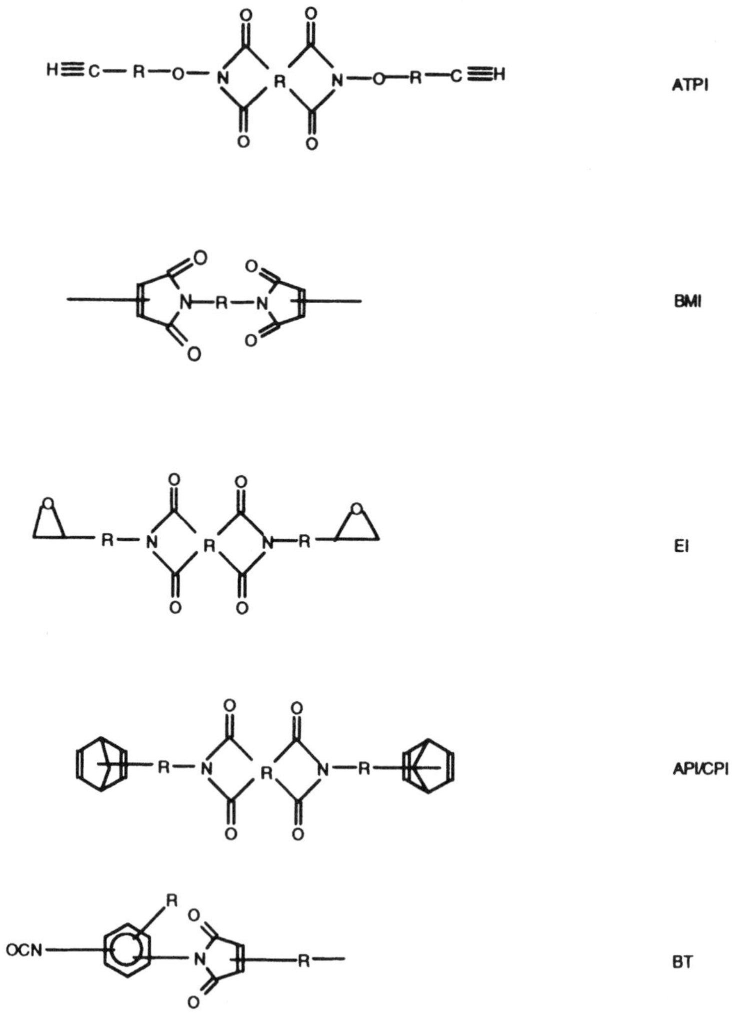

Figure 8.4 Idealized chemical structures for various addition polyimides.

Bismaleimides

The bismaleimides (BMIs) are probably the most widely used high-temperature thermosetting material in the electronics industry. Their operating temperature range is between 175 and 230°C. The primary reason for their popularity is their ability to be processed under epoxy resinlike conditions without the evolution of void-producing volatile by-products. The BMI resins are autoclavable under low pressure but do require postcure. A typical autoclave cycle consists of 1 h at 80°C from 75 to 100 lb/in^2 plus 4 h at 177°C from 75 to 100 lb/in^2 plus 4 h at 204°C from 75 to 100 lb/in^2. Postcuring from 4 to 24 h at 220 to 260°C is required to maximize properties. The fully cured bismaleimides are insoluble in all solvents, highly cross-linked, somewhat brittle, and dense; they weigh 1.35 to 1.40 g/cm^3, have high T_g values (250 to 300°C), have low strain to failure (1 to 2 percent), and generally are used in the temperature range of 150 to 200°C, although some compositions can be used at temperatures up to 230°C. The bismaleimides are produced as solids, e.g, powders, or in solvent solutions. Unmodified BMIs are brittle, and a number of options exist to toughen these polymers. One such option is the use of polybenzimidazole (PBI) particles (2 to 70 μm) suspended in the BMI matrix, which effectively doubles the fracture toughness of the BMI resin. Copolymerization is another option. There are BMI compositions too numerous to review, and the reader is directed to Refs. 4, 6, and 9 as well as the manufacturers' bulletins. The bismaleimide resins are used mainly in the production of electrical-grade laminates.

Acetylene-terminated polyimides

The acetylene-terminated polyimides (ATPIs) provide another route to low-void API matrix resin composites. Currently, National Starch and Chemical Corp. manufactures these materials under the trade name Thermid. The fully imidized structure is soluble only in N-methyl pyrrolidone (NMP) solvent. The Thermid resins have a narrow process window because of their high melt temperatures, although a fluoro-substituted material that melts at a lower temperature (160 to 180°C) has a wider process window. The ATPI resins have low dielectric constants (2.9) and have been used in a number of electrical applications, such as passivation coatings, alpha-particle shields, interlayer dielectrics, and encapsulants.

Bismaleimide-triazine (BT)

The BT resins are blends of bismaleimides and biscyanates and were developed by the Mitsubishi Gas Chemical Co. The latter materials are also known as triazine resins hence the acronym BT. These resins have

T_g values in excess of 250°C. These materials are processed in ketone solvents. These resins are usually blended with brominated epoxy resins to modify T_g and the flammability ratings.[7,8] The BT resins find use as the matrix resin for printed wiring boards.

Condensation and Addition Polyimide Hybrids

These materials are represented by a polymer structure that contains imidized structures along its chain length but also contains terminal functional groups capable of undergoing addition polymerization. Such a structure is represented by the PMR resins developed by TRW, Inc., in the late 1960s and further developed by NASA Lewis Research Center. These materials still evolved a low-molecular-weight alcohol during the prepregging operation, but during the final stages of cure the only volatiles evolved are solvent. The major disadvantage with these resins is the very high processing temperature needed (up to 320°C) to optimize the high-temperature properties of these polymers. Approaches were taken to lower the final cure temperature, but they resulted in lowering of the glass transition temperatures. These materials are used as matrix resins in structural composites and not for electrical applications. Additional information can be obtained in Ref. 4.

Polybenzimidazoles

Polybenzimidazoles are heterocyclic polymers characterized by having extremely high thermal stability and excellent hydrolytic dimensional and ablative stability. They have excellent chemical stability and are unaffected by solvents, acids, and bases. These polymers have the benzimidazole group as part of the repeating unit along the polymer chain. That group is shown here:

The polymers are produced by the melt condensation of *bis*-orthodiamines, e.g., tetraminobiphenyl and dicarboxylates such as diphenyisophthalate. The specific amine and carboxylate given are the starting materials for the commercial polymer sold today.

Polybenzimidazole has an excellent combination of toughness, thermal stability, chemical resistance, friction, and wear properties. Its compressive strength of 58,000 lb/in^2 is the highest among plastics, and it has a T_g of about 425°C. The polymer can withstand a temperature of 760°C for 3 min without a change in dimensions or mechanical properties.[9] Char yield is exceptionally high—about 85 percent. PBI is also highly resistant to deformation and has low hysteresis and high elastic recovery. PBI resins are attacked by polar aprotic solvents and strong oxidizing aqueous acids. PBI has an oxygen index of 58, does not burn in air, and evolves little to no toxic fumes. Thermal expansion is similar to that of aluminum. It is resistant to 200 Mrads of gamma radiation. The polybenzimidazoles have a host of applications in the aerospace, mining, electrical, and chemical process industries. In the electrical industry, it is used in switches, connectors, insulation, and other electrical component parts. Typical properties of PBI are shown in Table 8.7. Hoechst Celanese is the only manufaturer of polybenzimidazoles.

TABLE 8.7 Typical Properties of Unreinforced PBI

Property	Value
Specific gravity	1.3
H$_2$O absorption, 24 h, %	0.4
Tensile strength, lb/in^2 × 10^3	23
Elongation, %	3
Tensile modulus, lb/in^2 × 10^3	850
Flexural strength, lb/in^2 × 10^3	32
Flexural modulus, × 10^3	950
Compressive strength, lb/in^2 × 10^3	58
Compressive modulus, lb/in^2 × 10^3	850
Izod impact strength, ft · lb/in	0.5
Poisson's ratio	0.34
Heat deflection temperature at 264 lb/in^2, °C	815
Coefficient of linear thermal expansion, 10^{-5}/°F	1.3
Continuous-use temperature, °C	420
Thermal conductivity, Btu · in/(h · ft^2 · °F)	2.8
Dielectric strength, V/mil	550
Volume resistivity, Ω · cm	8 × 10^{14}
Dissipation factor	
1 kHz	00
10 kHz	0.003
0.1 MHz	0.034
Dielectric constant	
1 kHz	3.3
10 kHz	3.3
0.1 kHz	3.2
1 GHz	3.5
Arc resistance, s	186

Other High-Performance Polymers

The other classes of polymers that are considered high-performance materials are the aramids, fluoropolymers, liquid-crystal polymers, polyketones, polyphenylene sulfide, cyanate esters, benzocyclobutenes triazines, and the silicon-carbon SYCAR polymers. These materials are finding increased use in advanced microcircuits and electronic packaging such as in multichip module technology.[11]

See Chaps. 3 and 4 for information on some of these polymers. However, additional information on some is presented below.

Cyanate Esters

Although the basic chemistry and properties of these materials were detailed in Chap. 4, an interesting feature of these materials is their blending compatibility. A variety of flame-retardant resins can be obtained by blending a fluorocyanate ester resin with other nonfluorinated cyanate ester materials. At the same time, this blending also can be used to lower certain electrical properties. Dielectric constants as low as 2.7 and a dissipation factor of 0.002 at 1 MHz have been obtained by tailoring the blending of cyanate ester resins.[12] The cyanate ester resins are supplied by Rhone Poulenc. A variation of the cyanate ester resin technology is the resin system designated by the name *phenolic-triazine* developed by Allied Signal Inc. These resins have high glass transition temperatures (300°C), require no catalyst for curing, and evolve no volatile by-products during cure. These resins are presently used not in the electrical industry but rather in the high-performance structural composite industry.[13]

Silicon-Carbon Resins

Although these SYCAR resins contain a silicon-oxygen backbone, they are not members of the silicone resin family of polymers. The hydrocarbon groups (R) in the structure

$$\left(\!\!-O-\underset{\underset{R}{|}}{\overset{\overset{R'}{|}}{Si}}-O-\!\!\right)_{\!\!x_1}$$

$$\left(\!\!-O-\underset{\underset{R'}{|}}{\overset{\overset{R}{|}}{Si}}-O-\!\!\right)_{\!\!x_2}$$

are alkanes and hence provide functional groups of low polarity. The extent of cross-linking is achieved through a hydrosilation reaction. These resins have a rather unique combination of properties: exceptional moisture resistance and thermal stability, low dielectric constant, low dissipation factor, high T_g, process flexibility, excellent storage stability, and low ionic content. Some of the properties were described in Chap. 4, but more detail is presented here. As a printed-wiring board matrix resin, it had the following properties in an eight-ply 7628 E-glass laminate at 39 to 41 percent resin content: A dielectric constant of 3.7 to 3.9 at 1 MHz which remains flat with frequency from 100 MHz to 1 GHz.[14] A property comparison of the neat silicon-carbon-based resin with a number of other high-performance resins is given in Table 8.8. The SYCAR resins are available from Hercules Inc.

All the polymer types listed on page 166 of this chapter are commercially available. Continuing research in polymer chemistry unfolds new resin systems every year. Although they are not commercial, these new resin systems could represent the next generation of material with those unique properties that everyone is searching for. A brief description of these materials is given in this section.

Polyphenylene Rigid-Rod Polymers

Maxdem Inc., of San Diego, California, has developed a rigid-rod polyphenylene that is soluble in a number of solvents and is melt-processible. The polymers are designated Poly-X and are characterized by the following structural formula:

These materials are reported[16] to have very high modulus in the range of 2.5×10^6 lb/in^2.

Heterocyclic Polymers

Among the many other high-performance polymers are materials that can be grouped under the term *heterocyclic*. These materials are the polyquiniolines, polyphenylquinoxalines, polyquinoxalones, polybenzoxazoles, and polybenzothiazoles. These materials with their exotic-sounding names promise improved thermal, electrical, and mechanical properties over the best materials on the market today, and they rep-

TABLE 8.8 Comparative Resin System Data

Property	PCL-FR-511	Difunctional epoxy	Multifunctional epoxy	Polyimide	Cyanate ester
Dielectric constant of net resin	2.6	3.7	4.2	3.4	2.7–2.8
Dielectric constant at 1 MHz (8-ply 7628 E-glass)	3.9	4.6	4.5	4.6	3.9
Dissipation factor at 1 MHz	0.005	0.020	0.019	0.010	0.005
Volume resistivity, $\Omega \cdot cm$	1.0×10^{15}	1.0×10^{14}	3.8×10^{14}	2.1×10^{14}	
Surface resistivity, $\Omega \cdot cm$	1.0×10^{16}	1.1×10^{12}	2.7×10^{12}	3.7×10^{14}	
Arc resistance, s	180	128	128	180	
Dielectric breakdown, kV	60	>55	>55	>55	
Electric strength, V/mil	750	1200	1400	1600	800
Water absorption, %	0.02	0.11	0.09	0.35	0.7
Specific gravity	1.6	1.85	1.85	1.7	
Chemical resistance (solvents)	Excellent	Good	Excellent	Excellent	
Glass transition temperature by DMA (T_g), °C	190	125	135	280	245
Coefficient of thermal expansion z axis, [in/(in · °C)] $\times 10^{-5}$	50–190°C 8.5	25–125°C 5.5	25–135°C 4.4	25–275°C 3.5	25–245°C 8.1
Dimensional stability ($x = y$ axis), in/in	0.00036	0.0004	0.00035	0.0005	0.0003
Flexural strength (over 0.020 in), lb/in^2					
Lengthwise	50,000	68,000	70,000	80,000	
Crosswise	43,000	57,000	60,000	65,000	

SOURCE: Ref. 15.

resent a special area of materials research to develop advanced materials. Labadie and Hedrick[17] have synthesized soluble and processible versions of these polymers. The structure of these polymers is shown in Fig. 8.5, and some properties are reported in Table 8.9. An experimental sample of a polybenzoxazole fabric was obtained from the Dow Chemical Co. and impregnated with a cyanate ester resin and bismaleimide resin and converted to electrical laminates. Dielectric constants ranged from 2.5 to 2.9, and the dissipation factor ranged from 0.007 to 0.01.[18]

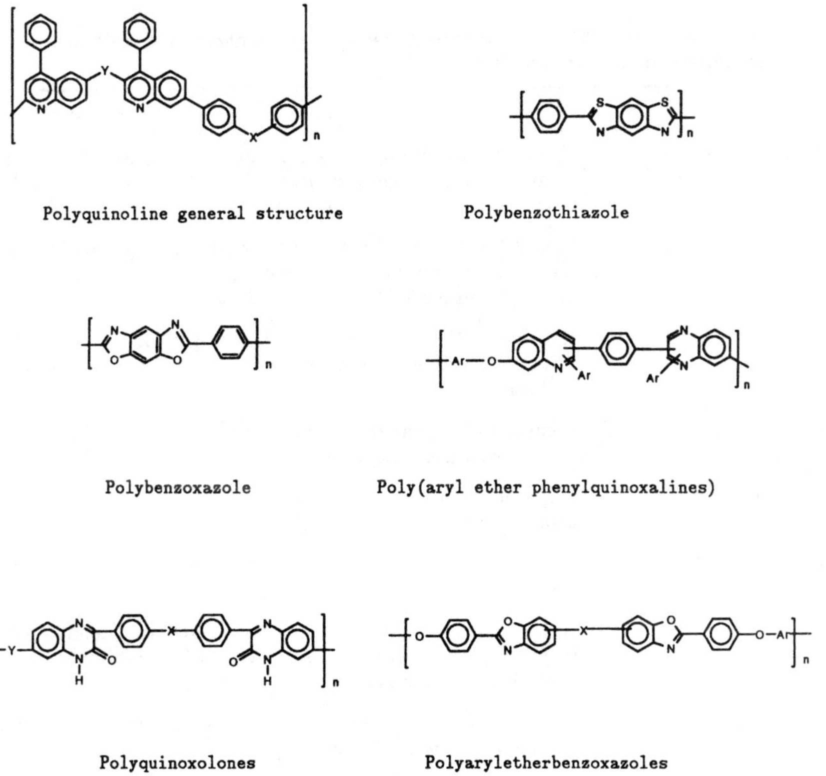

Polyquinoline general structure

Polybenzothiazole

Polybenzoxazole

Poly(aryl ether phenylquinoxalines)

Polyquinoxolones

Polyaryletherbenzoxazoles

Figure 8.5 Some heterocyclic high-performance polymers.

TABLE 8.9 Some Properties of High-Performance Heterocyclic Polymers

Polymer	T_g, °C	Solubility	PDT,* °C	Dielectric constant
Polyquinoline	288	NMP		2.5–2.6 at 10^6 kHz
Polyenzothiazole	>400	Acidic solvents	600	
Polybenzoxazole	>400	Acidic solvents	600	2.4 at 0.1 MHz
Polyaryletherphenyl quinoxaline	250–270	NMP	500	
Polyquinoxalone	250–450	NMP	510	
Polyaryletherbenzoxazole	213–259	NMP	505	

*Polymer decomposition temperature TGA, N_2 atmosphere.

TABLE 8.10 Industrial Fibers Used in High-Technology Nonwoven Materials, Textiles, and Structural Composites

Fiber	End uses
Alumina-boria-silica	Nextel flexible highly heat-resistant fiber for uses including ceramic reinforcement, furnace curtains, and wire and cable insulation
Aluminum oxide	Fiber FP metal reinforcement fiber adds strength and high-temperature stability to automotive components and is under development for aerospace applications
Aramid	Kevlar and Nomex fibers for high strength, flame-resistant filters, protective clothing, structural composites for aircraft and boats
Boron	Fiber used in epoxy resin and aluminum matrices for aerospace structures and sports equipment
Carbon	Fibers for structural composites in military, aerospace, and sports equipment; gears; bushings
Polybenzimidazole	Used for such markets as heat-protective apparel and aircraft seat fire blockers
PBZ	New family of heterocyclic rigid-rod and chain extended polymers has high tensile strength and thermal stability with possible applications in structural composites
Polytethylene	Spectra fibers for high-strength marine sailcloth and mooring ropes and protective clothing
Polyimide	Polyimide 2080 under license in fiber form to Austrian producer Lenzing. In rope form, fiber used to tie down insulating elements on NASA's space shuttle
Silicon carbide	Fiber used in metal matrix composites with aluminum and titanium for military, aerospace, and engine applications
Sulfar	Ryton heat- and chemical-resistant filtration fibers

SOURCES: Refs. 1 and 19.

High-Performance Fibers

While fibers are not used in their neat form for electrical applications, new developments have led to substantial improvements in their electrical, mechanical, and thermal properties. In the electrical industry these materials are converted into fabrics and impregnated with a variety of polymers for use as printed wiring board substrates. A list of fibers and their areas of application is shown in Table 8.10. Many of these fibers (e.g., aramids, PBZ, polyethylene, and polybenzimidazole) have application in the electrical industry.

Advanced High-Performance Thermoplastics

In addition to the thermoset plastics, there is another group of polymers in the thermoplastics category that can perform in the 175°C-plus environment. The demand for selected thermoplastics (polyphenylene sulfide, polysulfone, polyetherimide, polyethersulfone, polyarylate, liquid crystal polymers, and polyetheretherketone) is expected to increase to 13 percent of all plastics by the year 2000.[20] Electronics is the top user of high-performance thermoplastics. (See Fig. 8.6.) A brief description of these polymers follows.

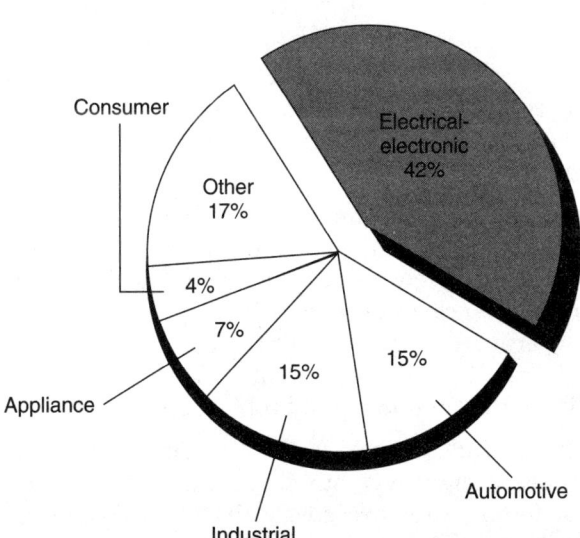

Figure 8.6 High-performance thermoplastic use. (*Source:* Ref. 20.)

Liquid crystal polymers (LCPs): Excellent chemical, radiation, weathering, and flame resistance. Heat deflection temperature of 356°C in some glass-filled grades. Producers: Amoco Chemical, Hoechst Celanese, du Pont.

Polyamide-imide (PAI): Amorphous material with good solvent, wear, and friction resistance. Heat deflection temperature above 277°C. Producer: Amoco.

Polyarylate: Aromatic polyester that exhibits excellent toughness, ultraviolet light and dimensional stability, and flame retardance. Heat deflection temperature up to 173°C. Producer: Amoco.

Polybenzimidazole: Heterocyclic polymer with excellent hydrolytic, compressive, and dimensional stability. Commercially available in stock shapes and finished parts. Heat deflection temperature of 433°C for unfilled molded material. Producer: Hoechst Celanese.

Polyetherimide (PEI): Broad chemical resistance, dimensional stability, and creep resistance. Heat deflection temperature of 215°C in glass-filled grades. Producer: GE Plastics.

Polyethersulfone (PES): Amorphous thermoplastic with good chemical resistance and hydrolytic stability. Heat deflection temperature of 204°C. Producers: Amoco, BASF.

Polyimide (PI): Fully reacted linear polymer with excellent toughness and electrical properties. Heat deflection temperature of 360°C in graphite-filled grades. Producers: du Pont, Lenzing, Ciba-Geigy.

Polyketones: Family of semicrystalline polymers with superior chemical resistance, strength, and stiffness. Heat deflection temperature of 325°C for glass-filled grades. Producers: Amoco (a proprietary polyketone), ICI (polyetheretherketone), BASF (polyetherketoneetherketoneketone).

Polyphenylene ether (PPE): Linear amorphous polymer frequently alloyed with polystyrene for processing. Heat deflection temperature of 165°C in glass-filled grades. New alloys with nylon and polyphenylene sulfide extend performance levels. Glass-filled nylon alloys have heat deflection temperatures up to 232°C, and blends with polyphenylene sulfide can extend range to 260°C. Producers: GE Plastics, Huls, BASF, Asahi, Sumitomo, Mitsubishi Gas.

Polyphenylene sulfide (PPS): Semicrystalline thermoplastic with good chemical resistance, dimensional stability, and electrical properties. Heat deflection temperature exceeds 260°C for glass- and carbon-filled grades. Producers: GE Plastics (produced for the company by Toso), Hoechst Celanese (in partnerhip with Kureha), Phillips Petroleum, Toray, DIC, Toso, Tophren.

Polyphenylsulfone: High degree of chemical resistance. Heat deflection temperature of 260°C in some glass-filled grades. Producer: Amoco.

Polyphthalamide (PPA): Semicrystalline polymer with excellent mechanial properties. Reinforced grade has a heat deflection temperature of 283°C. Producers: Amoco, Mitsui.

Polysulfone (PS): Rigid amorphous thermoplastic. Heat deflection temperature of 176°C for glass-filled grades. Producers: Amoco, BASF.

Summary

Polymers that exhibit special high-performance properties play an important role in industry today and will continue to play a critical role as technology advances. The materials discussed in this chapter represent the state of the art today. Some of these materials have been around for 10 years or more; others are short-lived. The longevity of these high-performance polymers depends on several factors: cost, processibility, and maintenance of property integrity in service. A list of the current high-performance polymers and some of their properties is shown in Table 8.11.

References

1. "High-Temp Polymers Still Hot," *Chemical Week Markets Newsletters,* February 10, 1993, pp. 29, 39.
2. S. Witzler, ed., "Composites News," *Advanced Composites,* March/April 1993, p. 8.
3. K. L. Mittal, ed., *Polyimides: Synthesis, Characterization and Applications,* vols. 1, 2, Plenum Press, New York, 1984.
4. D. Wilson, P. M. Hergenrother, and H. D. Stenzenberger, eds., *Polyimides,* Chapman and Hall, New York, 1990.
5. C. P. Wong, ed., *Polymers for Electronic and Photonic Applications,* Academic Press, New York, 1993.
6. A. J. Klein, "High Temperature Imides," *Advanced Composites,* July/August 1988, p. 45.
7. J. T. Gotro and B. K. Appelt, "Characterization of a Bis-maleimide Triazine Resin for Multilayer Printed Circuit Boards," *IBM Journal of Research and Development,* vol. 32, no. 5, September 1988.
8. David Wei Wang, "Advanced Materials for Printed Circuit Boards," in R. Jaccodine et al., eds., "Electronic Packaging Materials, Science III," *Proceedings of the Materials Research Society Symposium,* vol. 108, 1988.
9. L. DiSano, "High Temperature PBI Takes Shape," *Materials Engineering,* May 11, 1992.
10. R. Greene, ed., *Modern Plastics Encyclopedia,* McGraw-Hill, New York, 1991.
11. E. Sweetman, "Characteristics and Performance of PHP-92: AT&T's Triazine-Based Dielectric for Polyhic MCMs," *The International Journal of Microcircuits and Electronic Packaging,* vol. 15, no. 4, 1992.

TABLE 8.11 General Properties of Some High-Performance Polymers

Polymer	Designation	Supplier	T_g, °C	Tensile strength, $\times 10^3$ lb/in²	Tensile modulus, $\times 10^6$ lb/in²	Elongation, %	PDT, °C	Dielectric constant E'	tan δ	Dielectric strength, V/mil
Polyaramids	Kevlar	Du Pont	345	400	18	2.5	—	2.5	0.004	435
	Nomex	Du Pont	275	25*	—	11	—	2.1	0.0002	480
Fluoropolymers	Teflon	Du Pont	127	2–5	0.06	300	—	1.9	0.00007	539
	Teflon AF	Du Pont	160	3.9	0.2	30	—	—	—	—
	—	Shell	270	14	0.6	1.8	—	3.5	0.007	480
CE	—	Rohne Poulenc	289	13	0.47	3	—	2.9	0.005	450
PAI	Torlon	Amoco	275	27	0.7	15	>400	4.2	0.026	580
PI	Kapton	Du Pont	>360	25	0.43	75	500	3.5	0.002	7000
PI	LARC-TPI	Rogers	260	17	0.58	4	>499	—	—	—
PI	K-III	Du Pont	250	15	0.5	14	>400	—	—	—
PEI	Ultem	GE	~230	15	0.43	60	—	3.1	0.0013	830
PIS	Imide Sulfone	Hoecht Celanese	270	9	0.7	1	—	—	—	—
PBI	Celazole	Hoecht Celanese	425	23	0.85	3	—	3.3	0.000	550
PES	Ether Sulfone	Amoco	220	10	0.35	70	—	—	0.008	425
PPS	Ryton	Phillips 66	88	12	0.5	5	—	3.1	0.0004	380
PEK	—	ICI BASF	162	13	0.5	5	—	3.5	0.002	—
BCB	Cyclotene	Dow	350	—	0.34	6	350	2.7	0.0008	—
Silicon-carbon	SYCAR	Hercules	790	—	—	—	480	2.6	0.003	750
Silicone	—	Dow Corning	−123	5	—	500	—	—	—	400
LCP	LCP	Amoco Holdust	—	29	—	5	—	4	0.005	850
Phenolic triazine	PT	Allied Signal	300	—	—	—	440	—	—	—
Polyquinoline	PQ	—	288	—	—	—	—	2.6	—	—
—	—	—	>400	610	0.47	1.3	600	—	—	—
PBT	—	Dow	>400	840	0.53	1.6	600	2.4	—	—
PBO	—	—	250	—	0.33	23	500	—	—	—
Polyquinoxaline	—	—	270	—	—	—	—	—	—	—
Polyquinoxalone	—	—	250	—	0.57	15	510	—	—	—
	—	—	450	—	—	—	—	—	—	—

*lb/in.

12. W. M. Alvino and Z. N. Sanjana, unpublished results, Westinghouse internal report, October 23, 1990.
13. S. Das, D. C. Prevorsek, and B. T. DeBona, "Phenolic-Triazine Resins Yield High Performance Thermoset Composites," *Modern Plastics,* February 1990, p. 73.
14. *Electronic Materials Report,* vol. 8, no. 4, April 1992.
15. H. M. Enzien, J. D. Gagnon, D. E. Maurer, and E. M. O'Brien, "A New Low Dielectric Constant, Low Moisture Absorption, High T_g Material for Use in Multilayer PWB Fabrication," Nepon West Conference, 1992.
16. R. Baum, "Rigid Rod Polyphenylenes Made Processable," *Chemical and Engineering News,* February 1993, p. 27.
17. J. W. Labadie and J. L. Hedrick, "Recent Advances in High Temperature Polymers for Microelectronic Applications," *Journal of the Society for the Advancement of Material and Process Engineering,* vol. 25, no. 6, November/December 1989.
18. W. M. Alvino, Westinghouse internal report, unpublished results, 1990.
19. M. S. Reisch, "High Performance Fibers Find Expanding Military, Industrial Uses," *Chemical and Engineering News,* February 2, 1987, p. 10.
20. *Chemical and Engineering News,* August 30, 1993, p. 30.

Chapter 9

Organic Coatings

Overview

Organic coatings represent a very large segment of use in many industries. The term *organic coatings* comprises paints, varnishes, enamels, lacquers, water- and solvent-based coatings, powder coatings, organosols, plastisols, and solventless liquid coatings. These coating systems are more or less viscous liquids; however, some forms are solids such as in powder coatings. There are basically three components to a coating: the binder, which is the film-forming substance, and usually high-molecular-weight substances—a pigment and a volatile liquid. Pigments add color and opacity but are removed from the formulation if a clear coating is needed. The volatile liquid provides a packaged coating system with a practical viscosity for rheological control and does not become part of the final coating film after evaporation. In conventional paints, the binders can be unsaturated vegetable oils such as linseed, castor, and soya. In these paint systems, the solvent evaporates from the surface, and the paint film is converted to an insoluble form by oxidation. Varnishes are unpigmented paints using the same drying oils and resins as ordinary paints do. In addition, some synthetic polymers are used to formulate varnishes. Enamels are pigmented varnishes and generally are formulated with higher-molecular-weight polymers compared with those used for paints. In addition, enamels are lower in solids content than paints are. Lacquers differ from paints and enamels because they are formulated with higher-molecular-weight thermoplastic soluble polymers blended with plasticizers and other ingredients and dissolved in a suitable solvent. Lacquers produce a tough film after solvent evaporation which is not cross-linked and thus is soluble in organic solvents. Plastisols are dispersions of small-particle-size polymer in a plasticizer which is a solvent for the polymer at elevated temperatures. Powder coatings are solids, i.e., polymeric powders that are deposited on a heated metal part immersed in a fluidized bed or electrostatically sprayed onto the part.

Coatings are both functional and decorative, providing properties in the area of electrical performance and/or environmental protection. In the area of electrical performance, the significant properties of coatings that are of interest include the dielectric strength, dielectric constant, volume, and surface resistivity. How these properties are important or how much they contribute to coating performance depends on the end-use application. In the area of environmental protection, the coating provides a barrier to moisture, physical and mechanical abuse, chemical attack, corrosion, weathering, and the accumulation of dust and/or debris that might otherwise contribute to component failure.

In this chapter we will examine those coatings of importance to the electrical and electronic industries. A listing and description of all types of organic coatings are given in Ref. 1, and more detailed information about organic coatings in general can be found in Ref. 2. Many of the property data for plastics in general apply to their coatings as well and will not be repeated in this chapter except to reinforce a particular property where it significantly contributes to the component performance. Chapters 3, 4, and 5 provide much of the property information on polymers that can be used as coatings. For our purposes, Table 9.1 lists the various applications of coatings of importance to the electronics industries which will be discussed in the following sections of this chapter. A general listing of most types of polymers and their properties used as coatings in one form or another is given in Table 9.2 (pp. 180–181).[3]

TABLE 9.1 Coating Uses in Electrical Industry

Wire enamels and coil coatings
Conformal coatings
Encapsulants—potting and casting
Interlayer dielectrics
Passivation coatings
Resists
Fabric impregnation

Coating Types

Wire enamels and coil coatings

These coatings consist of high-molecular-weight polymers, either thermoplastic or thermoset, dissolved in a suitable solvent. The coatings electrically insulate the magnet wire coils in electric equipment such as motors, generators, transformers, solenoids, and relays. These coatings provide the primary insulation on the wire. It can be the only insulation, or the coil can be further coated with another coating to encapsulate and fix the wires in place for additional protection. The types of coatings used on wire are listed in Table 9.3 (pp. 182–185) along with some properties.

Plain enamel coatings are essentially varnishes from natural drying oils containing lots of unsaturation in the polymer which oxidize upon drying of the coating. These coatings are not used very much (noncritical applications) and have been essentially replaced by superior-performing products.

Acrylic enameled wire coatings are aqueous dispersions and when cured, yield tough, flexible films that are resistant to most solvents, mild acids, and alkalies. The cut-through resistance of these coatings is low.

Alkyds are glycerol-phthalate-based resins which are combined with drying oils. The coatings have good heat stability, dielectric properties, and resistance to oils. They are rated for 105°C use. These coatings rapidly air-dry or bake-dry to yield a hard, durable surface that provides environmental protection of the windings. The alkyd coatings have excellent arc resistance. They do not carbonize on the surface or sustain an electric arc and are compatible with nylon-, epoxy-, polyester-, and glass-covered wire. Two modifications of the alkyds yield better heat resistance. These materials are phenolic and silicone modified compositions.

Diallylphthalate coatings are solvent-based and are used for sealing, dipping, and laminating. After curing, these coatings have excellent resistance to moisture, chemicals, corrosive gases, and thermal aging. Diallylphthalate and isophthalate resins comprise these coating formulations and can operate at temperatures of 150 and 180°C, respectively.

Epoxies are available in forms ranging from solids to liquids at room temperature. One- and two-component systems are available. The resins are formulated with curing agents that cause the resin to form a cross-linked structure when cured, and the resins have excellent adhesion, electrical properties, moisture resistance, and process ease. Solvent solutions as well as solventless liquid formulations are available.

TABLE 9.2 General Properties of Polymers Used as Coating

	Dielectric strength, V/mil	Resistivity, $\Omega \cdot cm$	Water absorption, %[a]	Dielectric constant at 10^{10} Hz	Loss tangent at 10^{10} Hz
Thermoplastics					
Asphalt and tars	300	10^{10}	0.06	3.5	0.04
Fluorocarbon	450	10^{18}	0.00	2.1	0.0003
Polyethylene	500	10^{16}	<0.01	2.3	0.0005
Polystyrene	550	19^{18}	0.04	2.5	0.0003
Polyvinyl chloride	400	10^{15}	0.15	2.8	0.006
Wax	400	10^{17}	0.02	2.6	0.001
Silicone-polyimide	1500–2800	10^{15}–10^{17}	<1	3.0	0.007
Parylene	500–7000	10^{13}–10^{16}	0.03	2.8	0.01–0.00:
Polyketone	550	10^{15}		3.4	0.003
Polyetherketone	750	10^{15}		3.3	—
Polyetheretherketone	750	10^{14}		3.4	0.0015
Polyaryletherketone	590	10^{13}	0.5	3.5	0.001
Polysulfone	600	10^{13}	0.5	3.2	0.001
Polyethersulfone	500	10^{13}	0.5	3.5	0.001
Liquid-crystal polymers	1220–1700	10^{14}		3.0–4.5	0.005
Thermosets					
Alkyd	350	10^{14}	0.4	3.8	0.025
Allylester	400	10^{14}	0.7	—	—
Butadiene styrene	600	10^{16}	0.03	2.4	0.006
Epoxy	450	10^{14}	0.20	2.9	0.018
Phenolic	350	10^{12}	0.3	4.7	0.04
Polyester	350	10^{13}	0.4	3.5	0.05
Silicones	600	10^{13}	0.03	2.8	0.002
Polyimides	3400	10^{16}	—	3.6	0.002
Silicone-epoxy	246–338	10^{15}	0.1	3.6	0.004
Benzocyclobutene	10,000 / 750	10^{19} / 10^{15}	0.2^f / 0.02	2.6 / 2.6^d	0.0008 / 0.005^d
Cyanate ester	—	10^{19}	0.1	2.6–3.1	0.005
Bismaleimide	490	10^{16}	4^f	3.5	0.007
Elastomers					
Buna-S rubber	500	10^{14}	—	2.5	0.01
Chloro rubber	400	10^{12}	—	2.7	0.05
Natural rubber	500	10^{16}	—	2.1	0.03
Silicone rubber	600	10^{13}	—	3.0	0.05
Thioplast	150	10^{11}	—	14	0.15
Urethane	350	10^{11}	0.4	3.5	0.04

[a] 24-h immersion.
[b] 1 is best, 5 is poor.
[c] M: Rockwell M; SA: shore A; SD: shore D.
[d] 1 MHz.
[e] Units are in seconds.
[f] 500-h water boil.
SOURCE: Adapted from Ref. 3, p. 188.

Organic Coatings

Relative arc track resistance[b]	Distortion temp., °C	Safe-use temp., °C	Linear expansion × 10³/°C	Ultimate tensile strength, klb/in²	Ultimate elongation, %	Hardness[c]
5	55	70	8	0.6	5	SD 60
1	120	260	5.5	3	200	SD 60
3	—	115	9.5	4.4	1000	SD 65
3	80	85	4	7.3	1.5	M 80
3	65	100	3	3	100	SD 80
3	25	55	11	0.3	5	SD 30
2	150–240	400	3–10	2	200	—
2	280–400	120	3.5–4.0	10.2	200	—
2	186	260	5.7	13.8	150	—
2	160	—	4.4	15.2	50	M 105
2	160	250	4.7	14.5	5	M 99
2	245	250	4.0	10	5	M 86
2	200	200	3.0	12.2	6	M 69
2	205	235	3.0	13	7	M 88
2	250–280	180–240	0–2.5	20.3–34.8	1–7	M 60–100
2	105	120	4	8	—	SD 90
3	>90	100	4	5.8	—	M 70
3	125	245	5	4.4	4	SD k80
2	200	230	4–8	10.2	<1	M 90
4	80	80	4	7.3	1.5	M 126
3	90	165	6	8.0	<5	M 100
2	40	260	13–100	2.5	8	—
2[e]	>310	<430	0.3–80	14–20	80–110	—
2	—	<200	3–6	8	—	SD 60
2	300(N₂)	3.4	10.3	<1	—	—
—		180	8.5	—	—	—
—	249	180	3.4	13	1	—
—	270	180	3.4	12	1	—
4	—	120	6	0.3	400	SA 50
3	—	—	9	2.5	500	SA 70
4	—	65	4	3	700	SA 50
2	>230	260	—	0.58	100	SA 60
4	—	120	10	0.3	400	SA 40
4	>65	95	10	5.1	400	SA 60

TABLE 9.3 Wire Enamel Characteristics

Type/thermal class, °C	Advantages	Disadvantages
Plain enamel (oleoresinous) 105	Low cost Good film continuity and cut-through resistance and overload resistance Ease of stripping Tight dimensional tolerances	Low abrasion resistance Not for heavy-duty winding Limited compatibility with other varnishes
Solderable acrylic 105	Solderable Better winding, physical, and chemical properties than plain enamel Low cost Withstands solvents, mild acids, and alkalies without stress crazing after winding	Low cut-through resistance Soldering temperatures must be maintained at 455°C
Solderable nylon acrylic 105	Solderable Superior windability over straight acrylic—high abrasion resistance Better solvent resistance than straight acrylic Low cost	Dielectric strength lower than straight acrylic Not as resistant to moisture as straight acrylic Soldering temperatures must be maintained at 455°C Poor radiation stability
Modified polyvinyl formal resin 105	Excellent windability High adhesion rating Compatible with most varnishes Suitable for use with transformer oil Good overall chemical and physical properties	Must be stripped before soldering Crazes when exposed to varnish solvents unless stress-relieved first
Polyurethane (PUR)	Solderable at 360–425°C Good dielectric strength, solderability Moisture and corona resistance	Not for use where severe overloads may occur Low abrasion resistance Susceptible to softening under prolonged exposure Poor radiation stability
Polyurethane with nylon overcoat 130	Solderable with better windability than straight polyurethane Excellent abrasion resistance High heat shock resistance Compatible with most varnishes and encapsulants Excellent solvent resistance	Moisture absorption High-frequency losses

TABLE 9.3 Wire Enamel Characteristics (*Continued*)

Type/thermal class, °C	Advantages	Disadvantages
Polyester 155	Resistant to heat and solvent shocks Toughness Good overload characteristics Excellent dielectric properties	Hydrolyzes in moist sealed atmosphere Must be stripped before soldering
Polyester with acrylic overcoat 155	Low-cost class 155 70% of insulation is waterborne High cut-through resistance Excellent electrical properties	Must be stripped before soldering Not for use in enclosed equipment Poor radiation stability
Solderable polyester 155	Solderable at 482°C Long thermal endurance at 175°C Good heat shock resistance Good overload characteristics for solderable wire Good radiation stability	Not for use in oil-filled transformers Not for use with systems using amine-type catalysts
Solderable nylon polyester 155	Solderable at 482°C Better windability than unjacked grade Good resistance to heat and solvent shock	Not for use in oil-filled transformers Not for use with systems using amine-type catalysts Not as resistant to moisture as unjacketed wire
Isonel* polyester with polyamide-imide overcoat 200	Physical and electrical properties almost equal to Formvar Excellent cut-through resistance Excellent resistance to Freon Good solvent resistance Compatible with most varnishes and impregnating compounds Good radiation stability	Must be stripped before soldering Not for use in oil-filled transformers Not for use in systems containing chlorine compounds Not for use with systems using amine catalysts
Polyester with nylon overcoat 155–180	Excellent windability Resistant to heat and solvent shocks Excellent in flexibility, scrape abrasion, and film adhesion Good overload characteristics	Must be stripped before soldering Not for use in enclosed equipment Poor radiation stability

See p. 185 for footnotes.

TABLE 9.3 Wire Enamel Characteristics (*Continued*)

Type/thermal class, °C	Advantages	Disadvantages
Polyester/polyamide-imide 200	Polyester base coat with amide-imide top coat Good windability More resistant to heat and solvent shocks than conventional polyesters Excels regular polyesters in resistance to cut-through and abrasion resistance Highly resistant to Freon and solvents Excellent dielectric properties Suitable for use in hermetic motors	Must be stripped before soldering Not for use in enclosed equipment where moisture or chlorine compounds are present
Omega† polyesteramide-imide 200	Single film coating Excellent overload characteristics Compatible with epoxy casting and encapsulating compounds Compatible with most varnishes and impregnating compounds Long thermal endurance at 180–210°C For use in oil-filled transformers Withstands high-speed winding applications	Must be stripped before soldering High price
Isomid* polyester polyimide 180	Single film coating Excellent cut-through, flexibility, and adhesion properties Excellent wet and dry dielectric properties Highly resistant to Freon and solvents Improved resistance to heat shock Compatible with most varnishes and impregnating compounds	Must be stripped before soldering High price

TABLE 9.3 Wire Enamel Characteristics (*Continued*)

Type/thermal class, °C	Advantages	Disadvantages
Pyre-ML‡ polyimide 220	Single film coatings Retains high dielectric properties at operating temperatures of 220°C Highest heat shock resistance and thermal stability of all insulations Highest resistance to radiation High dielectric strength and low loss characteristics Compatible with most varnishes, impregnating compounds, and oil-filled transformers Highly resistant to attack by solvents and chemicals Highest overload rating	Very difficult to strip before soldering Subject to hydrolysis in sealed systems containing moisture Solvent will craze unless stress-relieved Highest-price film-coated wire May be used in hermetic motors
Nylon 105	Dielectric strength toughness, solderable	High moisture absorption High electrical loss
Polytetrafluoroethylene 188	High thermal stability High chemical stability High dielectric strength Low dielectric constant Low friction	Poor adhesion Gas permeability, cold flow
Butvar/Formvar	Bondable Good dielectric strength Good heat shock resistance	Low tolerance to vibration and high mechanical stress

*Schenectady Chemicals.
†Westinghouse.
‡Du Pont.

Oil modified-phenolic varnishes vary in properties depending on the kind of oil modification. Very hard, high-phenolic-content varnishes are used for electrical rotating equipment. Most varnishes in this group have good dielectric properties and chemical and moisture resistance. The thermal class of these coatings is 130°C.

Phenolic varnishes are resinous products of phenol and formaldehyde in alcohol or aromatic solvents. These coatings which have excellent chemical resistance, hardness, and adhesion are suitable for use in high-speed rotating equipment. The thermal class rating of these varnishes is 105°C.

A number of formulations comprise the high-temperature class of wire enamels: polyamide-imide, polyimide, polyester amide-imide, and polyester-imide. They are all rated for 200 to 220°C operation and have excellent dielectric properties, cut-through resistance, and abrasion resistance.

Polyurethane coatings are one-component systems derived from isocyanates and polyols. They have a thermal class rating of 130°C. They are used primarily because they permit soldering without stripping. These coatings are usually blended with polyvinyl formal to improve physical properties.

Polyester varnishes are derived from polyhydric alcohols and polybasic acids and have a thermal class rating up to 155°C depending on the formulation. These varnishes are available as very flexible materials or rigid, tough coatings. The varnishes have excellent durability, chemical and moisture resistance, and dielectric properties. Normal heating and cooling of an armature or stator puts heavy strain on the varnish, and polyester varnishes are elastic enough to withstand these stresses without cracking.

Silicone varnishes are complex mixtures of multifunctional polysilicones. The difunctional units are the flexible components, and polymers made solely of the units are fluids and gums. Rigid-type materials are obtained from trifunctional silicones although even these silicones are still somewhat flexible. The thermal class rating is 180°C. The silicone varnishes are excellent for impregnating coils, mica, glass fiber insulation, capacitors, and resistors. There are other silicone formulations that are designed for potting, embedding, and encapsulation of electronic equipment. The silicones provide a good balance of mechanical, electrical, and thermal properties and are excellent protective coatings.

The National Electrical Manufacturers Association (NEMA) lists approximately 46 wire enamels and combinations covering the thermal class range from 105 to 220°C.

Coated magnet wire is made by passing the clean wire through tanks containing the enamel solution. The wire exits the tanks and passes through first dies to regulate the coating thickness and then controlled-temperature ovens where solvent is removed and the coating is cured. The process can accommodate single- or multiple-pass operation for thickness control.

The properties of wire enamels are measured according to procedures described in ASTM D 3288 Federal Specification J-W-1177 and NEMA Standard MN 1000. These properties are adhesion and flexibility, heat shock, elongation, abrasion resistance, spring-back, cut-through resistance, dielectric strength, low- and high-voltage continuity, solubility, solderability, heat and solvent bonding, and completeness of cure. A description of these tests follows:

Adhesion on the wire is necessary to prevent cracks in the film as the magnet wire is stretched. Adhesion is determined by elongating a 10-in specimen, size 13 AWG and heavier, 25 to 30 percent at a rate of 12 in/min. Normal visual inspection should reveal no cracks. For sizes 14 AWG and finer, the wire is given a sudden jerk with minimum elongation of 15 to 20 percent (depending on the wire size) or to the breaking point, whichever is less. Magnification of 6 to 15 times (depending on the wire size) is used to determine cracks in the film for 31 AWG and finer wires.

Flexibility is the ability of magnet wire after preelongation of 15 to 30 percent (depending on the wire size) to be wrapped not more than 10 turns around a mandrel 1 to 5 times (depending on the wire size) the AWG size of the wire without film failure. The smaller the mandrel diameter without film failure, the higher the flexibility rating for a given size wire. Normal visual inspection is used to determine film failure for 30 AWG and heavier wires. Magnification of 6 to 15 times (depending on the wire size) is used for 31 AWG and finer wires.

The heat shock resistance indicates how well an insulation holds up when elongated magnet wire is wrapped not more than 10 times around a mandrel typically 1 to 3 times the AWG size of the wire. The specimen is then removed from the mandrel and placed in a circulating air oven for 0.5 h at 20°C above the specified thermal rating of the magnet wire. There should be no cracks on subsequent normal visual inspection for size 30 AWG and heavier wire or under magnification of 6 to 10 times for 31 AWG and finer wires.

The scrap or abrasion resistance indicates the mechanical abuse that a magnet wire will withstand. This property is determined by a procedure in which a special needle is rubbed on the coated wire at a specified rate with a constantly increasing load until dielectric failure occurs at a potential of 7.5 V between the needle and the wire. The load weight at failure measures the scrap resistance. An alternative method measures the number of strokes to failure with a constant load.

Spring-back is a measure of the tendency of a coil to spring open when it is removed from the form. A low spring-back indicates high formability. A specimen of film-coated wire is wound three turns around a mandrel having a specified diameter ranging from $\frac{3}{4}$ in for size 30 AWG wire to $3\frac{1}{4}$ in for size 14 AWG wire, under tension ranging from 2 oz for size AWG 30 wire to 16 oz for size AWG 14 wire. A special device measures the degree of spring-back when tension is removed.

Thermoplastic flow, or cut-through resistance, indicates the temperature at which the insulation film fails. Two lengths of the specimen positioned at right angles to each other are loaded with 2000 g for size 18 AWG or 100 g for size 36 AWG. The test equipment applies 115 V of alternating current (ac) through the crossover of the wires as the temperature is increased at a rate of less than 5°C/min to the point of dielectric failure.

Dielectric strength is an indication of the film thickness needed to withstand operating and surge voltages. The specimen may be either two layers of magnet wire or two twisted wires with a specified number of twists. The ac voltage applied between the wire layers or twisted wires is increased gradually so that breakdown does not occur in less than 5 s. The voltage at breakdown is the basis for determining the dielectric strength of the specimen.

Low-voltage continuity is a measure of the film defects of magnet wire sizes 31 to 56 AWG through dielectric failure at low voltage. Formerly, a 100-ft specimen of magnet wire was passed through a 1-in-long mercury bath at a speed of 100 ft/min. A direct-current (dc) potential of 20 to 75 V (depending on the wire size) was applied between the bath and the wire. A discontinuity-indicating device operated when the resistance between the bath and the wire was less than 5000 Ω, but did not operate when the resistance was 10,000 Ω or more. The number of faults per minute indicated the low-voltage continuity. This test is being revised to eliminate mercury with its potential health hazard.

High-voltage continuity is a measure of the film defects of magnet wire sizes 14 to 30 AWG through dielectric failure at high dc voltage. A 100-ft specimen of magnet wire is passed with contact over an energized sheave at a speed of 60 ft/min. Before and after the contact sheave is a similar grounded guide sheave. The sheaves are arranged so that the length of wire contacting the energized center sheave is 1 in. The specimen wire is grounded. The specified open-circuit voltage applied to the contact sheave ranges from 500 V for sizes 25 to 30 AWG single film to 2000 V for sizes 14 to 24 AWG triple film. The sensitivity of the fault detection device is such that the circuit is capable of detecting any fault having a resistance of less than 30 MΩ, but will not operate when the insulation resistance of the test wire film exceeds 180 MΩ. The number of faults per minute indicates the high-voltage continuity.

Solubility in certain liquids is tested by immersing stress-relieved specimens of magnet wire 12 in long at least 6 in in the liquids specified. After removal from the liquids, specimens are promptly drawn once between the folds of cheesecloth held firmly between the forefinger and the ball of the thumb. Removal of the coating indicates failure. Another test for solubility is performed on samples removed from liquids and placed in a device in which a needle scrapes the surface of the film at right angles to the lengths of wire. The needle is loaded with 580 g for testing copper magnet wire and 340 g for testing aluminum magnet wire. Film failure is indicated by short-circuiting in a circuit having a potential of 7.5 V between the needle and the conductors. Perhaps the simplest method is to immerse the magnet wire specimens in specified liquids for 24 h at 25°C. Following removal of specimens, the coating is rated pass (no effect), fair (some surface softening or crazing), or fail (severe surface softening or crazing).

Solderability without the necessity of stripping insulation is a highly desirable property which facilitates assembly operations. Soldering tests are made for wire sizes 14 to 23 AWG by forming a loop of a 12-in length of magnet wire and twisting the ends together for a distance of ¾ to 1 in with 5 to 10 turns. The loop is first immersed in a rosin-alcohol flux and then immersed in a soldering pot of 50/50 tin-lead solder for 8 to 10 s at 430°C. For sizes 24 to 46 AWG, a magnet wire specimen is wound with 5 to 10 turns for a distance of ½ to ¾ in around the end of a 6-in length of 20 AWG tinned, hot-dipped copper wire. The sample is then immersed in a rosin-alcohol flux and then in a soldering pot of 50/50 tin-lead solder at 360°C for 4 to 6 s. Magnification of 6 to 10 times is used to inspect sizes 37 to 46 AWG wire. Normal vision is used to examine larger sizes.

Heat and solvent bonding are methods used to bond coils of magnet wire coated with bonding agents, thus eliminating the necessity of subsequent varnish treatment or encapsulation. Specimens are prepared from sizes 18, 26, and 36 AWG wires. Size 18 and size 26 AWG wires are formed into a 3-in single-layer coil around mandrels with diameters of 0.250 and 0.157 in, respectively. Size 36 AWG wire is formed into a coil of 50 turns around a mandrel with diameter of 0.0394 in. Heat bonding is accomplished in a forced-air oven for 1 h at 150°C, after which specimens are cooled to room temperature. Solvent bonding is performed by dipping specimens in specified solvent for 5 s with subsequent drying for 1 h at room temperature. Bond strength is measured in a special device which applies a specified load or, alternatively, an increasing load to the center of the coil suspended at its ends.

Conformal coating

Conformal coatings are generally solventless liquid-resin formulations that are commonly applied to electrical components (e.g., printed-wiring boards and chip carriers) to protect them from a variety of environmental effects. These resins conform to the topography of the board and the components thereon and are cured to form a relatively thin (1 to 5 mils) protective coating. While the main function of the coating is to provide a moisture barrier for circuit traces and components, secondary benefits are also provided against contaminants in the process line (dust, chemicals), mechanical damage, corrosion of metallizations, and other environmental hazards. Conformal coatings provide a high level of protection against the ingress of moisture which can seriously degrade electrical properties, causing lower insulation resistance between conductors, premature high-voltage breakdown, corrosion of conductors, and even short circuits. There is, however, no coating that will totally resist the effects of environmental stresses, and so these coatings do have a finite time of protection. The coatings are designed to operate under the requirements of the system in which they are used. As a result, the coatings must have an excellent combination of mechanical, chemical, electrical, and thermal properties in addition to being easy to apply.

Because of the protective function of these coatings, they become a permanent part of the electronic package and must perform this function for the life of the product.

Types

Of the vast number of organic coatings available, only a few have been found to have the combination of properties required of a conformal coating. MIL Specification I-A6058 defines five classes of polymers for conformal coatings: acrylics, epoxies, polyurethanes, silicones, and paraxylylene polymers. Other polymer types include the diallylphthalate resins, polyimide resins, benzocyclobutenes, and new silicon-carbon (SYCAR) resins. Selected properties of these materials are shown in Table 9.4. A brief description of these resins is given in the following section. The chemistry of the resins has already been mentioned (see Chaps. 3 and 4 under thermoplastics and thermosets) and will not be discussed in this section.

Acrylic coatings

Acrylic coatings have excellent moisture resistance and dielectric properties. They have poor resistance to many chemicals. Their resistance to solvents is low. It is this property, though, that makes them easy to remove for circuit repair. Their resistance to mechanical abrasion is poor. They are easily applied and are fast-drying, making them suitable for high-volume, automated production. Chlorinated solvents are used to remove acrylic coatings. Other properties of the acrylic coatings are fungus resistance, long pot life (permitting a wide choice of application procedures), little or no exotherm during cure so that heat-sensitive parts are not damaged, and no shrinkage during cure.

Epoxy coatings

Epoxy materials provide excellent chemical resistance and mechanical properties. Cured epoxy coatings are hard. Their mechanical stiffness may require that stress-sensitive components be protected with a com-

TABLE 9.4 Typical Characteristics of Conformal Coating Encapsulant Materials

Properties	Acrylic	Urethane	Epoxy	Silicone	Polyimide	DAP	Parylenes	Benzocyclo-butene	Fluoro-polymers	Silicon-carbon
Volume resistivity (50% RH, 23°C), $\Omega \cdot$ cm	10^{15}	11×10^{14}	10^{12}–10^{17}	2×10^{16}	10^{16}	1.8×10^{16}	14×10^{16}	10^{19}	10^{15}	10^{16}
Dielectric strength, V/mil	3500	3500	2200	2000	7000		7000	2500	3000	
Dielectric constant										
60 Hz	3–4	5.4–7.6	3.5–5.0	2.7–3.1	3.4	3.6	2.65	—	—	—
1 kHz	2.5–3.5	5.5–7.6	3.5–4.5	—	3.4	3.6	2.65	—	—	—
1 MHz	2.2–3.2	4.2–5.1	3.3–4.0	2.6–2.7	3.4	3.4	2.65	2.65	2.7	2.6
Dissipation (power) factor										
60 Hz	0.02–0.04	0.015–0.048	0.002–0.010	0.007–0.001	—	0.010	0.0002	—	—	—
1 kHz	0.02–0.04	0.04–0.060	0.002–0.02	—	0.002	0.009	0.0002	—	—	—
1 MHz	2.5–3.5	0.05–0.07	0.030–0.050	0.001–0.002	0.005	0.011	0.0006	0.0008	0.016	0.002
Thermal conductivity, 10^{-4} cal/(s · cm³ · °C)	3–6	1.7–7.4	4–5	3.5–7.5	—	4–5	—	—	—	—
Thermal expansion, 10^{-5}/°C	5–9	10–20	4.5–6.5	6–9	4.0–5.0	—	3.5	3.4	—	8
Resistance to heat, continuous, °C	250	121	121	204	260	176	60–100 (O_2) 200 (N_2)	300	≤140	170
Effect of weak acids	None	Slight to dissolve	None	Little or none	Resistant	None	None	None	—	None
Effect of weak alkalies	None	Slight to dissolve	None	Little or none	Slow attack	None	None	—	None	None
Effect of organic solvents	Attacked by ketones, aromatics, and chlorinated hydrocarbons	Resists most	Generally resistant	Attacked by some	Very resistant	Resistant	Slight swelling in aromatics	None	Attacked	None

SOURCE: Adapted from Ref. 4.

pliant buffer coating between the component and the hard epoxy coating. Repair of an epoxy-coated PWB is difficult. Removal of the coating must be accomplished with heat or mechanical abrasion. Solvents that could remove epoxy coatings would also damage the printed-wiring board substrate. The epoxy systems that are used for electronic components are generally two-component systems. These coatings provide good humidity, abrasion, and chemical resistance.

Polyurethane coatings

The polyurethane materials are the most widely used for conformal coating applications. They are available as one- or two-component systems with the former requiring 3 to 10 days to cure at room temperature to reach optimum properties. Two-component systems cure in 1 to 3 h at elevated temperature to reach optimum properties.

They make tough coatings with good mechanical as well as moisture and chemical protection properties. The materials have good adhesion on components and PWBs that have been properly cleaned before coating. However, because their chemical resistance is high, repair or coated assembly is difficult. Polyurethane coatings can be removed with a heated tool or by microabrasive blasting or abrasive grinding. They are unsuitable for most high-frequency applications because the dielectric and loss factors vary considerably with frequency as well as temperature.

Silicone coatings

Silicone conformal coatings have excellent electrical properties. Their low dielectric constant remains substantially constant over the microwave frequencies. Silicones have high temperature resistance and are tough and flexible. The coefficient of thermal expansion is high. They are not resistant to hydrocarbons such as gasoline. A primer on the surface to be coated may be necessary before a silicone coating is applied to improve its adhesion. Repair has been difficult in the past due to a high chemical and heat resistance. The silicone resins used as conformal coatings are one of the few that are useful at service temperatures greater than 180°C. The Dow Corning Corporation is the major supplier of silicone coatings.

Parylene coatings

The parylenes are unique resins. First, they are vapor-deposited in a vacuum onto the component. Second, the coatings completely cover sharp edges and corners. The coatings and deposition process are noted for their ability to penetrate hard-to-reach places. The parylenes have excellent moisture, chemical, electrical, and mechanical properties. They form a very permeable resistant film and a 1-mil film will satisfy most moisture protection needs. It is difficult to repair a coated board. The coating can be removed by heating, abrasion, or microabrasive blasting. Union Carbide is the major supplier of the parylene materials.

Polyimide coatings

The polyimide coatings as a class of polymers are considered by many to be the top of the line in performance properties; i.e., they have the best combination of mechanical, electrical, and thermal properties, allowing them to be used at temperatures greater than 200°C for continuous operation. A variety of chemical compositions are available depending on the monomers used to prepare the polyimide; by varying the ingredients, the properties of the final product can be altered to meet specific end-use requirements. The polyimides used as coatings are supplied as low-solids (<25 percent), high-viscosity solvent solutions. The solvent is a high-boiling-temperature (<150°C) aprotic material such as N-methyl-pyrrolidone. As discussed in Chap. 3, the cure chemistry of the polyimides requires elevated temperatures (>250°C) to convert the amic-acid precursor to the final imide form. Photo-sensitive polyimide coatings are also available, but these materials, while sensitive to radiation, in order to affect chemical changes must also be thermally cured at elevated temperatures to convert to the imide structure. The cured aromatic polyimides possess excellent chemical resistance to most normal process solvents. The only material that dissolves the polyimide is antimony trichloride.[5] Suppliers of polyimide coating materials include du Pont, Ciba-Geigy, Hitachi, Toray, and Ashai.

Fluorocarbon polymer coating

The fluoropolymer conformal coatings are actually fluorine-containing acrylic polymers. The exact chemistry is proprietary, but the particular product described in this section (Fluorad FC-725 from 3M Co.) is reported to be a fluorinated terpolymer containing about 27 weight percent fluorine as side chain constituents on an acrylate polymer backbone. This particular material is a repairable coating and is qualified under MIL-I 46058. Some properties of FC-725 are listed in Table 9.5. The fluoropolymer acrylate coatings have very low surface energies and are difficult to adhere to. For example, the surface energies of these coatings are 11 to 14 dyn/cm compared to Teflon which is 18 to 20 dyn/cm, silicone at 24 dyn/cm, polyethylene at 31 dyn/cm, and epoxy, urethanes, and acrylate at 40 dyn/cm. This material is supplied by the 3M Co.

TABLE 9.5 Properties of FC-725 Fluoropolymer Conformal Coating

Property	Value
Dielectric strength, V/mil	2500
Dielectric constant at 1 MHz	2.7
Dissipation factor at 1 MHz	0.016
Volume resistivity at 50% RH, $\Omega \cdot cm$	10^{15}
Pot life	Excellent
Repairability	Excellent

Silicon-carbon coatings (SYCAR)

These coatings, while not specifically designated for use as a conformal coating, were developed by Hercules Inc. These materials are used as impregnating resins with various fabrics and as encapsulants. They have not been used as conformal coatings, but there is no reason why they could not be used for such an application. Both solvent and solventless compositions are available. The SYCAR resins have exceptional moisture resistance, low dielectric constant, high T_g, and good thermal stability. Curing of the resins requires several hours at 180°C. The material has been developed for semiconductor encapsulant applications. The resins have a long pot life and good overall combination of mechanical, electrical, and thermal properties. These are useful for encapsulation for chip-on-board tape-automated bonding, chip carriers, and multichip modules. Since these resins are thermoset and have excellent solvent resistance, their repairability would probably not be good.

Benzocyclobutenes

The benzocyclobutene (BCB) polymers, developed by the Dow Chemical Co., are a class of materials that possess excellent physical, chemical, and electrical properties. The BCB polymers have been used in many electronic packaging applications, e.g., as encapsulants and interlevel dielectrics.[7–10] Key properties of these materials are low water absorption, low dielectric constant and loss factor, and a high level of planarization. The coatings are supplied as solvent (mesitylene) solutions between 35 and 63 weight percent solids (see Table 9.6). The coating (1 mil or less) thermally polymerizes in nitrogen (1 h or less at 250°C). Rapid cure can be effected in 10 min.[11] The Dow Chemical Co. supplies this resin.

Allyl phthalate resins

These resins are considered part of a larger family of polyester resins. There are two important allyl resins: diallyl phthalate (DAP) and dial-

TABLE 9.6 **Typical Properties of Cyclotene 3022 Resins***

Solvent	Mesitylene
Resin content, %	35–63
Viscosity, cSt	14–870
Density, g/cm^3	0.93–0.99
Spin-coated thickness, μm	1–25
Dielectric constant, 1 MHz	2.7
Dissipation factor, 1 MHz	0.0008
Breakdown voltage, V/cm	3×10^6
Volume resistivity, $\Omega \cdot$ cm	1×10^{19}
Coefficient of thermal expansion, ppm	52
T_g, °C	>350
Elongation, %	6
Water absorption at 85% RH, wt %	0.2
Planarization, %	>90
Thermal stability, °C	350 (<1% wt. loss/h)
Refractive index, λ = 632.8 nm	1.56
Poisson's ratio	0.34
Tensile strength, lb/in^2	12,000

*Dow Chemical Co.

lyl isophthalate (DAIP). These resins have been around since the early 1950s, and they still hold a competitive position in the electronics market because of their outstanding heat and humidity resistance, which they return under long-term exposure. The properties have been described in Chap. 4 and will not be repeated here. The materials are used as potting or encapsulation resins for electrical components as well as molding resins. Major suppliers are also listed under DAP resins in Chap. 4.

Selection of coating and encapsulant

There are six properties one must consider in choosing a conformal coating from a manufacturing standpoint: pot life, viscosity, solids content, coating ingredients, application ease, and cure temperature. Of equal importance are the operational properties of the cured coating, and these include electrical properties, thermal properties, moisture and humidity resistance, repairability, and mechanical properties. All these properties should provide a balance between optimum properties and minimum cost. Table 9.7 summarizes the significance of these properties.

TABLE 9.7 Key Properties of Coatings and Significance

Property	Type of variable*	Significance
Pot life	M	Time a coating can remain open before use on assembly line. Affects planning operations
Viscosity	M	Proper viscosity allows easy flowing of the liquid around and under components. Affects coverage
Solids	M	Reflects amount of material converted to final film. Affects coating thickness. Solvents can be expelled into atmosphere
Cure time and temperature	M	Affects production rate. Determines optimum properties of coating
Application ease	M	Affects production rate
Coating ingredients	M	Components in coating must not adversely affect component (no swelling, softening, or corrosive effects)
Electrical	P	Affects insulation resistance, signal propagation speed, capacitance, attenuation, voltage breakdown
Mechanical	P	Affects component stress and long-term protection capabilities
Thermal	P	Affects heat-transfer capabilities, component stress, operating temperature limits
Moisture and humidity	P	Determines barrier protection properties. Affects electrical properties
Repairability	P	Easy repair should be goal. Affects cost

*M = manufacturing; P = performance.

Application methods

It is worthwhile to define the two terms as they relate to the coating of electronic devices, i.e., a conformal coating and/or an encapsulant. The purpose of both is to protect the electronic device by encasing or encapsulating it. They differ in that a conformally coated device is usually repairable; the encapsulated device is not.

Encapsulation often is considered an alternative to conformal coating for protection of printed-circuit assemblies. One advantage of encapsulation is improved component protection from shock and vibration. However, there are disadvantages, including the following:

Rigid encapsulation and potting compounds can mechanically stress circuit components unless a soft conformal coating is first applied as a cushion.

Encapsulation sometimes requires special molds compared to the easy application methods of conformal coating. Thus, encapsulation is more costly. Further, the process is inherently much slower and less suited to mass production.

Encapsulating compounds typically act to create a heat-transfer barrier, sharply reducing heat dissipation of circuit components. This may degrade component performance and reliability.

After encapsulation, components cannot be inspected, nor are they readily accessible. This means encapsulated printed-circuit boards are difficult or impossible to repair.

An encapsulated printed circuit is heavier, which can affect its suitability for certain weight-sensitive applications. The encapsulating material itself often outweighs the components that it is designed to protect.

Conformal coatings are generally thinner, although that is not absolute. One could consider a conformal coating as a form of encapsulation, although other forms of encapsulation include potting, molding, and glob-topped techniques which will be discussed in Chap. 12. In any case, an electronic device or component can be conformally coated or encapsulated by spray coating, dipping, brushing, or flow coating.

Spraying. Spray coating is widely used to apply coatings to protect all types of printed-wiring boards. It is a technique that is particularly adaptable for coating uneven surfaces. Proper coating viscosities are required for effective coating.

Dipping. Dip coating is an efficient method of coating a component provided the coating has the proper viscosity. The rates of immersion and withdrawal are critical factors in obtaining complete coverage of the assembly.

Brushing. Brush coating is not a very effective method for obtaining uniform coverage. It is useful if a small area needs to be repaired. It is also not very practical for coating large parts.

Flow coating. Flow coating consists of pouring the coating as a curtain as the device to be coated passes through the coating curtain. This method works well if the electronic assemblies have a minimum number of flat packs of ICs.

A summary of the advantages and disadvantages of conformal coatings and encapsulants is presented in Table 9.8.

TABLE 9.8 Review of Coating Advantages and Disadvantages

Coating material	Advantages	Disadvantages
Epoxy	Solvent resistance is good. Excellent mechanical properties, adhesion	Not repairable. Electrical properties good but not outstanding
Acrylics	Fast-drying. High-speed production. Good moisture and dielectric properties. Repairable	Poor chemical resistance. Low-temperature use only
Urethanes	Tough coatings. Good moisture and chemical resistance	Unsuitable for high-frequency applications. Repairable by heated tool or abrasive grinding
Silicones	Good electrical properties. Constant E over microwave frequencies. Good thermal cycling properties	High coefficient of thermal expansion. Low mechanical strength. Repair is difficult
Parylene	Good chemical resistance. Coats well	Thickness limitation. Not repairable
Polyimides	Good thermal stability. Good chemical resistance. Good overall combination of properties	High-temperature cure is needed. Not repairable
Fluoropolymers	Repairable. Good electrical properties and moisture resistance	Poor chemical resistance. Surface difficult to overcoat. Low surface energy
Silicon-carbon resin	Long pot life. Low moisture absorption	Repairability unknown
Benzocyclobutenes	Excellent combination of high-temperature and moisture resistance	High-temperature cure required. Not repairable
Allylphthalate	Excellent long-term moisture and heat resistance	High-temperature cure required. Not repairable

Other Coating Processes and Applications

In addition to conformal coatings, there are a number of other processes that the electronics engineer can use with the above-mentioned coating materials as well as other materials in order to provide enhanced protection of the electronic device. These processes appear to be similar, but there are small differences and the processes are defined to identify the coating processes available to the design engineer.

Encapsulation

Encapsulation is the encasement of a component in a resin coating by dipping, spraying, or embedding with or without a mold. Encapsulants can be designated as potting compounds, glob-topped, or molding compounds, all of which are used to encase electronic components.

Potting

Potting involves surrounding the component housed in a container with a liquid resin which is then cured in place. The container becomes an integral part of the system, and the interfacial adhesion between container and coating is a critical property to be maintained if a long-lasting, reliable package is to be obtained.

Casting or embedding

Casting is similar to potting except that the container or mold is removed from the part after the coating is cured.

The polymers used for encapsulation, potting, casting, or embedding are derived from four classes of materials: epoxy, polyester, polyurethane, and silicone resins. Some of the newer resins—at least the solventless formulations—include some bismaleimides, cyanate esters, and silicon-carbon (SYCAR) resins that could also function as encapsulants. A comparison of some characteristics of encapsulating resins is shown in Table 9.9.

TABLE 9.9 A Comparison of Encapsulating Resins

Resin	Dielectric properties	Overall adhesion	Shrinkage	Maximum-use temp., °C	Coefficient of thermal expansion	Chemical resistance
Epoxy	Excellent	Excellent	Low	150	Moderate	Good
Bismaleimide	Excellent	Good	Moderate	180	Low	Good
Cyanate ester	Excellent	Excellent	Low	175	Low	Excellent
Polyester	Good	Fair	High	175	Moderate	Poor
Silicone	Excellent	Low	Low	200	High	Good
Silicon-carbon (SYCAR)	Excellent	Good	Low	160	Moderate	Excellent

Glob-topped encapsulation

Glob-topped compounds are very viscous resin pastes that are usually applied to components already attached to a printed-wiring board or other electronic substrate. These resins are usually a highly filled composition of epoxy, silicones, and polyurethanes, but other resins can be used.

Molding

As part of the general category of encapsulation, molding resins are highly used as a packaging material for electronic devices because they offer high-performance protection at reasonable cost. The process involves injection of a thermoset molding resin into a mold containing many electronic devices; the molding resin flows around the devices and fills the mold. The coating is cured, and the encapsulated molded electronic device is removed. Molding compounds include the epoxy resins, specifically cresol-novolacs, silicones, liquid-crystal resins, and polyphenylene sulfide. An excellent review of plastic packaging of electronic devices is given in Ref. 12.

In all the above processes, it is critical for the designer to consider the properties of the coating and their impact on the performance of the electronic device. Some important properties are thermal expansion, thermal conductivity, modulus of elasticity, adhesion, moisture resistance, environmental stability, thermal shock resistance, flame resistance, and electrical properties.

Impregnation

This coating process is defined as the filling of the voids and interstices of a material with a liquid resin. These interstices can be those of an electrical or electronic component or device, such as coil impregnation in rotating electrical equipment, or it can be the impregnation of an organic or inorganic reinforcement (paper or fabric) that is used in printed-wiring board substrates.

Resin types

The types of resins used in impregnation are almost always low-viscosity liquids so as to ensure optimum impregnation and coverage of the part or device. The materials can be solventless or solvent solutions of resins, and all the resins mentioned in this chapter are candidates for use as impregnating resins. To review, those resins are epoxy, silicone, cyanate ester, polyester, polyimides, silicon-carbon resins, urethane, fluorocarbon, and bismaleimides. In addition, phenolic and melamine resins are considered to be impregnants because they are used to impregnate a variety of reinforcements for use in industrial, decorative, and electrical laminates.

Application methods

There are basically four impregnation methods: trickle, vacuum, vacuum-pressure impregnation, and dip coating.

Trickle impregnation. Trickle impregnating equipment operates on the principle of applying metered quantities of catalyzed resin (primarily polyester and epoxy resins) in a continuous stream to stator and armature coils in such a manner that the resin is applied only to the windings with no maskings or cleanup necessary. The units being impregnated are connected to a power supply, and resistance heating is used for preheating and curing of the resin, or, in some cases, radiant heating is used.

Vacuum impregnation. These completely assembled units consist of a tank—in most cases heated—which is connected to a vacuum pump and provided with an inlet connection. The resin mix and parts are held under vacuum (in separate chambers) for a period of time to remove moisture, gases, and other contaminants. Then the resin is admitted into the parts chamber and cured at proper temperature.

Vacuum-pressure impregnation. Motor and generator armatures and stators, coils, transformers, capacitors, condensers, and other electrical components are vacuum-pressure impregnated to eliminate corona-causing voids and to provide the superior internal insulation so important to system reliability.

Vacuum-pressure impregnators are generally cylindrical and equipped with quick-opening covers. Sizes range from 2 to 12 ft in diameter by any required depth. Most frequently, pressure ranges up to 100 lb/in^2 gauge and vacuum to the low millimeters. Operations may be manual, semiautomatic, or fully automatic.

Impregnants of various types are used, depending on the application, from simple oils and varnishes through epoxies and polyesters, and they include waxes, styrene, and asphalt.

Reinforcement impregnation. In this coating impregnation process, a reinforcing material (fabric or paper) is pulled through a resin solution to saturate the reinforcement. Excess resin is removed by passing the saturated material through wiper rolls. It is then passed through drying ovens to remove solvent and to partially advance the polymerization of the resin. This coated material, called a pre-preg, is wound and stored for conversion to a final form, usually by a heat and pressure lamination process.

The types of products and their properties that are produced through impregnation are covered in Chap. 12.

References

1. M. W. Hunt, ed., *Materials Engineering,* Penton Publications, 1992, p. 1563.
2. H. Mark, N. Bikales, C. Overberger, and G. Menges, eds., *Encyclopedia of Polymer Science and Engineering,* vol. 3, Wiley, New York, 1985.
3. C. P. Wong, ed., *Polymers for Electronic and Photonic Applications,* Academic Press, New York, 1993, p. 188.
4. C. F. Coombs, Jr., ed., *Printed Circuits Handbook,* McGraw-Hill, New York, 1988, p. 206.
5. L. M. Minges and members of the ASM International Handbook Committee, *Electronic Materials Handbook,* vol. 1: *Packaging,* ASM International, Materials Park, OH, 1989.
6. R. Kirchoff, U.S. Patent 4,540,763, 1985.
7. D. Burdeaux, P. Towsend, J. Carr, and P. Garrou, "Benzocyclobutene (BCB) Dielectrics for the Fabrication of High Density, Thin Film Multichip Modules," *Journal of Electronic Materials,* vol. 19, no. 12, 1990, p. 1357.
8. E. S. Moyer, E. Rutter, M. Bernius, P. Harris, H. Pranjoto, and D. Denton, "Photodefinable Benzocyclobutene Formulations for Thin Film Microelectronic Applications, Part II" (preprint), *International Electronics Packaging Society Proceedings,* Austin, TX, September 1992.
9. H. Pranjoto and D. D. Denton, "Moisture Uptake of BCB Films for Electronic Packaging," *Mat. Res. Soc. Symp. Proc.,* vol. 203, 1991, p. 295.
10. R. H. Heistand, II, R. DeVellis, T. Manial, A. Kennedy, T. Stokich, P. Towsend, P. Garrou, T. Takahashi, G. Aderna, M. Berry, and I. Turlik, "Advances in MCM Fabrication with Benzocyclobutene Dielectric," *International Journal of Microcircuits and Electronic Packaging.,* vol. 15, no. 4, 4th quarter 1992.
11. Dow Chemical Co. bulletin, *Typical Properties of Cyclotene 3022 Resins,* Midland, MI, 1993.
12. L. T. Manjoine, *Plastic Packaging of Microelectronic Devices,* Van Nostrand Reinhold, New York, 1990.

Part 2

Design Considerations for Plastics

Chapter

10

Plastic Properties and Testing

In addition to characterizing polymers on the basis of fundamental properties such as molecular weight, molecular weight distribution, and refractive index, the plastic material itself can be subjected to a variety of chemical, mechanical, thermal, electrical, environmental, and optical tests to provide some measure of the engineering properties of these materials for practical use. The proper choice of a plastic material for a particular application is not a trivial task: a great deal of analysis is necessary. A whole spectrum of properties (as listed above) have to be considered as well as the influences of the manufacturing process and the actual design of the part, before a particular plastic is chosen for a given application. Plastics offer a number of outstanding features such as high strength and stiffness per unit weight; excellent electrical, thermal, and wear resistance properties; and these factors combined with an ease of processing and fabricating make these materials one of the most versatile in industry today. The properties of plastics can be further altered by mixing them with an almost infinite number of fillers and other reinforcements such as fibers, fabrics, and paper. It is the goal of this chapter to acquaint the reader with information on what factors influence the properties of plastics, what are those properties, how are they measured, and what actual values one can expect from a given class of polymers. Note that there are a wide range of property values for a particular plastic (see Table 10.1) and that specific property values for a given plastic formulation should be obtained from the material data sheets supplied by the manufacturer. For those interested in more detailed theoretical and fundamental discussion of polymer properties, see Refs. 1 to 5. Reference 6 contains the ASTM test methods for the evaluation of the properties of polymers.

TABLE 10.1 General Properties of Plastics and Elastomers

General properties	Acrylics	Acetals	ABS	Thermoplastics Fluoro-plastics	Polyamides	High-temperature polyamides
Tensile strength $\times 10^3$ lb/in^2	5–11	6–16	3–7	2–9	5–26	25–535*
Tensile modulus $\times 10^3$ lb/in^2	280–450	400–1300	200–400	38–300	200–2000	11–20
Elongation, %	2–30	2–300	5–200	10–300	5–350	—
Flexural strength $\times 10^3$ lb/in^2	8–17	9–24	5–12	2–3.5	3–34	—
Flexural modulus $\times 10^3$ lb/in^2	310–50	200–1050	200–420	60–250	150–1500	—
Compressive strength 2% offset $\times 10^3$ lb/in^2	7–18	4–18	7–11	0.7–14	2–20	—
Izod impact, notched, ft · lb/in	0.2–2.3	0.8–17	2–13	2.5–no break	1–20	—
Thermal conductivity, Btu/(h · ft^2 · °F · in)	1.44–1.56	1.56–1.92	0.12–2.4	1.44–1.74	1.2–3.3	0.7–1.21
Specific heat, Btu/(lb · °F)	0.35	0.35	—	0.22–0.33	0.28–0.58	—
Thermal expansion coeff, 10^{-5} in/(in · °F)	3–6	2–7	3.2–6	4–11	1.2–8.4	1.12†
Heat deflection temp., °C, 264 lb/in^2 load	65–120	115–210	107–154	100–140	47–260	—
Brittleness temp., °C	—	—	<–40	—	—	—
Glass transition temp., °C	–24 to +115	—	—	140–327	40–50	375
Melt temp., °C	<100 to >200	160–180	—	—	200–265	Does not melt
Volume resistivity, Ω · cm	10^{14}–10^{16}	10^{14}–10^{16}	10^{16}	10^{14}–10^{18}	10^{13}–10^{15}	10^{16}–10^{17}
Dielectric strength, V/mil	400–530	400–600	300–600	260–2000	300–840	200–730
Dielectric constant						
60 Hz	3.4–4.5	3.7–4.0	2.5–3.5	2–10	3.2–5.6	1.2–3.5
1 MHz	2.5–3.2	3.1–4.0	2.4–3.2	2–7.5	3.0–5.4	—
Dissipation factor						
60 Hz	0.026–0.06	0.0047	0.003–0.04	0.0002–0.05	0.007–0.06	0.003–0.02
1 MHz	0.01–0.03	0.0036–0.009	0.005–0.016	0.0002–0.184	0.01–0.1	—
Arc resistance, s	No track	120–240	85–130	50 to >360	81–135	—

Thermoplastics

General properties	Polycarbonates	Polyethylene	Polypropylene	Polyimides	Polyamide-imides	Polysulfones	Polyketones
Tensile strength × 10^3 lb/in²	9–20	1.6–9.0	3–14	7–25	22–32	9–30	3.5–33
Tensile modulus × 10^3 lb/in²	340–3000	4–900	130–240	3–600	650–2110	300–2600	520–3300
Elongation, %	1–150	80–1000	3–200	1–70	6–18	2–100	1–150
Flexural strength × 10^3 lb/in²	10–36	—	5–24	19–28	27–48	12–38	16–48
Flexural modulus × 10^3 lb/in²	240–2900	7–150	170–610	430–700	600–2600	300–2600	440–3300
Compressive strength 2% offset × 10^3 lb/in²	10–31	—	4.4–6.5	13–40	31–40	10–40	18
Izod impact, notched, ft·lb/in	1–16	1.8–30	0.4–12	0.5–15	2–12	1.2–19	1.6–2.7
Thermal conductivity, Btu/(h·ft²·°F·in)	1.35–1.53	2.28	1.21–1.72	2.5–6	—	1.4–2.0	2.9–6.2
Specific heat, Btu/(lb·°F)	0.30	0.46–0.55	0.45–0.48	0.27–0.31	—	0.26	—
Thermal expansion coeff, 10^{-5} in/(in·°F)	0.9–5.1	8–17	3.8–5.9	2.3–2.8	1.7	1.2–3.0	1.4–2.4
Heat deflection temp., °C, 264 lb/in² load	125–152	—	49–60	290–360	274	150–216	165–320
Brittleness temp., °C	−200	−4 — −148	—	—	—	—	—
Glass transition temp., °C	150	−25	−1 — −13	360–410	275	190–225	143–177
Melt temp., °C	—	83–135	160–168	None	—	—	330
Volume resistivity, Ω·cm	10^{16}	10^{15}–10^{19}	10^{17}	10^{17}	10^{16}–10^{17}	10^{15}–10^{17}	10^{16}–10^{17}
Dielectric strength, V/mil	380–450	480	450–700	300–7700	600	340–480	400–480
Dielectric constant							
60 Hz	3.0–3.5	2.3	2.2	3.4–4.1	—	3.0–3.5	3.5
1 MHz	2.96–3.48	—	2.2	3.5–3.9	3.5	3.0–3.5	—
Dissipation factor							
60 Hz	0.0008–0.0012	<0.0005	0.0005	0.002–0.003	—	0.0007–0.007	0.002
1 MHz	0.0067–0.024	—	0.0002	0.004–0.011	0.02	0.003–0.007	—
Arc resistance, s	120	—	123–140	150–230	125	41–122	—

*Nomex paper, lb/in.
†After heat setting at 260°C.

TABLE 10.1 General Properties of Plastics and Elastomers (*Continued*)

	Thermoplastics						
General properties	Vinyl chlorides	Poly-etherimide	Polyphenylene sulfide	Polyesters	Polyphenylene oxide	Styrenes	PBI
Tensile strength × 10^3 lb/in^2	6–12	14–28	7–29	8–25	6–18	4–13	16–29
Tensile modulus × 10^3 lb/in^2	300–600	430–3300	480–2200	240–3500	340–1000	330–1750	850–3100
Elongation, %	1–450	2–60	1–4	1–300	2–60	1–50	1–3
Flexural strength × 10^3 lb/in^2	6–22	22–33	14–43	10–30	8–23	4–20	24–46
Flexural modulus × 10^3 lb/in^2	120–450	480–1300	550–2400	260–2700	300–500	230–1450	810–2900
Compressive strength 2% offset × 10^3 lb/in^2	10–85	20–24	16–32	8–25	12–18	2–19	18–58
Izod impact, notched, ft · lb/in	0.5–10	1–2	0.5–4.8	0.7–9	2–7	0.2–4.5	0.5
Thermal conductivity, Btu/(h · ft^2 · °F · in)	0.6–1.2	0.12	1.1–6.3	1.1–2.2	1.1–1.5	0.29–1.40	2.8
Specific heat, Btu/(lb · °F)	—	—	—	0.2–0.5	—	0.25–0.35	1.3
Thermal expansion coeff., 10^{-5} in/(in · °F)	1–8	1.1–3.1	2.9–5.4	1–5	1–4.5	1.6–5.6	—
Heat deflection temp., °C, 264 lb/in^2 load	55–110	199–215	100–260	50–250	8.8–150	70–100	435
Brittleness temp., °C	−70–+20	—	—	—	—	—	—
Glass transition temp., °C	75–110	215	88–90	70–90	100–190	74–120	427
Melt temp., °C	—	—	285	220–285	—	—	—
Volume resistivity, Ω · cm	10^{11}–10^{16}	10^{16}	10^{15}–10^{16}	10^{15}	10^{17}	10^{16}	10^{14}
Dielectric strength, V/mil	24–500	750–800	340–520	350–570	400–1000	300–520	550
Dielectric constant							
60 Hz	3–8	—	—	3.1–4.3	2.6	2.4–4.7	—
1 MHz	—	3.1	—	2.9–3.9	2.6	2.4–4.0	3.3
Dissipation factor							
60 Hz	0.02–0.11	0.0013	—	0.002–0.006	0.0007	0.0001–0.006	—
1 MHz	—	—	—	0.01–0.03	0.0024	0.0001–0.01	—
Arc resistance, s	—	85–126	34–200	50–190	75	20–150	186

General properties	Thermosets					
	Alkyds	Allyls	Epoxies	Amino resins	Phenolics	Unsaturated polyesters
Tensile strength $\times 10^3$ lb/in^2	3–9.5	6–11	4–27	5–10	3–17	1–100
Tensile modulus $\times 10^3$ lb/in^2	500–3000	600–2200	50–500	100–2000	800–3300	1–6000
Elongation, %	—	3–5	1–60	3–350	0.4–1.0	—
Flexural strength $\times 10^3$ lb/in^2	6–26	8.5–20	1–70	6–24	5–36	8–80
Flexural modulus $\times 10^3$ lb/in^2	2000	1000–1500	36–2500	—	4.5–3300	1–300
Compressive strength 2% offset $\times 10^3$ lb/in^2	12–38	18–31	17–50	25–45	10–38	1–37
Izod impact, notched, ft · lb/in	0.3–18	0.3–15	0.2–2	0.2–0.3	0.2–10	0.2–18
Thermal conductivity, Btu/(h · ft^2 · °F · in)	4–7	—	1–6	4–4.5	6	1.2–5
Specific heat, Btu/(lb · °F)	0.3	—	0.4	—	0.28–0.40	0.2–0.55
Thermal expansion coeff., 10^{-5} in/(in · °F)	0.2–0.55	2–5	1–3.5	—	0.8–2.2	1–6
Heat deflection temp., °C, 264 lb/in^2 load	120–230	116–260	33–275	127–200	105–260	65–200
Brittleness temp., °C	—	—	—	—	—	—
Glass transition temp., °C	—	—	—	—	—	—
Melt temp., °C	—	—	—	—	—	—
Volume resistivity, Ω · cm	10^{14}	10^9–10^{12}	10^9–10^{16}	10^{11}–10^{12}	10^9–10^{12}	10^{12}–10^{15}
Dielectric strength, V/mil	290–350	250–450	280–440	70–400	200–475	200–440
Dielectric constant						
60 Hz	5–7.5	3.3–5.2	3.3–5.7	6–12	5–16	2.8–7.0
1 MHz	4.5–7	3.4–6.5	2.7–4.8	—	4–7	2.8–6.1
Dissipation factor						
60 Hz	0.02–0.04	0.008–0.05	0.004–0.1	0.02–0.15	0.05–0.6	0.003–0.18
1 MHz	0.008–0.2	0.009–0.15	0.013–0.06	—	0.02–0.1	0.006–0.06
Arc resistance, s	180–240	85–140	75–185	70–186	5–200	80–180

TABLE 10.1 General Properties of Plastics and Elastomers (*Continued*)

General properties	Silicones	Urethanes	Thermosets Cyanate esters	Bismale-imides	Benzocyclo-butene
Tensile strength × 10^3 lb/in²	0.35–1.5	0.175–11	10–13	11–14	12
Tensile modulus × 10^3 lb/in²	9	10–100	420–460	560–620	290
Elongation, %	20–800	3–1000	2.5–4	2–3	6
Flexural strength × 10^3 lb/in²	—	0.7–19	18–25	24–27	—
Flexural modulus × 10^3 lb/in²	0.1–0.25	10–610	420–480	580–590	475
Compressive strength 2% offset × 10^3 lb/in²	0.1–15	20	—	—	—
Izod impact, notched, ft · lb/in	—	0.4–>25	0.7–1.1	—	—
Thermal conductivity, Btu/(h · ft² · °F · in)	1.3–1.5	1.4	—	—	—
Specific heat, Btu/(lb · °F)	—	—	—	—	—
Thermal expansion coeff., 10^{-5} in/(in · °F)	45	—	3–4	—	5
Heat deflection temp., °C, 264 lb/in² load	−260	—	—	—	—
Brittleness temp., °C	—	—	—	—	—
Glass transition temp., °C	—	—	—	—	—
Melt temp., °C	—	—	—	—	—
Volume resistivity, Ω · cm	10^{14}–10^{15}	10^{11}–10^{15}	10^{16}	10^{13}	10^{13}
Dielectric strength, V/mil	125–550	300–500	1600–2100	—	7500
Dielectric constant					
60 Hz	2.7–5	4–7.5	—	4.41	—
1 MHz	2.6–4	6.5–7.1	2.6–2.9	—	2.7
Dissipation factor					
60 Hz	0.001–0.030	0.015–0.017	—	—	—
1 MHz	0.001–0.020	—	0.001–0.005	0.01	0.0008
Arc resistance, s	115–250	0.1–0.6	—	—	—

	Thermosets		
General properties	Silicon-carbon resins	Parylene	Polyimides
Tensile strength × 10^3 lb/in^2			4–23
Tensile modulus × 10^3 lb/in^2			460–4650
Elongation, %			1
Flexural strength × 10^3 lb/in^2	9		6.5–50
Flexural modulus × 10^3 lb/in^2	300		422–3000
Compressive strength 2% offset × 10^3 lb/in^2	—		19–32
Izod impact, notched, ft · lb/in	—		0.6–15
Thermal conductivity, Btu/(h · ft^2 · °F · in)	—		5–12‡
Specific heat, Btu/(lb · °F)	—		—
Thermal expansion coeff., 10^{-5} in/(in · °F)	36		13–50§
Heat deflection temp., °C, 264 lb/in^2 load	—		300–350
Brittleness temp., °C	—		—
Glass transition temp., °C	140–190		—
Melt temp., °C	—		—
Volume resistivity, Ω · cm	10^{15}		10^{15}–10^{16}
Dielectric strength, V/mil	750		300–560
Dielectric constant			
60 Hz	2.6		—
1 MHz	—		3.5–4.7
Dissipation factor			
60 Hz	0.002		—
1 MHz	—		0.004–0.011
Arc resistance, s	180	50–230	—

‡10^{-4} cal · cm/(s · cm^2 · °C)
§10^{-6} in/(in · °C)

TABLE 10.1 General Properties of Plastics and Elastomers (*Continued*)

		Elastomers				
General properties	Butyl	Chlorosulfonated polyethylene	Epichlorohydrin	EPM/EPDM	Chloroprene	Fluoroelastomers
Tensile strength × 10^3 lb/in^2	2	0.5–3.5	2–3	0.5–3.5	0.5–3.5	1.5–2
Tensile modulus × 10^3 lb/in^2	50–500¶	100–3000	150–2000	100–3000	100–3000	200–2000
Elongation, %	300–800	100–700	200–800	100–700	100–800	150–450
Resilience	Fair	Good	Good	Good	Excellent	Fair
Thermal conductivity, Btu/(h · ft^2 · °F · in)	0.053	0.065	—	0.15	0.11	0.06–1.3
Thermal expansion coeff., 10^{-5} in/(in · °F)	32	27	—	32	34	—
Brittleness temp., °C	−50	−45	−20 – −60	−55	−45	−15 – −50
Glass transition temp., °C	—	—	—	—	−109	—
Volume resistivity, Ω · cm	10^{16}	10^{14}	—	10^{16}–10^{17}	10^{13}	10^{13}
Dielectric strength, V/mil	600–700	650	—	500–1000	400–600	500
Dielectric constant						
60 Hz	2.3	7	—	2.2–3.0	8	5–10
1 MHz	2.2	6	—	2.2–2.8	6.7	—
Durometer hardness	30A–100A	40A–95A	30A–95A	30A–90A	30A–95A	55A–95A
Compression set	25	35–80	20	20–60	20–60	15–30

General properties	Elastomers					
	Natural rubber	Nitrile	Ethylene acrylics	SBR	Polysulfides	Chlorinated polyethylene
Tensile strength × 10^3 lb/in^2	2.4–4.6	1–3.5	1.9	1.8–3	0.5–1.5	0.9–3
Tensile modulus × 10^3 lb/in^2	480–850	490	800	300–1500	—	700–2200
Elongation, %	300–750	400–600	450	450–500	210–450	100–700
Resilience	Outstanding	Good	Fair	Good	Fair	—
Thermal conductivity, Btu/(h · ft^2 · °F · in)	0.082	0.14	—	0.14	—	—
Thermal expansion coeff., 10^{-5} in/(in · °F)	37	39	—	37	—	—
Heat deflection temp., °C, 264 psi load	—	—	—	—	—	—
Brittleness temp., °C	−60	−50	−34	−60	−45	−50
Glass transition temp., °C	−109	—	—	—	−55	—
Melt temp., °C	—	—	—	—	—	—
Volume resistivity, Ω · cm	$10^{1\ddagger}$	10^{10}	10^{12}	10^{13}	10^{13}	—
Dielectric strength, V/mil	600–800	250	730	600–800	—	—
Dielectric constant						
60 Hz	—	—	—	—	7.3	—
1 MHz	—	—	—	—	6.8	—
Durometer hardness	30A–100A	30A–100A	64A	30A–90D	20A–80A	50A–95A
Compression set	10–30	5–20	—	5–30	29–38	5–30

‡10^{-4} cal · cm/(s · cm^2 · °C)
¶100% tensile modulus.

TABLE 10.1 General Properties of Plastics and Elastomers (*Continued*)

General properties	Elastomers		
	Poly-butadienes	Silicones	Urethanes
Tensile strength × 10^3 lb/in^2	2–2.5	1.5	0.8–8
Tensile modulus × 10^3 lb/in^2	300–1500	—	25–5000
Elongation, %	450	100–800	250–800
Resilience	Excellent	Poor–excellent	High–damp
Thermal expansion coeff., 10^{-5} in/(in · °F)	37	45	5–25
Heat deflection temp., °C, 264 psi load	—	—	—
Brittleness temp., °C	−100	−110	−50
Glass transition temp., °C	−108 – −14	—	—
Volume resistivity, Ω · cm	—	10^{14}–10^{16}	10^{10}–10^{13}
Dielectric strength, V/mil	400–600	400–700	330–700
Dielectric constant			
60 Hz	—	2.9–4	4.7–9.5
1 MHz	3.3	2.9–4	5.9–8.5
Durometer hardness	40A–80A	20A–90A	10A–80D
Compression set	10–30	10–30	10–45

Mechanical Properties

Tensile strength

Tensile strength is defined as the amount of stress applied to stretch a material to its breaking point. Tensile strength increases with the molecular weight of the polymer, reaching a maximum and leveling off. It is also dependent on cross-linking in the polymer, the nature of the chemical groups present within the polymer structure, and the type and quantity of ingredients mixed with the polymer. Linear polymers exhibit high tensile strengths because the polymer chains uncoil, stretch, and align themselves as the polymer is elongated. Elastomers have low tensile strengths while thermoplastics and thermosets have medium to high tensile strengths. Plastics exhibit different responses to a tensile force and have been generally classified according to their response. For example, polytetrafluoroethylene is a soft and weak material that is characterized by a low modulus, low yield stress, and medium elongation at break. A hard and brittle material such as a phenolic is characterized by a high modulus and low elongation. Polyacetal, e.g., is a hard and strong material because it has a high modulus, high yield stress, low elongation, and high tensile strength at break. Polycarbonate is a hard and tough material because of its high modulus, yield stress, high elongation, and tensile strength at break. Polyethylene, on the other hand, exhibits multiple behaviors; in fact, it can be soft and tough, hard and tough, and even hard and strong depending on the type of polyethylene. Highly cross-linked polymers below their T_g exhibit hard and brittle behavior although the polymers can still be very strong and tough. Above the T_g of a polymer, the tensile properties begin to decrease in value. Crystalline plastics are generally harder than amorphous ones.

Elongation

Elongation is the increase in length of a plastic sample, when stretched to its break point and expressed in terms of the percentage of the original length. Linear polymers have greater elongations than cross-linked polymers. Elongation measurements can be used to roughly compare the ductility of materials.

Modulus of elasticity

The elastic modulus is the slope of the initial stress-strain curve and represents the stress-strain ratio within the area of the elastic limit of the material. It is measured only within this area and is a measure of the stiffness or rigidity of a material. Perturbations in the stress-strain behavior of plastics are due to the deviations from Hooke's law (stress versus strain). This causes deformation under load with time and gives rise to such phenomena as cold flow and creep, which is observed in plastics. Cold flow is plastic deformation under a constant rate of deformation under small applied stress; creep is the same, but the rate of deformation may also change during the applied load with time and extension of the polymer. Increasing temperatures will markedly affect plastic flow in thermoplastics (increase) but will have an insignificant effect on plastic flow in thermoset polymers.

Flexural strength

The flexural strength of a polymer is the resistance to breaking of the material as it is bent across its main axis. The characteristic of a polymer associated with flexural strength is called *fatigue*. Fatigue is defined as the resistance of the plastic to repeated flexing of the material through a known arc. The flexural strength of polymers normally is correlated with the macromolecule size and the degree of cross-linking of the polymer chains. Long, linear macromolecular chains are free to move with relative ease and have a high degree of flexibility. This leads to good fatigue resistance, and the modulus in flexure is low. High flexural moduli are obtained when there is a high degree of cross-linking in the polymer which results in a decrease in fatigue resistance.

Compressive strength

The compressive strength of a polymer is closely allied to the stress-strain characteristics in tension. The difference lies in the nature of the test piece. Not all polymers have a definite compressive strength. A brittle plastic in tension generally behaves as a ductile material in compression because the microcracks that form in tension do not form in compression. Not all polymers fail in compression, and a compression strength value may be reported as a certain percentage of deformation.

Impact resistance

The impact resistance of a plastic is the ability of the material to absorb the energy of a rapidly applied load. The resistance of a polymer to a sudden impact is an important property for many applications. The ability of a plastic to withstand this sudden impact is related to the toughness of the plastic and can be approximated from the area under the tensile stress-strain curve which is directly proportional to the impact strength. The two tests used to measure impact strength are the Izod and Charpy tests. Tests are run on notched and unnotched samples. In general, the impact strength increases with the molecular weight. Increased cross-linked density lowers impact strength. Thermoplastics have higher impact strength than thermoset materials. Elastomers are not characterized by impact strength because their structural features (elasticity) are different from those of plastics.

Creep resistance

The creep resistance of a plastic is characterized in terms of its behavior and response to an applied stress (tensile, compressive shear, flexure) over a long period. The deformations that occur in plastics under continuous loading over a longer period (days to years) can be significant, and creep resistance is a very important plastics property. The creep resistance of plastics depends on the chemical structure; i.e., if the structure allows movement of the molecular chains, then the creep will be greater. Thermoplastics exhibit larger creep than thermosets, and amorphous plastics exhibit larger creep than crystalline ones. Temperature increases creep so that above the glass transition temperature T_g of plastics one expects higher creep values than below T_g. Because creep is the result of the viscous component of polymers (polymers are viscoelastic materials), the effect over time (short or long) of small applied stresses is just as severe.

Hardness

The hardness of a plastic is characterized by a combination of properties which can be measured: scratch resistance, abrasion resistance, and indentation resistance. Scratch resistance is measured by moving the plastic under a loaded diamond point. Abrasion resistance is usually measured by a loss of weight or change in optical properties of the abraded surface. Hardness is measured in terms of hardness scales as follows: Shore A and D for soft to relatively hard polymers, respectively, Barcol hardness for just above the Shore D range, and Rockwell M for very hard polymers. Crystalline polymers tend to be harder than amorphous ones, and thermosets are harder than thermoplastics. There are exceptions. The cross-link density also affects hardness.

Electrical Properties

Practically all polymers find some use in the electrical industry. Plastics are inherently nonconductive and are known for their superior insulating properties. Of course, one can alter these properties through blending with a variety of fillers, plasticizers, lubricants, etc., to obtain a range of resistivities in the final product. That range of resistivities is shown in Fig. 10.1; and it is this versatility of polymers that makes them so useful in a variety of applications both within the

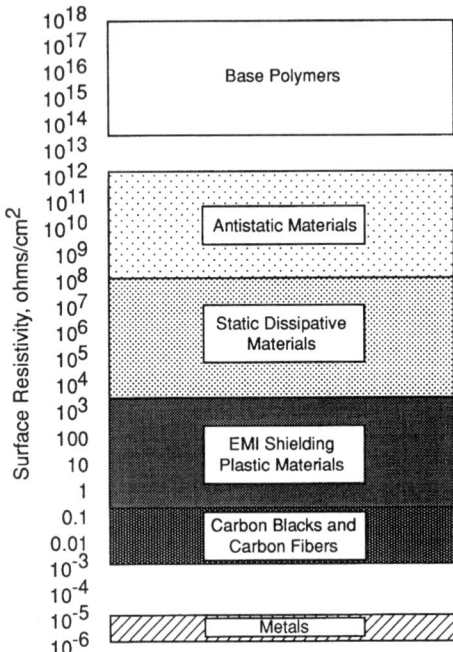

Figure 10.1 Surface resistivities of various materials. (*Source:* Ref. 7.)

electrical industry and outside that industry. In fact, the remarkable advances in the polymer industry have contributed to the growth of the electrical industry. Plastics are indeed one of the basic components of most electrical systems.

Dielectric constant

The dielectric constant is the ratio of the capacitance formed by two plates with a material between them to the capacitance of the same plates with air or vacuum between them. The dielectric constant is a measure of the ability of a polymer to store electrostatic field energy. The dielectric constant ϵ', k, or K is a dimensionless quantity. Dry air has a value of 1 while polymers have dielectric constants in the range of 2 to 10. The value of the dielectric constant varies with frequency, chemical structure (nonpolar, polar, symmetric, and unsymmetric molecules), and temperature. The chemical structure of polymers determines whether the dielectric constant is a function of electronic or dipole polarization.

In the former case, only electrons within the polymer molecule move whereas in the latter case either all or parts of the polymer molecule move. For nonpolar symmetric polymers, the ϵ' values are less than 2.7

TABLE 10.2 Propagation Delays for Insulators

Insulator	Dielectric constant	Propagation delay, ns/ft
Vacuum	1.0	1.0
Teflon, glass laminate	2.2	1.5
Epoxy, Kevlar laminate	3.6	1.9
Polyimide, quartz laminate	4.0	2.0
Epoxy, glass laminate	4.7	2.2
Alumina	10	3.2

SOURCE: Ref. 8.

while polar polymers have values greater than 3. The incorporation of additives will significantly affect the dielectric constant. In polar polymers the dielectric constant is also affected by whether the polar groups are attached to the main polymer chain or a side chain. In general, the lower the dielectric constant, the better the polymer as a dielectric for high-frequency electrical transmission applications. The value of ϵ' of an insulator is related to the signal speed of an electric circuit. Table 10.2 shows the propagation delays of several insulators as a function of the dielectric constant. Low dielectric constants are better for high-frequency applications to minimize power loss while high dielectric constants are favored for capacitance applications.

Dissipation factor

The dissipation factor (DF) of a material is expressed numerically as the tangent of the loss angle. In an ideal dielectric, the current flows 90° out of phase with the voltage; and in a nonideal dielectric, the current leads the voltage by some angle which is less than 90°, and the angle by which the current is less than 90° is called the *loss angle*. Other names are *loss tangent* and tan δ. The power factor is similar to DF but represents the amount of power absorbed by the insulation material from the alternating-current (ac) field. It is the ratio of the power dissipated in watts in a dielectric to the product of voltage and current (watts). When the dissipation is less than 0.1, the power factor differs from the DF by less than 0.5 percent. In a perfect insulator, there is no power loss (i.e., the material would not absorb power) so that those materials that have values close to zero would be better dielectrics in this respect. Such materials include polyethylene, polystyrene, PPS, and polypropylene. Keep in mind that the dissipation factor is affected by the polymer structure (polar versus nonpolar groups), temperature, moisture content, compounding ingredients, degree of oxidation, and frequency. This property is important because it represents the efficiency of an insulation to dissipate heat. Heat generation with a polymer can cause serious degradation of the material if the temperature is uncontrolled. Low values of the dissipation factor are preferred for the insulation to function in an efficient manner.

Resistivity

Resistivity is the resistance to the conduction of electricity. Plastics are excellent insulators, some better than others. Plastics can be made to be conductive by compounding with suitable ingredients such as powdered metals and carbon black or by changing the chemical structure of the plastic to make it more suitable for electrons to flow. In polymers, three main resistance properties are generally used to characterize these materials: volume resistivity, surface resistivity, and insulation resistance.

Volume resistivity. Polymers are valuable because of their general resistance to the passage of an electric current. The volume resistivity is the reciprocal of the current that will pass between two electrodes on opposite faces of a 1-cm cube of a material when a unit potential gradient exists between the electrodes. The volume resistivity represents the extent to which current can flow through the bulk of the material. The volume resistivity of a good insulator should be about 10^{15}. It is dependent on the type of plastic, imperfections in the plastic, i.e., conducting path, impurities (ionic or metallic) defects in the structure, and moisture content. It is interesting to note that in metals the resistivity increases linearly with increasing temperature but in plastics the resistivity decreases exponentially. The units are ohm-centimeters ($\Omega \cdot$ cm).

Surface resistivity. The surface resistivity is the resistance to current flow over the surface of a 1-cm^2 plastic. The units are ohms. The values obtained are more a function of the nature of the plastic surface than some intrinsic property of the material. The surface resistivity of a material is affected by moisture, temperature, and sample contour. Moisture is probably the single biggest item that will lower the resistivity. Both volume and surface resistance measurements are often used as a quality control check on the uniformity of a plastic material.

Insulation resistance. The insulation resistance of a plastic is the ratio of the applied voltage on the electrodes to the total current between them. It is expressed in ohms and is a composite of both volume and surface resistance measurements.

Dielectric strength

The dielectric strength of a plastic is the maximum voltage gradient which must be applied to cause an arc or spark discharge through the bulk material. It represents the resistance of a material to increasing voltage which can be applied stepwise or continuously. It is measured as the breakdown voltage per thickness and is expressed in volts per mil. Two tests are used in practice to measure the dielectric strength. In a short-time test, voltage is applied continuously from zero to breakdown. The value obtained is known as the *short-term dielectric breakdown strength*. In the other test, voltage is applied in incremental amounts (~100 or 500 V) and held at each voltage for a specific time to see if the material breaks down. This latter test is known as the *step-by-step test*. Higher values are usually obtained from the short-term test. All insulation will break down at some specific voltage; this breakdown voltage is influenced by the duration and rate of the applied voltage, thickness of the sample, temperature, electrode dimensions, surrounding medium, and frequency of the applied voltage. As the thickness increases, the voltage necessary to cause breakdown also increases; but this relationship is not linear, and the voltage gradient at breakdown decreases with increasing sample thickness. A $\frac{1}{8}$-in-thick plastic sample may break down at 400 V/mil, but a 1- to 3-mil sample of the same material may require a considerably higher voltage, say, 1000 V/mil before the sample fails.

Arc and tracking resistance

The arc and tracking resistance of a plastic is a measure of the ease with which a conductive path can form on the surface of a plastic when subjected to an electric arc. In testing for arc resistance, two point electrodes are placed on the same surface of a plastic, and an arc is passed between the two electrodes. Initially the arc travels through the air, but with increased time the material breaks down to form carbonaceous tracks along its surface, and then the arc follows these tracks. The length of time for this to happen is the arc resistance and is measured in seconds. The longer the time, the better the arc resistance. The value of the arc resistance depends on the ease of decomposition of the plastic and the nature of the decomposition products.

Corona

Corona, known also as *partial discharge,* occurs when an electrical stress is sufficiently high to cause a breakdown of the surrounding medium, usually air or oil, into a conductive path. This results in discharges to the surface of the plastic and in turn can lead to deterioration of the insulation. These discharges manifest themselves as humming or sizzling sounds emanating from the insulation system and are most prominent at high-moisture locations. The corona inception voltage is a measure of the onset of discharge. Polymers that oxidize readily are more prone to corona and include the polyolefins and many of the elastomers. Silicones and polyimides are quite resistant to corona.

Thermal Properties

Temperature affects the properties of all materials. Both physical and chemical changes can and will occur with the exposure of the material to different temperatures regardless of whether the temperature is -200 or $+200°C$. Plastics, unlike metals or ceramics which have a higher temperature exposure limit, have a considerably lower upper-temperature exposure limit, and this restricts the temperature range over which plastics can be utilized. On the lower end of the temperature range, plastics are comparable to if not better than other materials including metals and ceramics. Although the effect of temperature on thermoplastics and on thermosets is the same, the magnitude of that effect will differ for metals, ceramics, thermosets, thermoplastics, and elastomers. A number of properties are used to characterize the thermal nature of plastics, and they are discussed in the following sections.

Thermal conductivity

Thermal conductivity is defined as the quantity of heat per unit time passing through a unit area and unit thickness of a sheet of material when exposed to a unit temperature gradient. It is defined by

$$K = \frac{Q/A}{dT/dx}$$

where $\frac{Q}{A}$ = heat flow across unit area A

$\frac{dT}{dx}$ = temperature difference T across thickness x

The thermal diffusivity d is related to K by

$$d = \frac{K}{\rho C_p}$$

where ρ = density and C_p = heat capacity at constant pressure.

As a general rule, neat plastics have low thermal conductivities. The thermal conductivity and thermal diffusivity of plastics are essential to the appropriate processing of the particular plastic into its final form and to the selection of a plastic for a given application. Low conductivity values are not favorable in the processing of plastics. If heat conduction is poor, plastic melt flow is affected and part uniformity is poor. Structural changes in polymers that result in an increase in the frequency of contact of adjacent polymer chains will increase thermal conductivity. Factors that increase the free volume between chains decrease the thermal conductivity. For example, crystallinity in polymers usually increases conductivity because of improved packing of the polymer chains. Branching in polymers decreases their heat conduction. Increased cross-linked density in thermoset polymers increases heat conduction.[9] And last but not least, the type and concentration of filler affect the thermal conductivity of plastics and elastomers. The thermal conductivities of various materials are compared in Fig. 10.2.

Specific heat

As a crystalline solid is heated, its temperature rises, becomes constant at its melting point, and rises again when it is fully converted to its liquid state. If one plots the heat content per unit mass (enthalpy) against temperature, the change in enthalpy can be obtained; and at the melting point, it is the latent heat. Such a plot for crystalline and amorphous

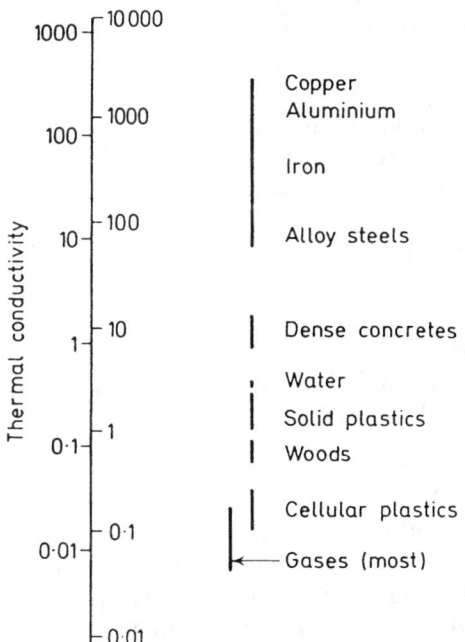

Figure 10.2 Thermal conductivities of various materials. (*Source:* Ref. 10.)

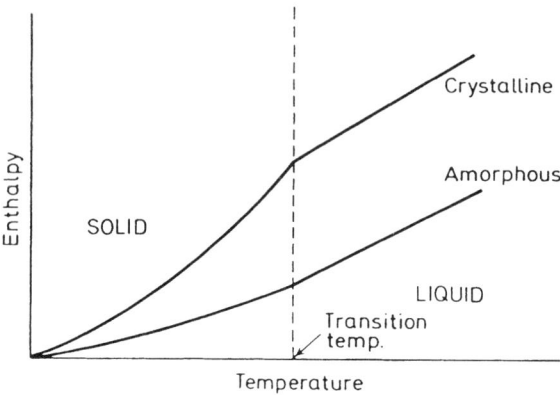

Figure 10.3 Effect of temperature on heat content of crystalline and amorphous polymers. (*Source:* Ref. 10.)

polymers is shown in Fig. 10.3. The slope of the initial and final parts of the curve gives the heat content per unit mass per unit temperature rise and is known as the *specific heat*. It is also defined as the amount of heat necessary to raise the temperature of 1 g of a substance 1°C.

The specific heats of plastics are much higher than those of metals. The lower the specific heat of a material, the lower the heating costs; and specific heats are very important in plastics processing. The specific heat of crystalline polymers tends to increase at a faster rate below its glass transition temperature T_g than that of an amorphous polymer. Above T_g the specific heat of both amorphous and crystalline polymers remains relatively constant with temperature, but the absolute value for the crystalline plastic is higher than that for the amorphous material. The specific heat of polymers is important in processes such as quenching, melting, and crystallization. Fillers and/or reinforcements affect the specific heat of plastics. The specific heat of polymers is dependent upon the degrees of freedom with which the polymer chain molecules can take part as the temperature is raised. In thermoset resins, the heat capacities have values that fall within the range of linear thermoplastics [0.40 to 0.55 cal/(g · °C)].

Glass transition temperature

All properties of plastics are affected in some way by temperature, but the magnitude of that effect is different for each polymer. The fundamental property of a polymer that relates its behavior to temperature is called the *glass transition temperature T_g*. This point manifests itself in a plastic as a change in properties from a rigid, hard, brittle, or glassy behavior to a soft, flexible, or rubbery behavior. This change in behavior is brought about by a change in volume in the polymer caused by movement of the polymer chains. If we plot density or specific volume versus temperature, a slope change signifies the T_g. Glass transition effects are greatest in amorphous polymers. The glass transition temperatures of cross-linked polymers are generally higher than those of non-cross-linked polymers. The T_g values of thermosets are generally higher than those of ordinary thermoplastics. There are exceptions; e.g., a thermosetting silicone has a much lower T_g than a thermoplastic polystyrene (-120 versus $+100°C$, respectively). Other properties of polymers that change at the glass transition temperature include the modulus, thermal expansion, refractive index, heat capacity, impact resistance, mechanical damping, and dielectric constant. The T_g value has been shown to increase with increasing molecular weight, reaching a limiting value.[11] Other factors affecting the T_g of polymers include the chemical structure, composition, plasticization, copolymer addition, and cross-linking.

Coefficient of thermal expansion

The coefficient of thermal expansion (CTE) is the change in length of a material with temperature (either linear or cubic). Plastics have CTE values several times larger than metals or glass does. The CTEs of polymers are generally several times greater above the T_g than below it. Crystallinity in polymers generally lowers the CTE, although as a class of materials the crystalline polymers generally have higher CTEs than their amorphous counterparts. Fillers and reinforcements significantly decrease the expansion of polymers.

Heat deflection temperature

The heat deflection temperature represents the lowest temperature at which a material being tested yields a specified distance under a given load. It is a measure of the temperature response of a material to a variety of stress conditions. Thermoplastics generally have lower values than thermosets, and cross-linked polymers generally have higher deflection temperatures. Fillers and reinforcements significantly affect the heat deflection temperature of both thermoplastics and thermosets.

Flammability

Some plastics will show a great tendency to burn while others are slow to burn; some burn with great difficulty, and others will not burn at all. The flammability of plastic materials is affected by their elemental composition and whether they contain various ingredients that retard or inhibit flammability. Some of these ingredients include compounds containing phosphorous, bromine, chlorine, fluorine, antimony, alumina trihydrate, nitrogen, and boron. Often polymers which are not flammable are blended with flammable polymers to reduce their burning characteristics.

There are a number of tests available to characterize the flammability of a plastic, and they are listed in Table 10.3. Flammability tests are divided into ignition and burning tests, and the former ascertains how readily and under what conditions a material will ignite while the latter measures the rate of burning of the substance once it is ignited. A listing of some of these properties for various polymers is given in Table 10.4.

TABLE 10.3 Flammability Tests

Test	Source	Comment
Smolder susceptibility	California Bureau of Home Furnishings	Tendency of material to support smoldering combustion within itself. Applied to fabrics and furnishings
Ignitability	ASTM D-1929, D-229	Temperature at which a material will ignite in presence of a pilot flame
Flash-fire propensity	Douglas Aircraft Co.	Tendency of material to produce a fire that spreads rapidly
Flame spread	ASTM E-84, UL-773, E-648, E-162	Rate of travel of a flame front
Heat release	Ohio State University	Heat released by a given weight or volume
Fire endurance	ASTM E-119, UL-181	Length of time for flame to penetrate a material
Ease of extinguishment	ASTM D-2863, Oxygen	Minimum % oxygen in an O_2-N_2 atmosphere that will support burning
Smoke evolution	ASTM E-84, E-662	Degree of visual obstruction produced by smoke
Toxic gas evolution	DIN 53436 (German Standard)	Identification of toxic volatiles

TABLE 10.4 Flammability Properties of Various Polymers

Polymer	Flash ignition temp., °C	Self-ignition temp., °C	Approximate oxygen index
Epoxy	—	—	20–30
Cotton	230–266	254	17
Paper	230	230	18
Wool	200	—	25
Polyethylene	341–357	349	17
Polypropylene, fiber	—	570	18
Teflon	—	530	95
PVC	391	454	40
PVC-acetate	320–340	435–577	—
Polyvinylidene chloride	532	532	60
Polystyrene	345–360	488–496	18
ABS	—	466	19
SAN	366	454	19
PMMA	280–300	450–462	17
Polyamide	421	424	26
Polycarbonate	467	580	25
Phenolic glass	520–540	571–580	40–60
Melamine glass	475–500	623–645	—
Polyester glass	346–399	483–488	20–60
Silicone glass	490–527	550–564	30–40
Polyurethane foam, rigid	310	516	—
Nomex	—	—	42
PBI	—	—	40
Polyethylene terephthate	—	—	20
Polyimide	—	—	30
Polysulfone	—	—	38
PPS	—	—	>40
PPO	—	—	29

Environmental Properties

Plastics are subject to the same environmental concerns as metals with the possible exception of electrochemical corrosion. However, with the advent of conductive plastics, this exception may no longer be true. Plastics are affected by chemicals, temperature, radiation (ultraviolet and gamma radiation), stress, and biological deterioration. It is important for the design engineer to be aware of what affects plastics, how they are affected, and what tests measure these effects, in order to choose the right plastic for a given environment.

The effect of the environment on plastics is an exceedingly complex problem because it is dependent on the chemical nature of each individual plastic. Molecular weight, chemical structure, crystalline and amorphous content, cross-linking, the nature of the intermolecular forces, and bond energies of the covalent bonds comprising the polymer molecule as well as additives in the polymer can all affect the degree of resistance of the plastic to environmental conditions.

Certain chemical structures within a plastic are more susceptible to attack than others. For example, the more unsaturation (double-bond character) a polymer molecule contains, the more susceptible it is to oxidation. Polyisoprene has lots of double bonds and is easily destroyed by ozone while polyisobutylene is very resistant due to its low unsaturation. Ozone will also attack polyolefins, but after the initial attack a barrier forms which prevents further attack.[12] As the molecular weight of the polymer increases, the polymer becomes less soluble. A phenomenon known as *stress cracking* is related to molecular weight and solubility. Polymers will crack by selectively absorbing certain solvents. Both solubility and stress crack resistance are reciprocally related to molecular weight, chain branching, and cross-linking. Temperature accelerates polymer deterioration. The rate of chemical attack is lower below the T_g of the plastic. The rate of chemical attack is also dependent on the morphology of the polymer. Thermoplastics are generally less resistant to chemicals than thermosets. The former will swell but not dissolve.

Chemical resistance

The effects of chemicals on plastics are manifested by physical and/or chemical changes such as swelling, dissolution, stress cracking, hydrolysis, and oxidation.

The following sections contain specific information about the environmental effects on polymers.

Thermoplastics

ABS resins

The chemical resistance of the acrylonitrile-butadiene-styrene (ABS) resins increases with the level of acrylonitrile concentration. Strong acids such as 65 percent nitric acid, perchloric acid, concentrated sulfuric acid above 50 percent, and oleum attack ABS. Acetic, chloracetic, and butyric acids degrade ABS. ABS has good resistance to most alkalies up to 65°C. Ammonium hydroxide does attack ABS at 65°C. Acid salts, alkaline salts, and neutral salt solutions are completely acceptable up to the maximum working temperature of the resin system. ABS is unaffected by most dry and wet gases, except wet sulfur dioxide, which has a very significant effect. The ABS resins have very poor chemical resistance to most organic solvents and are readily attacked by aromatics (benzene, toluene, xylene), ethers, esters, chlorinated aromatics, and chlorinated aliphatics, amines, ketones, and hot alcohols.

Acetal resins

The acetal homopolymer is not resistant to strong alkali, but the copolymer is resistant. The acetals have excellent resistance to most organic solvents, including aliphatic and aromatic hydrocarbons, alcohols, ketones, esters, glycols, and chlorinated solvents. Acetals are unaffected by vegetable and mineral oils. They are completely resistant to all inorganic salt solutions within the pH range of 4 to 14.

The acetal copolymer is not resistant to weak acids above 65°C and cannot be used with strong acids at ambient temperature. It is extremely resistant to distilled water and is completely unaffected by hot (180°F) water over long periods of exposure. Acetal pipe placed underground is unaffected by fungi, insects, or rodents.

Acrylics

The acrylic resins are resistant to weak acids, alkalies, inorganic salt solutions, oils, and water. PMMA is more resistant than the other acrylic esters. Strong oxidizing acids (HNO_3) and strong alkali degrade PMMA; and esters, phenols, aromatic hydrocarbons, ketones, and chlorinated hyrocarbons will soften and dissolve PMMA.

Fluorocarbons

As a class of polymers, the fluorocarbons are probably the most chemically resistant of all. TFE and FEP are not compatible with certain environments. The fluorocarbon resins will react with molten alkali metals (such as metallic sodium), fluorine, strong fluorinating agents (such as chlorine trifluoride and oxygen difluoride), and molten sodium hydroxide at temperatures above 300°C.

PFA has similar chemical resistance properties to TFE and FEP. CTFE is less resistant and is swelled by chlorinated solvents at elevated temperature. Polyvinylidene fluoride (PVF_2) is not as chemically resistant as the other fluorocarbon resins. It is, however, resistant to most inorganic acids and bases. Fuming nitric, oleum, other sulfonating agents, and concentrated caustic solutions show some effect. The fluorocarbons are resistant to common aliphatic and aromatic solvents.

Nylons

The nylons have very good resistance to most chemical compounds and to hydrolysis. They are biologically inactive and resistant to fungal and bacterial attack. They have good solvent resistance to most of the common organic solvents including aromatic and aliphatic hydrocarbons, ketones, and esters. They are essentially unaffected by lubricants, oils, fuel, many phosphate esters, and refrigerants. The nylons show very poor chemical resistance to strong mineral acids, oxidizing agents, and certain salts such as potassium thiocyanate, calcium chloride, and zinc chloride. Nylon 6/6 is generally considered to have the best chemical resistance properties of the nylons. They are unaffected by alkalies and most salt solutions. Polyamides dissolve in phenol and can be hydrolyzed in hot water.

Polycarbonate

The unstressed carbonate resin is unaffected by dilute mineral and organic acids. It is insoluble in aliphatic hydrocarbons, petroleum ether, and most alcohols. It is partially soluble in aromatic hydrocarbons, such as benzene and toluene, and ketones and esters. It is extremely soluble in chlorinated hydrocarbons, such as methylene chloride, and materials such as dioxan 1,4. Ammonia and amines readily attack it, and it is slowly decomposed by strong alkaline solutions.

Molded polycarbonate items are prone to stress cracking or crazing when exposed to certain chemical compounds or vapors; and this appears to be due to residual stresses locked in during the molding operations being relieved when contacted by compounds exhibiting partial solvency. This susceptibility to stress cracking and crazing can be eliminated or minimized by proper mold design or special annealing treatments in the temperature range of 120 to 150°C.

Polyolefins

The polyolefins have excellent chemical resistance to many chemicals and solvents (see Table 10.3). At ambient temperature, they are resistant to acids and alkalies except for oxidizing acids such as nitric, chlorosulfonic, and fuming sulfuric. They are unaffected by hydrofluoric acid. The polyolefins are generally insoluble in organic solvents at temperatures below 50°C. However, at higher temperatures they are soluble to varying degrees in hydrocarbons and halogenated hydrocarbons. They are appreciably affected by chlorinated solvents, aliphatic and aromatic hydrocarbons, certain esters, and oils. Some polyethylenes can be dissolved in hot (70°C) toluene, xylene, amyl acetate, trichloroethylene, petroleum ether, paraffin, turpentine, and lubricating oils.

Polyimides

The polyimides are resistant to mild concentrations of inorganic or organic acids and are unaffected by grease and oils and aliphatic and aromatic solvents. Concentrated mineral acids cause severe embrittlement of polyimide parts. They are severely attacked by strong bases such as aqueous ammonia, hydrazine, nitrogen dioxide, and primary or secondary amines. Exposure to steam or hot water at 100°C produces a decrease in tensile and flexural strength. However, most of the reduced tensile values can be restored by drying, which indicates the reduced properties are not due to chemical changes. At high temperatures some solvents containing functional groups such as m-cresol and nitrobenzene can cause swelling of polyimides without substantially reducing their mechanical strength.

Polyphenylene oxide resins

The polyphenylene oxide (PPO) resins show excellent chemical resistance to dilute acids, alkalies, and aqueous salt solutions. They are attacked by strong oxidizing acids. Polyphenylene oxide resins under stressed conditions will crack when exposed to some organic liquid environments such as aliphatic hydrocarbons, ketones, esters, and petroleum derivatives; stress-relieved polyphenylene oxide is unaffected in such environments. It is readily dissolved by chlorinated aliphatics and aromatics such as benzene and toluene.

The resistance of the polyphenylene oxide resin to hydrolytic breakdown is demonstrated by its excellent performance when exposed to steam at 132°C for long periods.

Polyphenylene sulfide (PPS)

The cured polymer is essentially unaffected by aliphatic hydrocarbons, alcohols, ketones, chlorinated aliphatic compounds, esters, aliphatic ethers, toluene, liquid ammonia, ammonium hyroxide, aqueous sodium hydroxide, organic acids, and most inorganic salt solutions. Dilute hydrochloric and nitric acids, concentrated sulfuric acid, and strong oxidizing agents such as chromic acid, bromine water, and sodium hypochlorite attack PPS, particularly at elevated temperatures. No change in weight was noted after 4 months of exposure in xylene, dimethyl formamide, chlorobenzene, 10 percent nitric acid, concentrated hydrochloric acid, 48 percent sulfuric acid, and concentrated sulfuric acid at ambient temperature. There are no known solvents for polyphenylene sulfide below 190°C.

Polystyrene

Polystyrene is dissolved by a number of hydrocarbons such as benzene, toluene, and ethylbenzene and by chlorinated hydrocarbons such as carbon tetrachloride, chlorobenzene, chloroform, and methylene chloride. It is attacked by ketones (with the exception of acetone) and esters. Certain other materials such as acids, alcohols, oils, cosmetic creams, and foodstuffs promote crazing and cracking. Polystyrene has good resistance to many ordinary chemicals such as weak acids, all concentrations of alkalies, and aqueous solutions of many salts. It is readily attacked by oxidizing agents. Polystyrene has a low moisture absorptivity. It also has good resistance to water vapor transmission.

Polysulfones

Polysulfone is highly resistant to mineral acids, alkalies, and salt solutions. Resistance to detergents and hydrocarbon oils is good, even at elevated temperatures under moderate levels of stress. It is unaffected by aliphatic hydrocarbons. Polysulfones will be attacked by polar organic solvents such as esters and ketones, chlorinated hydrocarbons, and aromatic hydrocarbons. Polysulfone is not attacked by water through chemical hydrolysis. Products made from polysulfone can be steam-sterilized repeatedly without any significant degradation of physical properties or chemical resistance. Polysulfones can stress-crack when exposed to certain solvents.

Vinyl chloride polymers

The vinyl chloride polymers are resistant to inorganic acids (such as hydrochloric, nitric, phosphoric, and sulfuric), organic and inorganic salts, glacial acetic acid, ketones, esters, and ethers, chlorinated and aromatic hydrocarbons soften and/or dissolve these polymers. Alcohols and aliphatic hydrocarbons have no effect. Acetic and formic acid show some attack on PVC.

The polyvinylidene chloride copolymers have better chemical resistance than PVC. However, the vinylidene chloride copolymers are attacked by ammonium hydroxide, dioxane, cyclohexanone, dimethyl formamide, and chlorinated hydrocarbons.

Polyether ketone polymers

The polyether ketone resins are resistant to nonoxidizing acids (HCl), alkalies, salts, and all organic solvents. These resins are attacked by sulfuric acid, nitric acid, and chlorine.

Polybenzimidazole

Polybenzimidazole is unaffected by xylene, toluene, gasoline, kerosene, ketones, chlorinated solvents, strong acids, and bases.

Polyamide-imide

These polymers are resistant to most acids and to aliphatic, aromatic, and halogenated hydrocarbons. At elevated temperatures some acids and strong bases will attack these materials.

Polyetherimide

The polyetherimide polymer is resistant to aliphatic hydrocarbons, mineral acids, salt solutions, and dilute base. It is attacked by halogenated solvents and strong bases.

Thermoplastic polyesters (PBT, PET, PCT)

The PET resin is resistant to all known chemicals and solvents; PCT is resistant to most, and PBT is resistant to most chemicals except strong acids and bases. PBT is swollen by ethylene dichloride.

Liquid-crystal polymers

The LCPs are highly resistant to most chemicals at both room and elevated temperatures. Boiling caustic will attack these materials.

Thermosets

Amino resins

The amino resins include melamine and urea formaldehyde materials. These materials are attacked by strong oxidizing acids and strong alkalies. Weak acids and bases show mild attack. Industrial-strength bleach (sodium hypochlorite) solutions attack the resins. The amino resins are unaffected by aliphatic aromatic solvents and chlorinated solvents.

Allyl resins

The allyl polymers are unaffected by weak acids, alkalies, and organic solvents. Strong acids and alkalies show a mild effect.

Epoxy resins

There are three main classes of epoxy resins: the standard bisphenol A type, the epoxy novolacs, and the cycloaliphatics. The standard epoxy resins are resistant to aliphatic and aromatic solvents, water, and alkalies. Oxidizing agents and strong acids will attack these materials. The epoxy novalacs and cycloaliphatic epoxy resins have characteristics similar to those of the standard epoxy resins but are not resistant to acids and oxidizing agents. The epoxy resins as a class of materials are not attacked by weak acids, salt solutions, or sodium hypochlorite solutions. Acetic, chromic, nitric, and sulfuric acids and oleum will attack epoxy resins. Amine-cured epoxy resins have good resistance to dilute bases while anhydride-cured resins have better resistance to dilute acids.

Phenolic resins

Phenol formaldehyde resins are unaffected by inorganic acids. Strong oxidizing acids such as nitric acid and strong bases will attack these resins. The resins are resistant to aliphatic, aromatic, and halogenated solvents.

Unsaturated polyesters

The unsaturated polyesters have good resistance to nonoxidizing acids, solvents, and weak bases. The resins are attacked by strong acids, strong alkalies, and organic solvents.

Silicone resins

The silicone resins are attacked by hot solutions of acids and alkalies. They have poor resistance to aliphatic and aromatic hydrocarbons. They have only fair resistance to halogenated solvents.

Urethanes

Polyurethanes in general are resistant to aliphatic solvents, alcohols, and ethers. They are attacked by strong acids and bases, hot water, and polar solvents. Large variations in polyurethane formulations give rise to a range of chemical resistance properties.

Cyanate esters

The cyanate ester resins as a class of polymers absorb less water at saturation than epoxy, bismaleimide, and polyimide resins. The resins are quite resistant to hydrolysis (i.e., the cyanurate ester linkage in these polymers) and withstand thousands of hours of immersion in boiling water. All the cyanate ester resins are resistant to aqueous organic and mineral acids, aqueous alkalies, bleach (NaOCl), alcohols, ketones, and hydrocarbon solvents. Some grades are attacked by alkalies at 93°C.

Benzocyclobutenes

Benzocyclobutene polymers are resistant to acids, alkalies, and hydrocarbons. After long exposure times at elevated temperature, these resins are attacked by nitric acid, phosphoric acid, and strong alkalies. In addition, benzocyclobutene resins have very low water absorption. The maximum steady-state moisture uptake is reported to be 0.2 percent.[13]

Bismaleimides

The bismaleimide resins are resistant to acids and common organic solvents but are attacked by strong alkalies and methylene chloride.

Silicon-carbon (SYCAR) resins

These silicon-carbon resins have very low moisture absorption (0.05 percent after 24-h immersion in water). They have excellent resistance to acids and bases and fair resistance to ketones. They swell in aromatic and aliphatic hydrocarbons. They are resistant to methylene chloride in the circuit board laminate test (MIL Specification 13949).

Polyphthalamide

The polyphthalamide resins, like their aliphatic counterparts, absorb about 6 percent moisture. The resins have excellent resistance to aliphatic, aromatic, and chlorinated hydrocarbons as well as to esters, ketones, and alcohols. The resins are attacked by strong acids and oxidizing acids and are dissolved by phenol and cresol.

Elastomers

Butyl

Butyl rubber has very low permeability to oxygen and nitrogen. It is resistant to acids and alkalies as well as animal and vegetable oils. Aliphatic and aromatic solvents attack butyl rubber.

Chlorosulfonated polyethylene

Chlorosulfonated polyethylene is resistant to concentrated sulfuric acid, strong and weak alkalies, alcohols, oils, and greases. It has good resistance to sodium hypochlorites and poor resistance to aliphatic, aromatic, and chlorinated aldehydes and ketone solvents.

Epichlorohydrins

These elastomers have excellent oil resistance. They are attacked by ketones, esters, aldehydes, and chlorinated and nitro-substituted hydrocarbon solvents.

Ethylene propylene elastomers

The ethylene propylene elastomers have good resistance to acids and alkalies and are unaffected by alcohols. They are resistant to phosphate-based hydraulic fluids, but diester-type synthetic lubricants will attack these elastomers. They are attacked by aromatic, aliphatic, and chlorinated solvents.

Chloroprene

This elastomer has excellent gas permeability resistance. It is resistant to aliphatic hydrocarbons, alcohols, fluorinated hydrocarbons, dilute mineral acids, concentrated alkalies, and inorganic aqueous salt solutions. It is attacked by chlorinated solvents, esters, ketones, phenols, aromatic hydrocarbons, nitric and sulfuric acids, potassium dichromate, and peroxides.

Fluorinated rubbers

These elastomers as a class of materials are quite resistant to chemicals. They are attacked by ketones, esters, nitro compounds, aldehydes, and ethers. The fluoroelastomers are resistant to aliphatic hydrocarbons, chlorinated solvents, and dilute acids and bases.

Silicone elastomers

The silicone elastomers are resistant to dilute acids and bases, alcohols, and animal and vegetable oils. They are attacked by aromatic and chlorinated solvents and gasoline. Steam at elevated temperature degrades the silicones. They are attacked by strong acids and bases. The fluorosilicones have improved resistance compared to the neat silicone elastomers.

Nitrile elastomers

The nitrile rubbers have good resistance to aliphatic hydrocarbons, dilute acids, and bases. Strong oxidizing agents such as nitric and sulfuric acids attack the nitrile elastomers. The nitrile rubbers are not resistant to ketones, esters, and ethers.

Ethylene acrylic elastomers

These elastomers exhibit good resistance to aliphatic, aromatic, and chlorinated solvents. They show poor to fair resistance to concentrated acids. They are resistant to dilute acid.

Styrene-butadiene elastomers

These elastomers are not very resistant to oils and solvents. They are attacked by ketones, esters, and aromatic and aliphatic hydrocarbons. They are resistant to dilute acid and alkali.

Polysulfide elastomers

The polysulfide rubbers are not significantly affected by alcohols, ketones, esters, acids, bases, or aliphatic or aromatic solvents. They have low permeability to water, gases, and organic solvents. They do swell in benzene.

Aging of Polymers

Deterioration affects all categories of materials and is a critical problem in those materials whose purpose is to provide a critical operation or protective function in equipment, such as the maintenance of mechanical, chemical, electrical, thermal, radiation, and corrosion integrity of materials used in the equipment.

Polymeric materials rely on their chain nature for their bulk properties. Anything that causes that chain structure to be impacted can have a profound effect on the performance of that material. Materials exposed to environmental conditions are subjected to complex forces. The damage mechanisms associated with an environment cover a wide range and tend to be material-specific.

The treatment of the deterioration process of nonmetallic materials as a chemical rate phenomenon was first introduced by Dakin.[14] Use was made of the Arrhenius equation to study the degradation of organic materials. The equation shows the temperature dependence of the reaction rate and expresses the life of a material as an exponential function of the reciprocal of absolute temperature:

$$\frac{dP}{dt} = A \exp\left(\frac{-E}{RT}\right) \quad (1)$$

In this equation, dP/dt is the reduction in property with respect to time, A is a constant, R is the gas constant, T is the absolute temperature, and E is the activation energy of the aging reaction. Integration of Eq. (1) gives

$$\ln t = \frac{E}{R}\frac{1}{T} + B \quad (2)$$

where E represents the heat of activation or the energy required to convert unreactive molecules to "active ones" and is a measure of the activation energy for the material degradation. A plot of ln t versus $1/T$ yields a straight line with slope E/R, and this is known as the *Arrhenius plot*. In using the Arrhenius equation to predict service life or storageability of a material, one must first decide on the point which will be considered the end of the product's useful life. This is an absolutely critical choice because the property and failure modes determine the eventual life of the material. Then the time to reach the endpoint at a series of constant temperatures is determined and is plotted as mentioned above. Extrapolation of the line to lower temperatures allows one to predict the longer time required to reach the same physical state of the material. Although the development of the test methods was originally confined to electrical insulation, it has since been found to be a general relationship applicable to most organic materials.[15–17] The equation expresses the life of a material as an exponential function of the reciprocal of absolute temperature.

Temperature accelerates many chemical and physical processes that lead to failure of a material and/or component. Other factors that also accelerate the aging process are radiation, stress, moisture, fluid immersion, and gaseous atmosphere. Thus it is necessary to consider both material function and environment when service-life predictions are developed. The problem of aging and prediction of service lifetimes is exacerbated when there is a multistimulus environment. And care must be exercised in the prediction of service lifetimes in a multicomponent environment.

In using this equation in accelerated testing, one must be cognizant of the fact that multiple degradation mechanisms will adversely affect the long-term prediction capabilities. The Arrhenius equation determines this deviation (a change in slope of ln t versus $1/T$). It may be that a different mechanism of degradation operates as temperature alone is changed; e.g., aging at lower temperatures sometimes will produce a different degradation mechanism than that observed at higher temperatures. In addition, a combination of environmental parameters could produce a change in the degradation mechanism.

The author and many others have done considerable work in the area of polymer aging; see Refs. 18 to 29 for additional information on specific plastics, properties, and environmental conditions.

Nature of degradation processes

We have said that material deterioration occurs in many ways: Oxidative and thermal degradation of polymers is an autocatalytic process that can lead to chain scission and/or cross-linking. Polymer stability is a complex function of the chemical structure of the polymer and depends on the aliphatic and aromatic content and the arrangement of the atoms within the polymer. Certain arrangements of atoms are more susceptible than others to thermal and oxidative degradation. Radiation degradation involves wavelengths lower than 300 nm because they are sufficiently energetic to excite electrons along the polymer chain. Photodegradation is the most common type among polymers because polymers contain chemical groups that absorb radiation in this region. Ionizing radiation is also very damaging because it interacts with the nucleus of the atoms, generating free radicals along the polymer chain. In mechanical degradation, stress is applied to a polymer in a variety of ways (grinding, milling, stretching, ultrasound) which can cause bond cleavage with the formation of radical species which then undergo degradative reactions. Biological degradation occurs in polymers when microoganisms attack the material. In this case, enzymes preferentially attack the amorphous regions in the polymer. A summary of the resistance of polymers to temperature, uv radiation, and microorganisms is given in Table 10.5. Both Underwriters Laboratories (UL) and the Institute of Electrical and Electronic Engineers (IEEE) have standards dealing with the evaluation of the thermal performance of polymers and their use at elevated temperatures. See Chap. 14 for information about these organizations. The UL standards that apply to plastics for electrical applications are UL-746A, B, C and 1146 and the following:

IEEE Publication 1	*General Principles for Temperature Limits in the Rating of Electric Equipment and for the Evaluation of Electrical Insulation*
IEEE Publication 57	*Test Procedure for Evaluation of the Thermal Stability of Enameled Wire in Air*
IEEE Publication 98	*Standard for the Preparation of Test Procedures for the Thermal Evaluation of Solid Electrical Insulating Materials*
IEEE Publication 99	*Recommended Practice for the Preparation of Test Procedures for the Thermal Evaluation of Insulation Systems for Electric Equipment*
IEEE Publication 101	*Guide for the Statistical Analysis of Thermal Life Test Data*

IEEE Publication 1 also lists the thermal classification of materials as follows: Class 0 (90°C), class A (105°C), class B (130°C), class F (155°C), class H (180°C), class (220°C), and class C (over 220°C).

Design Considerations for Plastics

TABLE 10.5 Ultraviolet Thermal and Biological Resistance of Plastics

Polymer	Service temp. range, °C	Ultraviolet resistance	Fungus resistance
Acrylics	60–90	Excellent	Not applicable
Acetals	90–104	Fair	Excellent
Cellulosics	49–77	Good to excellent	Poor
PTFE	−80–+250	Excellent	Excellent
FEP	−80–+200	Excellent	Excellent
PFA	260	Excellent	—
PCTFE	−80–+200	Excellent	—
E-CTFE	−80–+170	Excellent	—
PUDF	150	Excellent	Excellent
PVF	−40–+110	Excellent	Poor
Ionomer	<70	Poor	—
PEK	260	Poor	—
PEEK	260	Poor	—
PEKEKK	260	Poor	—
XYDAR	−190–+250	Excellent	—
VECTRA	−190–+250	Excellent	—
Nylons	−50–+150	Poor	Excellent
HT nylons	220	—	Excellent
Nomex			
Kevlar	>+220	—	Excellent
Polyamide-imides	250	Fair	—
Polyimides	−190–+300	Fair to good	Excellent
Polyetherimide	170	—	Excellent
Polyarylate	160	—	—
PBT	140	—	—
PET	140	—	Excellent
PCT	150	—	—
Polycarbonates	120	Excellent	Excellent
Polyethylene	75–100	Poor	Poor
Polypropylene	110	Poor	Excellent
Polyphenylene oxide	−170–+135	Good	—
Polyphenylene sulfide	230	Good	—
Polystyrene	75	Fair	Excellent
ABS	100	—	Poor
SAN	90	Good	Excellent
Polysulfone	170	Fair	Excellent
Polyarylsulfone	190	Fair	Excellent
Polyethersulfone	185	Fair	Excellent
Polyphenylsulfone	190	Fair	Excellent
PVC	55	Poor	Poor
CPVC	55	Poor	Poor
PVDC	55	Poor	Poor
Allyl	150–230	Good	Excellent
Bismaleimide	177	—	—
Epoxy, aromatic	100–150	Fair to good	Poor

TABLE 10.5 Ultraviolet Thermal and Biological Resistance of Plastics (*Continued*)

Polymer	Service temp. range, °C	Ultraviolet resistance	Fungus resistance
Epoxy, cycloaliphatic	100–150	Excellent	Poor
Phenolic	120–230	—	Poor
Unsaturated polyesters	65–150	Fair	—
Polyurethanes	100–130	Fair	Poor
Silicones	260	Excellent	Excellent
Cyanate esters	180	—	—
Benzocyclobutenes	200–250	—	—
Polyxylylene	150	—	—
Urea and melamine	77–200	Poor	Poor
SYCAR	Up to 180	—	—
PBI	−190–+400	—	—
Butyl rubber	−45–+150	Excellent	Excellent
Chlorosulfonated polyethylene	−45–+135	Very good	—
Epichlorohydrin	−60–+160	Excellent	Poor
Ethylene-propylene	−55–+150	Outstanding	Poor
Chloroprene	−50–+125	Very good	Poor
Fluoroelastomers	−50–+260	Excellent	Poor
Fluorosilicones	−55–+200	Excellent	Poor
Natural rubber	−55–+70	Fair	Poor
Nitrile	−50–+120	Good	Poor
Acrylates	−40–+175	Excellent	Poor
Ethylene-acrylics	−35–+175	Excellent	Poor
Styrene-butadiene	−60–+120	Fair	Poor
Chlorinated polyethylene	−50–+150	Good	Poor
Polybutadiene	−100–+130	Fair	Poor
Polysulfides	−45–+125	Excellent	Poor
Silicones	−117–+260	Excellent	Poor
Urethanes	−50–+120	Good	Poor

Nuclear Radiation Effects[19]

Organic insulating and dielectric materials experience both temporary and permanent changes in characteristics when they are subjected to a radiation environment, such as that found in space and in the field of a nuclear reactor or a radioisotope source. Data indicate that the temporary effects are generally rate-sensitive with a saturation of the effect at the higher radiation levels. The enhancement of the electrical conductivity is the most important of the temporary effects, with increases of several orders of magnitude being observed. The magnitude of the increase is dependent upon several factors, including the material being irradiated, ambient temperature, and the radiation rate.

Absorption of energy, excitation of charge carriers from nonconducting states, and the return of these carriers from conducting to nonconducting states are considered responsible for the induced conductivity.

The cumulative results of the temporary effects, as they pertain to the electrical parameters of insulating materials, are a reduction in breakdown and flashover voltages as well as an increase in leakage current or conductance; the latter also is identified as a decrease in the material's insulation resistance. However, these temporary changes in electrical characteristics are often not large enough to prevent the use of organic insulators and dielectrics in a radiation environment. This is especially true if the designer considers these changes and makes allowances to minimize their effects. When the designer is under severe space limitations or the application includes a high radiation exposure rate, it may be necessary to limit insulating material considerations to the inorganics.

Permanent effects of radiation on organic insulating and dielectric materials are normally associated with a chemical change in the material. Most important among these chemical reactions that occur are chain scission and cross-linking. These chemical reactions or changes modify the physical properties of the material. For example, a softening of the material, decreases in tensile strength and melting point, and a greater solubility could be the result of chain scission. Cross-linking results in hardening; an increase in strength, density, and melting point; and a decrease in solubility. Thus, the permanent effects of radiation on organic materials are chemical changes that result in a change in the physical properties. This physical degradation may also be disastrous to the electrical characteristics of a component part, such as printed-circuit boards, wire insulation, and connectors. Radiation induced embrittlement of insulating structures, such as these, where the insulation cracks or flakes could in turn cause a circuit to fail electrically through an open or short circuit. This type of failure often occurs when an insulator or a dielectric material fails in a radiation environment, i.e., physical degradation followed by the failure of electrical properties. Changes in the dielectric loss or dissipation factor and insulation resistance also are the result of permanent effects from exposure to a radiation environment. These changes are often quite small, and it would be the more uncommon application where they would pose any problem.

A secondary reaction that may occur when an organic insulator or dielectric is irradiated is gas evolution. It is unlikely that the volume of gas would be of serious concern except in the case of organic fluids, where sufficient pressure may be produced to distort or rupture a sealed enclosure. Gas evolution from solid organic polymers is less than that for liquids because of a greater possibility of recombination and limited diffusion. Another problem with some evolved gas species is that they are corrosive. This is true of the gases produced during the irradiation of halogenated hydrocarbons, such as polytetrafluoroethylene (Teflon) and Kel-F. Failure from other causes is likely to occur before the corrosion could become a problem, but some care in this area may be advisable when one is selecting sealed parts, such as miniature relays that contain electrical contacts.

Environmental conditions other than radiation contribute to the degradation of organic insulators and dielectrics. Temperature, atmosphere, and/or humidity may be important for some materials. For example, the absence of oxygen is known to increase the tolerance of tetrafluoroethylene to radiation by 1 to 2 orders of magnitude. This effect could be an important factor in consideration of its possible use in a radiation environment.

The relative radiation resistance of representative organic insulators is given in Fig. 10.4.

Elastomers

The effects of radiation on elastomers are variable because of the influence of the various compounding agents. Several methods can be used to improve the radiation resistance of elastomers. These include the use of fillers, the addition of radiation-resistant resins, and the use of organic additives called *antirads*. In general, carbon-black fillers are superior to mineral fillers. Resins have been used to improve the radiation resistance of gum stocks but appear to have little effect on black stocks.

Natural rubber. Irradiation of natural rubber induces cross-linking and tends to decrease the elastic properties and increase the hardness. This is similar to the effects of overvulcanization, whereby natural rubber acquires a rigidity comparable to that of glass. Natural rubber is unaffected by radiation up to about 1×10^6 rads. Tensile strength is not affected until the rubber is exposed to 2.4×10^8 rads. Elongation and set at break are not affected up to about 5.5×10^6 rads.

Styrene-butadiene rubber. Styrene-butadiene rubber (SBR) resists radiation better than any of the synthetic hydrocarbon rubbers, but it is not equal to natural rubber in radiation resistance. Threshold damage is reached at 2×10^6 rads. The tensile strength of the material changes less rapidly than that of natural rubber. The rate of property changes for both hot SBR (polymerized at 50°C or higher) and cold SBR (polymerized at 5°C) is about the same under irradiation. Cold SBR has better initial physical properties than hot SBR, and this superiority is evident after irradiation.

244 Design Considerations for Plastics

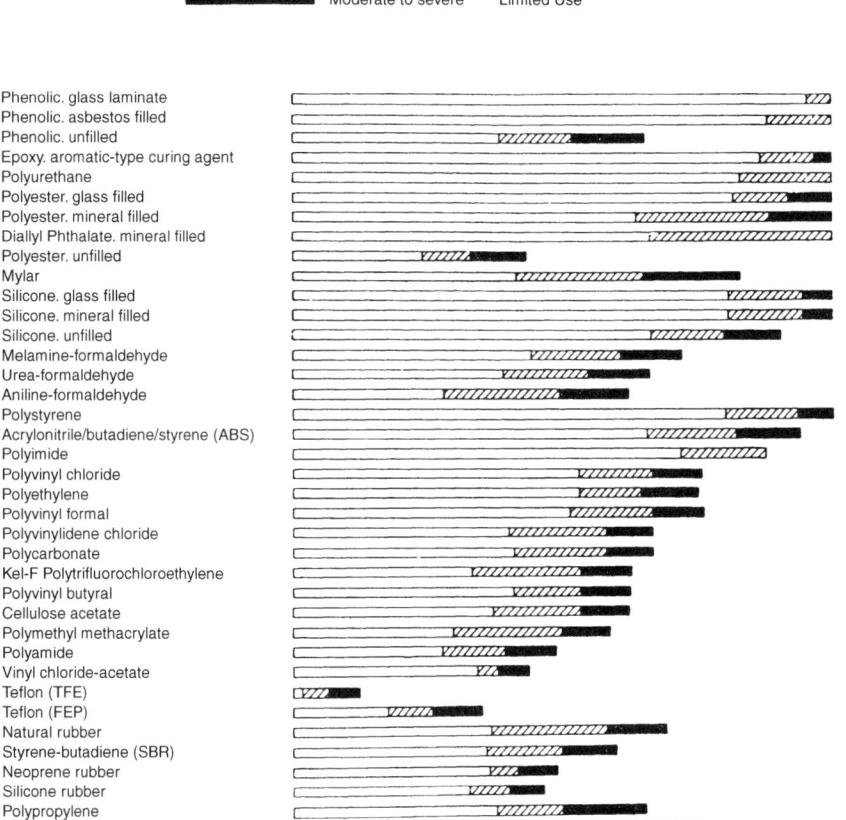

Figure 10.4 Relative radiation resistance of organic insulating materials (based on changes in physical properties).

Butyl rubber. In general, butyl rubber appears suitable for use only at relatively low radiation doses. The tensile strength of the material decreases with increasing irradiation. A damage level of 25 percent is reached for hardnesses at about 5×10^7 rads and for tensile strength and elongation at about 10^7 rads. The material shows no evidence of stress cracking after irradiation.

Hypalon (chlorosulfonated polyethylene) rubber. It is difficult to predict the tensile strength of Hypalon, because the material exhibits different trends during irradiation. In some cases, the material retains nearly its original value up to 9×10^7 rads, after which it starts to increase with higher dosages. In others cases, the tensile strength increases at low doses, drops considerably at about 4.5×10^7 rads, and then starts to increase with continued exposure. Continued exposure tends to increase the hardness and decrease the elongation of Hypalon. There is evidence that stability can be improved by adding aromatic plasticizers.

Neoprene rubber. In general, the properties of neoprene (polychloroprene) rubber are similar to those of nitrile rubber after irradiation.

Nitrile rubber. Nitrile rubber (NBR) is a copolymer of acrylonitrile and butadiene. Formulations with a high acrylonitrile content have about average radiation stability compared to other elastomers. Compression set degrades by about 25 percent at 7×10^6 rads. Tensile strength is variable after irradiation and increases by 25 percent at 1.5×10^8 rads.

Acrylic rubber. Acrylic, or polyacrylate, rubbers are based on polymers of butyl or ethyl acrylate. Both types of polymers behave similarly when irradiated. They undergo a slight amount of cross-linking and chain cleavage when exposed to 9×10^7 rads. Their hardness increases with increased exposure. However, their tensile strength behaves erratically. It increases or decreases after short exposure, remains relatively unchanged for intermediate exposure, and drops and essentially increases with prolonged exposure.

Polysulfide rubber. The hardness of polysulfide rubber does not change significantly up to 2.5×10^8 rads. However, tests conducted at 4.5×10^8 rads have caused such great damage that the hardness could not be measured. Also, at the higher exposure, both elongation and tensile strength were reduced to zero, suggesting that the material may undergo chain cleavage. No stress cracking was observed at either exposure.

Fluorocarbon rubber. Fluorocarbon rubbers undergo both cross-linking and chain cleavage when irradiated. Fluorocarbon rubbers exposed to 9×10^7 rads increase about 25 percent in hardness and lose about 80 percent of their elongation. The tensile strength also varies. The stability of the fluorocarbon rubbers is quite dependent on the environment. Thus, the stability of the rubbers in air cannot be used to predict their stability in other media.

Silicone rubber. In general, the radiation resistance of silicone rubbers is below the average of other elastomers. Physical properties are damaged by 25 percent at exposures less than 10^7 rads.

Polyurethane rubber. Polyurethane rubber is not affected up to 9×10^5 rads. In general, the urethane elastomers tend to soften at exposures up to 4×10^8 rads and then become increasingly harder. The materials also tend to decrease in both tensile strength and elongation after irradiation. Compounding appears to have little effect on the radiation resistance for most types.

Plastics

Plastics showing the greatest stability with respect to tensile strength are polystyrene, asbestos-filled phenolics, furane resin, and polyvinyl carbazole. They show practically no change up to 10^{10} rads.

Rigid plastics have very low initial elongations, in most cases less than 2 percent. Irradiation has little effect on this property for most rigid plastics. Those plastics which do not appreciably change at 10^9 rads include aniline formaldehyde, phenolic resin with asbestos and with asbestos-fabric laminate, polystyrene with mineral filler, polystyrene, polystyrene with white pigment filler, and polyvinyl carbazole.

Plastic materials showing the greatest radiation stability with respect to elastic modulus include aniline formaldehyde, melamine formaldehyde with cellulose filler, polyethylene, phenolic resin with graphite filler, polyester with mineral filler, polystyrene, polystyrene with black or white pigment fiber, and polyvinyl carbazole.

Polymers having the least change in shear strength when irradiated include aniline formaldehyde, phenolic resin with asbestos fiber or graphite, nylon, mineral-filled polyester, allyl diglycol carbonate, polystyrene with black or white pigment filler, and polyvinyl carbazole.

Melamine formaldehyde with cellulose filler, phenolic resin with asbestos or graphite filler, mineral-filled polyester, allyl diglycol carbonate, polystyrene, polystyrene with white pigment filler, polyvinyl carbazole, and vinyl chloride-acetate show good radiation stability with respect to impact strength.

The best plastics are unaffected by radiation up to doses of 10^7 to 10^9 rads and are changed by 25 percent by doses of 10^8 to 10^9 rads. These include polystyrene, mineral-filled phenolic, polyester, mylar, polyvinyl chloride, and polyethylene. Plastics with poor radiation resistance include cellulosics, polyamides, unfilled polyesters, and Teflon.

Polytetrafluoroethylene. Polytetrafluoroethylene (PTFE) has shown a rather high susceptibility to radiation damage. The rapid degradation of properties by ionizing radiation is primarily attributed to main chain scission by liberated fluorine atoms and the production of entrapped fluorocarbon gases. Tensile strength and ultimate elongation decrease, and the material becomes embrittled through this damage mechanism.

The embrittlement becomes severe with extended irradiation (10^7 rads), and the polytetrafluoroethylene crumbles to a powder. The approximate damage threshold and the 25 percent damage dose are 1.7×10^4 rads and 3.4×10^4 rads, respectively.

There is evidence that the damage observed when PTFE is irradiated is a function of several factors. These include the various types of polytetrafluoroethylene such as TFE and the copolymer FEP/P, the ambient atmosphere, and the test temperature. It has been demonstrated that Teflon-FEP is more radiation-resistant than TFE.

The effect of elevated temperature in combination with irradiation is to accelerate the degradation of PTFE's physical properties.

Polytetrafluoroethylene also experiences changes in electrical properties when it is subjected to a radiation environment. The electrical parameters that have shown a sensitivity to radiation include the dissipation factor or loss tangent, volume resistivity, dielectric constant, and dielectric strength. The changes observed are often significant in many practical applications as long as the material's mechanical integrity is maintained. Therefore, even though changes in electrical properties do occur, the degradation of physical properties is the criterion often used in determining the acceptability of this material for use in a specific application.

The volume resistivity of PTFE decreases 2 or 3 orders of magnitude from initial values of 5×10^{17} to $1 \times 10^{18}\ \Omega \cdot$ cm or greater when irradiated under vacuum conditions to total doses of 10^6 rads and higher. The degradation may continue after the radiation exposure is terminated with an additional decrease of 1 or 2 orders of magnitude over a period of several days. Recovery may also occur with the volume resistivity approaching its preirradiation value several weeks after the irradiation.

Dielectric-constant measurements of PTFE during and following exposure to a radiation environment have shown increases of less than 15 percent when PTFE is irradiated in air or vacuum to respective doses of 8×10^6 and 10^8 rads. Recovery is essentially complete within a day or two after the irradiation.

Polychlorotrifluoroethylene (Kel-F). Polychlorotrifluoroethylene, another fluoroethylene polymer, also suffers severe degradation of its physical properties when exposed to a radiation environment. It is reported to have a damage threshold of 1.3×10^6 rads and a 25 percent damage dose of 2×10^7 rads. The elongation of this material increased 47 percent and the impact strength decreased 16 percent, when it was subjected to a total dose of approximately 2.4×10^7 rads. The ultimate tensile strength was unaffected.

Electron irradiation with a total of 3.67×10^{16} e/cm^2 at 60°C seriously degraded a sample of polychlorotrifluoroethylene so that its physical and electrical properties could not be measured.

The degradation of the electrical properties of polychlorotrifluoroethylene from exposure to radiation results in a reduction in volume and surface resistivity.

Polyethylene. In some respects, polyethylene improves with exposure to radiation in that its softening point temperature increases for exposures of less than 10^7 rads. In addition, the tensile strength also increases until approximately 10^8 rads, after which it decreases and is 25 percent below the initial value at approximately 10^{10} rads. The damage threshold is greater than 10^7 rads.

A study where polyethylene of low and high densities and another which was carbon-black-filled were exposed to an electron dose of 5.8×10^{16} e/cm^2 at 60°C illustrates the differences these factors make. The hardness and stiffness in flexure of the high-density material decreased as a result of the irradiation while the low-density and carbon-filled polyethylene experienced increases in these properties. The high-density polyethylene also increased in tensile strength, and the others decreased.

The electrical properties of polyethylene also degrade when this material is exposed to a radiation environment. Volume resistivity, surface resistivity, and insulation resistance decrease by up to 3 orders of magnitude during irradiation, with permanent decreases of 1 order of magnitude. The dissipation factor at 1 kHz increases 1 to 2 orders of magnitude as a result of irradiation, and the dielectric constant changes less than ±5 percent.

Polystyrene. Irradiation studies of polystyrene have shown it to be one of the most radiation-resistant plastics. It has a damage threshold of 10^8 rads and does not experience 25 percent damage to its physical properties below 4×10^9 rads. Polystyrene is subject to post-irradiation oxidation that continues for several weeks; however, oxidation plays little or no part in the radiation damage that occurs.

Electron irradiation to a total dose of 5.8×10^{16} e/cm^2 at 60°C has resulted in decreases of approximately 50 percent in the tensile strength and ultimate elongation. The hardness and the stiffness in flexure also decreased 1 percent and 13 percent, respectively, during this same study. These results indicate that polystyrene becomes more flexible and softer as a result of the irradiation.

The insulating quality of polystyrene appears to be the only electrical property that is affected by exposure to radiation. Permanent decreases of 1 and 2 orders of magnitude have been observed in the volume resistivity and insulation resistance of this material following doses as low as 4.5×10^6 rads and as high as 10^8 rads. Other electrical parameters, such as the dielectric constant and dissipation factor, have shown little or no change from exposure to a radiation environment within this range of total dose.

Polyethylene terephthalate. Polyethylene terephthalate (Mylar) has shown improvement in its physical properties when exposed to limited radiation doses with very little degradation in electrical properties. Radiation exposure to doses of 10^8 rads and above causes severe embrittlement of polyethylene terephthalate to a degree that properties are unmeasurable.

Degradation of the electrical properties of polyethylene terephthalate within the doses described above of 10^6 to 10^7 rads is insignificant. Changes in the insulation resistance, volume resistivity, and surface resistivity as a result of irradiation are limited to approximately 1 decade. The dielectric constant and dissipation factor remain essentially unchanged.

Polyamide. Polyamide (nylon) sheet or film insulation experiences changes in both physical and electrical properties when subjected to a radiation environment. This material experiences threshold damage at a dose of 8.6×10^5 rads and 25 percent damage at 4.7×10^6 rads. These doses are based upon losses in ultimate elongation and impact strength. Another property of polyamide that deteriorates from radiation exposure is stiffness in flexure, which has increased between 52 and 181 percent, depending upon the nylon type, after exposure to an electron dose of 5.8×10^{16} e/cm^2 at 60°C. This same exposure improved the tensile strength by 49 to 107 percent. This agrees with other radiation studies which have shown increases in tensile strength of 25 percent at doses over 10^9 rads.

Information on the effects of radiation on the electrical properties of polyamide is limited to results of the electron irradiation mentioned above. Exposure to this radiation environment produced an increase of approximately 1 order of magnitude in the insulation resistance and a decrease of less than an order of magnitude for the dissipation factor. A decrease in dielectric constant was insignificant at 1 MHz and varied between 5 and 32 percent at 1 kHz, depending on the polyamide type.

Diallyl phthalate. Diallyl phthalate with various fillers such as glass or orlon has shown exceptional radiation tolerance. Little or no permanent degradation of physical or electrical properties has been observed with radiation exposures to doses of 10^8 to 10^{10} rads. Insignificant changes are observed in the hardness and flexibility of this material when it is irradiated to these total doses. The ultimate elongation and tensile strength of orlon-filled diallyl phthalate actually increased or improved with exposure to an electron dose of 5.8×10^{16} e/cm^2 at 60°C. The electrical properties of diallyl phthalate such as dielectric constant, dissipation factor, and insulation resistance are little affected by exposure to a radiation environment such as described above.

Polypropylene. Polypropylene, when exposed to a radiation environment, suffers a severe loss in physical properties. Above a total dose of 10^7 rads this material becomes embrittled. Decreases in tensile and impact strengths that approach 60 and 75 percent, respectively, at doses of 5×10^7 rads. An electron fluence of 5.8×10^{16} e/cm^2 at 60°C results in a decrease of 87 to 96 percent in ultimate elongation and tensile strength. The permanent change in electrical properties that occurs when polypropylene is irradiated to the above doses is negligible. The dielectric constant decreases slightly, and the insulation resistance decreases less than an order of magnitude. Measurements of ac loss such as power factor and dissipation factor at 1 kHz to MHz have varied from no observable change to an increase from between 0.0005 and 0.0008 to between 0.002 and 0.003.

Polyurethane. Polyurethane has shown good stability in both physical and electrical properties when it is exposed to a radiation environment. Irradiation in doses up to 7×10^8 rads has caused very little change in flexural modulus. At exposures of an electron fluence of 5.8×10^{16} e/cm^2 at 60°C, serious deterioration of physical properties occurred and included a 67 percent and a 176 percent increase in hardness and stiffness in flexure, respectively. A 59 percent decrease in tensile strength and a 99 percent decrease in ultimate elongation were also noted following irradiation to this electron fluence.

The effect of radiation on the electrical properties of polyurethane is negligible. There are insignificant permanent changes in the insulating properties, volume resistivity, or insulation resistance. The dissipation factor at 1 MHz was essentially unchanged while that at 1 kHz increased approximately 30 percent in the reactor study (6.0 to 7.4) and doubled in the electron irradiation study (0.02 to 0.04).

Polyvinylidene fluoride. Polyvinylidene fluoride shows higher radiation tolerance than other fluorocarbons such as Teflon and Kel-F. It withstands irradiation to a dose of 10^7 rads in air or vacuum with no indication of degradation in physical properties except color change. An order-of-magnitude increase in the radiation dose to 10^8 rads and above causes embrittlement and loss of flexibility and tensile strength. Low temperature increases the radiation tolerance of polyvinylidene fluoride in that doses of this magnitude, 10^8 rads, at cryogenic temperatures do not reach the damage threshold.

Changes in the electrical properties of polyvinylidene fluoride include decreases of 2 to 3 orders of magnitude in volume resistivity during and after irradiation at doses up to 2.1×10^7 and 6.6×10^7 rads in air and a vacuum cryotemperature environment, respectively. A decrease of approximately 5 orders of magnitude occurred with a dose of 2.1×10^8 rads in the air atmosphere. The dissipation factor increased less than 1 order of magnitude, and the dielectric constant was essentially unaffected by the irradiation.

Acrylics. Chemically, polymethyl methacrylate is very unstable to irradiation, undergoing chain scission and evolving gas readily. When the plastic is irradiated, it suffers main chain cleavage, and the side chains are ruptured at the same rate, causing gas formation. The internally trapped gas can expand the material 5 to 10 times when it is heated to its softening point of about 132°C. The gases evolved include hydrogen, carbon monoxide, carbon dioxide, and methane. Its physical properties are unaffected to 8.2×10^5 rads. Tensile strength and elongation are decreased at 1.1×10^7 rads, and the properties deteriorate above this dose, making the plastic very brittle. Its light transmission drops typically from 91 to 56 percent at 5.5×10^6 rads, although there is only a slight increase in haze. The plastic turns progressively brown with little loss in transparency with increasing dose.

Amino resins. Urea-formaldehyde-type plastics are, in general, about average for plastics in radiation resistance. They are unaffected at doses up to 8.3×10^6 rads and are 25 percent damaged at a dose of 5.1×10^7 rads.

Melamine-formaldehyde-type plastics are slightly more resistant than the urea formaldehyde types. They are unaffected at doses up to 7.4×10^6 rads and are 25 percent damaged at a dose of 1.1×10^8 rads.

Both the urea formaldehyde and the melamine formaldehyde type of plastics, when filled with cellulosic materials, become brittle, blister, swell, and crumble in a radiation field.

Aromatic amide-imide resins. These are the most radiation-resistant resins, exceeding the oriented polyethylene terephthalate (Mylar) by a factor of about 10.

Cellulose derivatives. Cellulosic polymers, such as cellulose acetate, cellulose acetate butyrate, cellulose nitrite, cellulose propionate, and ethyl cellulose, are among the least radiation-resistant plastic materials. Their physical properties, except breakdown voltage, deteriorate rapidly under gamma radiation. Cellulose acetate, considered to be one of the more radiation-resistant cellulosics, suffers 25 percent damage at a dose of 3×10^7 rads. Its electrical properties remain acceptable up to about 10^8 rads.

Fiberglass laminates. Fiberglass laminates, polyester cured with triallylcyanurate, polyester cured with styrene, epoxy (excepting resins cured with dicyandiamide), epoxy-phenolic, and silicone types are essentially unchanged in dielectric constant and loss tangent by radiation. In general, the mechanical properties are not significantly affected at dosages up to 8.7×10^8 rads. The weave in the glass fabric does not noticeably affect the durability of the finished laminate to environmental conditions. The catalyst used for the polyester system is not critical to the radiation resistance. On the other hand, the curing agent used for the epoxies constitutes a significant portion of the resin. Dicyandiamide is the common curing agent found to be deleterious to radiation resistance. There is some evidence that aromatic amines and anhydrides improve the radiation resistance of the epoxies.

Phenolics. Unfilled phenolics have relatively poor radiation resistance; under irradiation they swell, become brittle, and crumble. Damage of 25 percent occurs at 10^7 rads, and tensile and impact strengths of the materials drop off about 50 percent at 3×10^8 nrads. Also, a soluble product is formed during irradiation which causes the materials to disintegrate in water. Radiation stability can be increased by adding fillers, particularly mineral fillers. Phenol formaldehyde with an asbestos filler shows excellent stability and is one of the more radiation-resistant plastics, possibly because of its relatively high heat stability. It is unaffected by radiation exposures of more than 4×10^8 rads. Phenolic laminates have been exposed to 8.5×10^9 rads at room temperature without reaching threshold damage. Laminates irradiated to 2×10^7 rads at 900°F for 50 h have shown a flexural strength equivalent to unirradiated laminates also exposed at 900°F for 50 h.

Polycarbonate resins. Polycarbonate resins undergo a slow chain breakdown on irradiation. Data show that after a dosage of 50×10^6 rads, the tensile strength remains unchanged although the elongation is reduced 25 percent.

Polyesters. Unfilled polyesters have poor radiation stability; they harden and develop small cracks under irradiation. Although the stability of various polyesters will vary somewhat, their properties begin to change at approximately 10^5 to 10^6 rads. Tensile strength and impact strength decrease, although the tensile strength may increase at first. The addition of mineral fillers increases the radiation stability of polyesters by approximately 100-fold.

Upon radiation, polyethylene terephthalate (Mylar) fibers lose strength and become powder. Crystallinity does not change, which gives evidence that irradiation of Mylar does not induce cross-linking. Rather the effect of radiation is to cause chain cleavage. Mylar film does not change in tensile strength and elongation up to a dose of 3×10^7 rads. The film becomes brittle and darkens with large doses (above 10^8 rads). There is definitely some gas formation upon irradiation of Mylar. After a dosage of 7×10^6 rads, the permanent volume resistivity of a 0.002-in Mylar film remained unchanged. No significant change in dielectric constant has been found after doses of 10^7 to 10^9 rads on a 0.002-in Mylar sheet. Investigation of the post-irradiation effects on the electrical properties of Mylar shows that the dielectric constant, after a dose of 10^9 rads, decreases 4 percent immediately after irradiation, increasing rapidly for a day or two, leveling off in about 240 h to a value approximately 6 percent above the value measured immediately following irradiation. No effect was found when Mylar was irradiated to 10^8 rads and subjected to thermal aging at temperatures of 160, 180, and 200°C. At 10^9 rads, an effect was observed indicating that thermal aging becomes important between 10^8 and 10^9 rads.

Polyoxymethylene. Polyoxymethylene is rapidly depolymerized by radiation and loses strength very quickly. The rate of radiolytic degradation apparently increases considerably with increasing temperature. Electrical failure occurs when mechanical degradation is complete, at about 1×10^7 rads.

Vinyl polymers and copolymers. Vinyl polymers and copolymers range from very radiation-resistant materials such as polyvinyl carbazole, which is unaffected by radiation doses up to 8.8×10^7 rads, to vinyl chloride acetate, which has a threshold damage of 1.4×10^6 rads.

Polyvinyl carbazole is one of the most radiation-resistant plastics. It has a threshold damage dose of 8.8×10^7 rads. However, the material is brittle, and therefore its applications can be limited.

Polyvinyl butyral is unaffected by radiation to a dose of 4.7×10^6 rads and is damaged by 25 percent at a dose of 1.9×10^7 rads, which is below average for polymeric materials. The tensile strength decreases rapidly after 10^7 rads. Upon irradiation the material first softens and then becomes more brittle.

Polyvinyl chloride (PVC) is equivalent to polyethylene in its radiation stability. Its properties begin to change at a dose of 1.9×10^7 rads, while it is 25 percent damaged at a dose of 1.1×10^8 rads. The tensile strength of PVC is not affected until it is given a dose higher than that which affects polyethylene. The tensile strength of polyethylene first increases and then decreases. PVC decreases more rapidly than polyethylene in tensile strength, but its elongation does not decrease as rapidly as that of polyethylene. The liberation of the more highly corrosive hydrogen chloride when PVC is irradiated makes this material unsuitable for many applications in a nuclear environment.

Polyvinyl formal is similar to polyvinyl butyral in its properties. However, its radiation resistance is better than that of butyral and slightly poorer than that of PVC. Its threshold damage dose is 1.6×10^7 rads, and 25 percent damage occurs at a dose of 8.2×10^7 rads.

Polyvinylidine chloride is between polyvinyl formal and polyvinyl butyral in radiation stability. It is approximately equal to urea formaldehyde and is about average for plastics. The threshold damage dose is 4.1×10^6 rads, and 25 percent damage occurs at 4.5×10^7 rads. It softens, blackens, evolves hydrogen chloride, and decreases in tensile strength when irradiated.

Vinyl chloride acetate behaves similarly to polyvinylidiene chloride when irradiated. The threshold damage is reached at 1.4×10^6 rads, and 25 percent damage occurs at 2.8×10^6 rads. It also evolves HCl readily under irradiation and turns black after a very short period of irradiation. It softens even before showing any appreciable darkening, and the elongation increases by over 500 percent before a dose of 5×10^6 rads has been reached.

Silicones. Silicones show a wide variation in radiation stability. The presence of phenyl groups in the silicone chain increases the radiation stability, and the presence of the methyl groups increases the flexibility. Since silicone resins generally have a phenyl component, they have reasonably good radiation resistance. The radiation stability of silicone resins is improved when they are reinforced with glass fibers. Glass-reinforced silicone laminates reach threshold damage at about 10^9 rads at room temperature. The tensile strength does not drop off until 2.5×10^9 rads. Asbestos-silicone laminates show no apparent change in properties up to 1 to 2×10^8 rads.

Nomex. This aromatic polyamide has outstanding radiation resistance. Nomex is essentially unaffected by 800 Mrads of ionizing radiation. At this dose, polyester laminates crumble. After exposure to 6400 Mrads of ionizing radiation, Nomex still retains useful mechanical and electrical properties.

Polybenzimidazole. This polymer has outstanding resistance to gamma radiation. No noticeable changes in the material have occurred after 200-Mrads exposure to cobalt 60.

Cyanate esters, bismaleimides, and benzocyclobutenes. No data exist on the gamma radiation resistance of these materials as of this writing. However, because of the high aromatic nature of these materials, their radiation resistance should be better than 10^8 rads before deterioration begins in some properties.

Miscellaneous organics. Radiation-effects information is available on organic bulk, sheet, and/or film materials other than those discussed on the preceding page. The information is limited to results from only one radiation-effects test of each material. These results are limited to those shown in Tables 10.6 and 10.7. Table 10.6 lists materials that were so seriously degraded that their physical and electrical properties could not be tested or measured. Table 10.7 lists those materials that survived exposure to the radiation environment and includes some of the particulars concerning changes observed in their physical and electrical properties.

TABLE 10.6 Miscellaneous Organic Bulk, Sheet, and/or Film Materials which Limited Information Indicates as Unsatisfactory at the Radiation Dose Indicated

Material	Total integrated exposure, $\times 10^{16}$ e/cm² at 60°C
Acetal resin	1.22
Acrylic plastic, molding grade (rubber-modified)	5.80
Allyl carbonate plastic, cast	4.10
Cellulose acetate	5.80
Cellulose butyrate	4.10
Cellulose propionate	4.10
Chlorinated polyether	2.90
Polycarbonate	5.80
Polyfluoroethylenepropylene, Teflon FEP (copolymer)	3.67
Polymethyl methacrylate, cast	1.22
Polymethyl methacrylate, molding grade	4.10
Styrene acrylic copolymer	2.90
Polyvinyl chloride, DOP plasticized	3.67
Polyvinyl chloride, rigid	4.10

TABLE 10.7 Radiation Effects on Miscellaneous Organic Bulk, Sheet, and/or Film Materials

Material	Total integrated exposure	Remarks
Acrylonitrile-butadiene-styrene	5.8×10^{16} e/cm² at 60°C	Hardness increased 13 percent; flexibility, tensile strength, and ultimate elongation decreased 49, 58, and 93%, respectively. Dielectric constant increased 1.5% and dissipation factor (DF) decreased slightly. Insulation resistance (IR) increased
Styrene-acrylonitrile copolymer	5.8×10^{16} e/cm² at 60°C	Tensile strength and ultimate elongation decreased 34 and 47%, respectively. Hardness was unchanged, and flexibility increased 5%. Dielectric constant increased 4 to 6%. DF increased to 0.01 at 1 kHz and 0.40 at 1 MHz. IR decreased one decade
Styrene-butadiene (high-impact styrene)	5.8×10^{16} e/cm² at 60°C	Flexibility and ultimate elongation decreased more than 90%, and tensile strength decreased 35%. Hardness increased. Dielectric constant increased slightly while DF increased 50%. IR increased
Styrene-divinyl benzene	5.8×10^{16} e/cm² at 60°C	Changes in physical properties were of no practical significance
Polyvinyl chloride acetate	5.8×10^{16} e/cm² at 60°C	Serious degradation prevented measurement of physical degradation. Dielectric constant decreased 7%, and DF increased one decade. IR did not change
Polyvinyl fluoride	5.8×10^{16} e/cm² at 60°C	Serious degradation prevented measurement of physical degradation. Dielectric constant decreased 7%, and DF increased one decade. Insulation resistance did not change
Epoxy-glass laminate	2×10^{13} n/cm² (E 0.1 MeV) 1×10^{8} rads	Unaffected
Epoxy-glass laminate, copper-clad	0.86×10^{6} rads	No induced conductivity and no change in ac loss properties
Polyester-glass laminate	2.5×10^{6} rads	Volume and surface resistivity decreased three decades. DF increased from 0.003 and 0.006 to 0.019 and 0.010. No change in dielectric constant
Silicone-glass laminate	5.0×10^{13} n/cm² at 200°C 1.0×10^{8} rads	A 49% loss in flexure strength; slight change in color, thickness, and weight

Organic fluids

Lubricants. The radiation tolerance of gas-turbine lubricants ranges from 10^7 to 10^9 rads. Although polyphenyl ethers show very good radiation resistance, it appears that hydrocarbon fluids, such as highly refined mineral oils and alkyl aromatics, may offer the best compromise among stability (radiation, oxidation, and thermal), lubricity, and low-temperature performance. From available data, it appears that for each fluid there exists a critical gamma exposure below which radiation exerts little or no effect. With some hydrocarbon fluids this exposure may be greater than 10^8 rads.

Hydraulic fluids. Polyphenyl ethers show outstanding stability at 10^9 rads for high-temperature applications where the pour point is not critical. Alkyl diphenyl ethers have slightly lower pour points but are somewhat less stable than the unsubstituted polyphenyl ethers. As with lubricants, alkyl aromatics may offer the best balance in pour point and stability.

Heat-transfer fluids. The terphenyls show the highest radiation resistance and are recommended over other heat-transfer materials for high radiation exposures. Of the terphenyls, the *para* isomers show the greatest resistance at low exposure. At higher exposures the three isomers approach one another in stability. Moniso-propylbiphenyl, biphenyl ether, and silicate esters also appear to possess good radiation stability. Ethylene glycol, chlorinated diphenyls, DC 710 silicone, and phosphate esters are resistant to about 8.5×10^7 rads.

References

1. T. Alfrey, Jr., "Mechanical Properties of Polymers," in H. Mark, (ed.), *High Polymers,* vol. 6, Interscience, New York, 1968.
2. F. Brecke, *Physical Properties of Polymers,* Interscience, New York, 1962.
3. L. E. Nielsen, *Mechanical Properties of Polymers and Composites,* vols. 1 and 2, Marcel Dekker, New York, 1974.
4. J. V. Schmitz, ed., *Testing of Polymers,* vols. 1 and 2, Wiley-Interscience, New York, 1965.
5. C. C. Ku and R. Liepins, *Electrical Properties of Polymers,* Hanser Publishers, New York, 1987.
6. *Plastics, Rubber,* and *Electrical Insulation and Electronics,* vols. 8.01 to 10.05, *ASTM Book of Standards,* American Society for Testing and Materials, Philadelphia, 1992.
7. A. M. Litman and N. E. Fowler, "Electrical Properties," in *Engineered Materials Handbook,* vol. 2, ASM International, Metals Park, OH, 1988.
8. C. A. Harper, ed., *Handbook of Plastic, Elastomers and Composites,* 2d ed., McGraw-Hill, New York, 1992, Chap. 2.
9. S. W. Shalaby and P. May, "Thermal Properties," in *Engineered Materials Handbook,* vol. 2, ASM International, Metals Park, OH, 1988.
10. G. C. Ives, J. Mead, and M. Riley, *Handbook of Plastics Test Methods,* Butterworth & Co., London, 1971.

11. T. G. Fox and P. J. Flory, "Second-Order Transition Temperatures and Related Properties of Polystyrene," *Journal of Applied Physics*, vol. 21, 1950, p. 581.
12. F. H. Winslow, in R. K. Eby, ed., *Durability of Macromolecular Materials*, "Ozone Attack in Polyolefins," ACS Symposium Series No. 95, American Chemical Society, 1979.
13. H. Pranjoto and D. D. Denton, "Moisture Uptake of BCB Films for Electronic Packaging," *Matl. Res. Society Symposium Proceedings*, vol. 302, 1991, p. 295.
14. T. W. Dakin, "Electrical Insulation Deterioration Treated as a Chemical Rate Phenomenon," *AIEE Transactions*, vol. 67, 1948, p. 113.
15. T. W. Dakin, "Electrical Insulation Deterioration," *Electrotechnology*, vol. 124, December 1960.
16. F. H. Steiger, "The Arrhenius Equation in Accelerated Aging Studies," *Amer. Dyestuff Reporter*, vol. 287, May 5, 1958.
17. IEEE Standards for life testing of electrical insulation, nos. 1, 98, 99, and 101.
18. M. F. Rose, "Electrical Insulation and Dielectrics in the Space Environment," *IEEE Transactions on Electrical Insulation*, vol. EI-22, no. 5, October 1987.
19. *Thermal-Radiation Materials Application Data Manual, 1971–1984*, Westinghouse Electric Corp., PGH.
20. W. M. Alvino, "Aging Behavior of Polyarylsulfone Films," *J. Appl. Poly. Sci.*, vol. 15, 1971, p. 2521.
21. W. M. Alvino, "Thermal Properties of Poly[N,N'-(4,4'-Diphenyl Ether)4-Amidophthalimide] Film," *J. Appl. Poly. Sci.*, vol. 19, 1975, p. 673.
22. W. M. Alvino, "Ultraviolet Stability of Polyimides and Polyamide-imides," *J. Appl. Poly. Sci.*, vol. 15, 1971, p. 2123.
23. W. M. Alvino, G. Bower, L. Frost, and L. Scala, "Evaluation of Elastomers for Use in Thermal Solar Collectors," *I&EC Product Research and Development*, vol. 21, 1982, p. 691.
24. L. G. Colark, L. Kinard, J. Carter, Jr., and J. Jones, Jr., "Long Duration Exposure Facility (LDEF)," NASA, SP-473, 1984.
25. J. G. Funk, J. W. Strickland, and J. M. Davis, *The Preliminary Long Duration Exposure Facility (LDEF) Materials Data Base*, NASA TM 10762, October 1992.
26. C. G. Collins and V. P. Calkins, *Radiation Damage to Elastomers, Plastics and Organic Liquids*, General Electric Co., Contract No. AF 33(038)-21102, September 1956.
27. F. L. Bouquet, R. B. Somoano, and P. O. Frickland, *Effects of Radiation on Capacitor Dielectrics*, JPL Caltech, Contract no. NAS 7-100, Pasadena, CA, February 1987.
28. F. L. Bouquet and J. W. Winslow, *Radiation Effects on Polymer Properties*, JPL Caltech, Pasadena, CA, January 1987.
29. W. W. Parkinson and O. Sisman, "Use of Plastics and Elastomers in a Radiation Environment," *Nuclear Engineering and Design*, vol. 17, 1971.

Chapter 11

Design Considerations

The selection of a plastic for a particular application is not a trivial task even for the knowledgeable materials engineer. It is not trivial because there are so many factors that influence the selection. The key to any good material selection is a logical and systematic approach which usually involves the following: A clear definition of the application must be given and understood as well as the most probable mode of failure. This is critical because it affects plastic selection and test evaluation design. There should be a description of the property requirements such as engineering properties (physical, mechanical, electrical) and environmental properties (thermal, radiation, chemical resistance). Material processing requirements as dictated by part size should also be considered along with assembly, machining, decorating, and finishing, if applicable, in the material selection process. And last but not least, material cost and alternatives must be taken into account.

It is the purpose of this chapter to make the reader aware of those factors that are important in the selection and use of plastics. It is highly recommended that the user or designer work closely with an experienced materials engineer and with the technical support people provided by the industry which supplies the plastic materials.

A Brief Review

Polymers are high-molecular-weight materials made by the joining of lower-molecular-weight ones. Polymers can be conveniently classified as thermoplastics, thermosets, and elastomers. A fourth classification can also be defined as *reinforced polymers*. Thermoplastics can be repeatedly softened and hardened by heating and cooling, respectively. Thermoset polymers cure into a permanent infusible mass, although when first heated, they do flow for a short time. Reinforced polymers can be thermoplastics, thermosets, or elastomers and are polymers containing generally up to 50 percent by weight of fine fibers (organic or inorganic). The fiber length, distribution, orientation, type, and weight percent strongly influence the properties of the reinforced polymer material.

Polymers offer some rather unique benefits (economical and noneconomical) compared to other materials, and it is the combination of these benefits that makes this group of materials uniquely qualified for electronic applications as well as various engineering applications. This combination of properties often results in significantly improved product performance. Most polymers including reinforcements and fillers are natural electrical insulators and provide excellent thermal insulation properties compared to metals and some nonmetals. Polymers in general have excellent chemical resistance and in some cases provide superior resistance to environments where other materials fail. Polymers are lightweight and offer superior strength-to-weight and stiffness-to-weight ratios compared to most other materials. Polymers are considerably easier (handling, tooling energy requirements, equipment size, etc.) to process and fabricate than metals. In addition, plastics offer a broad range of optical properties, colorability, and toughness. The decision to use a particular material in electrical and electronic applications is not a choice between metal and plastic but rather between plastic and plastic.

Engineering Properties

The engineering properties include the physical, mechanical, and electrical properties of polymers. The properties of polymers in general and of individual polymer compositions within a given class of polymers can vary by several orders of magnitude. This is especially true when the polymers are modified with fillers, fabrics, fibers, and blending with other polymers. The range of property values for all classes of polymers was shown in Table 10.1 and provides a comparative database for all polymers. This information can be used as a guide and should be supplemented with the property data information generated by both the material supplier and customer in-house testing. In examining the effects of the above type of additives on the properties of polymers, consideration must be given to surface treatment of the fiber, which affects adhesion of the polymer to the fiber. In general, addition of reinforcements increases the tensile strength, tensile modulus, and impact strength; reduces shrinkage; can cause part properties to be anisotropic; and can increase wear and coefficient of friction properties.

In an isotropic material, the properties at any given point in the material are the same, independent of the direction in which they are measured. In an anisotropic material, the mechanical properties depend on the direction in which they are measured. A few examples will help you visualize these effects: Cast metals and plastics are generally isotropic. Rolled metals and extruded plastics are anisotropic. Composite plastics which contain reinforcements (such as fillers, fibers and fabrics) can have a high degree of property orientation depending on how the reinforcement is oriented, and generally these reinforced plastics exhibit anisotropy. Wood is a highly anisotropic material. The isotropic-anisotropic behavior of plastics also affects the way that a material shrinks. Mechanical properties are important because virtually all end-use applications (electronic, structural, etc.) involve some degree of mechanical loading. In practical applications, plastics are seldom subjected to a single steady deformation without the presence of other adverse factors, such as temperature and radiation. So it is necessary to consider both mechanical and other properties when plastics are used for electrical and electronic applications. Tensile strength, modulus, elongation, and impact strength are generally the most often selected properties that are considered for a variety of applications and are normally provided in data sheets by the material suppliers. However, often emphasis is placed too much on comparing the published values of different types and grades of materials rather than on determining the critical values relative to end-use requirements. A detailed description of the mechanical properties of plastics is beyond the scope of this chapter, but some basic information on these properties is provided in Chap. 10. However, mechanical properties that are important in the selection of a plastic for a given application are listed in Table 11.1, and the interested reader should see Refs. 1 to 4 for more detailed information.

As far as electrical properties are concerned, all engineering plastics are excellent insulators, but their electrical properties vary significantly depending on the material (see Table 10.1). The volume resistivity of a polymer is affected by temperature, relative humidity, test voltage, and how the sample is conditioned. Unlike volume resistivity, which is a property of the particular polymer, surface resistivity is a measure of the material's susceptibility to surface contamination, particularly moisture. As a result, the data are subject to large errors, and so large safety factors should be built into the design. Surface resistivity data are used to provide information on surface leakage. Dielectric strength is not a linear function of thickness and is also affected by temperature, conditioning environment, rate of voltage increase, and test duration. Any contamination or internal voids within the sample can cause premature failure of the material in this test. The dielectric constant of a plastic is important when the material is used in high-frequency applications because the constant affects the rate of signal propagation. The dielectric constant is affected by temperature, mois-

TABLE 11.1 Mechanical Property Considerations in Plastics

Normal stress behavior (yield strength, ultimate strength)

Normal strain behavior

Modulus of elasticity (proportional limit, yield point elastic limit, secant modulus)

Poisson's ratio

Shear stress

Shear modulus

Relationship of material constants (tensile modulus, Poisson's ratio, and shear modulus)

Direct shear

True stress

Compression strength and modulus

Bending strength and modulus

Rate/time dependence of mechanical properties

Creep

Stress relaxation

Impact loading

Fatigue endurance

---

ture, frequency, and part thickness. The dissipation factor (DF) provides information on the ability of the plastic to dissipate heat, and low values are important when plastics are used as dielectrics in high-frequency applications such as in radar and microwaves. The arc resistance of a plastic is a measure of the time it takes to form a conductive path on its surface. This conductive path is usually the result of decomposition products from the plastic, and the mechanism of formation can be very different for thermoplastics and for thermosets. High values of arc resistance are advantageous in applications where arcing is a possibility, such as in switches, circuit breakers, automotive ignition components, and high-voltage apparatus.

Environmental Properties

The way in which the environment affects the properties of plastics is another important aspect in choosing a particular plastic for a given application. These environmental factors include thermal, radiation, and chemical resistance of plastics and have a direct influence on the mechanical, electrical, and physical properties of the plastics. The thermal characteristics that are important are the thermal conductivity, coefficient of thermal expansion (CTE), heat deflection temperature

(HDT), and aging. Thermal conductivity is important in the operation of a device in order to prevent heat buildup. If heat were not removed, the life of the plastic used in the part would be adversely affected. The thermal conductivity does not vary significantly among plastics. Plastics are basically very good insulators, and fillers are generally added to improve the thermal conductivity. The heat deflection temperature is useful to the design engineer because it gives insight into the relative ability of plastic materials to perform at elevated temperatures while supporting loads. The CTE of plastics is very important because of the wider variations within this group of materials than in other materials. Plastics are affected more by temperature than many other materials because changes in their properties are effected at lower temperatures. Stresses are created whenever one material is connected to or encapsulated in another material. The CTE of plastics is affected by reinforcement, chemical structure, and the anisotropic-isotropic nature of the plastic part; and the extent of these effects and variations must be known for successful material selection and part design. The glass transition temperature T_g of a plastic is extremely important because it significantly affects the CTE. The CTE will be higher above than below its T_g. Above T_g, polymers tend to expand isotropically. The service life of a polymer depends a great deal on temperature as well as other environmental effects. Excluding softening of plastics, the most common effect of elevated-temperature exposure is oxidation of the polymer. This oxidation can have significant adverse effects on the polymer's properties and usually results in cross-linking and chain scission of the polymer chains. When this occurs, first physical strength is usually reduced, followed by a reduction in other properties and finally a reduction in electrical properties. It is because of this that the long-term resistance properties are often rated differently for various applications, and it is extremely important in determining the service life of a polymer that the mode of failure be exactly known. The effects of temperature on plastics are typically measured by aging at different temperatures for extended periods and plotting the decrease in the measured property as a function of time. The aging of plastics has been successfully treated as a chemical rate phenomenon[5] (see Chap. 10, page 237 for details), and aging at elevated temperatures is used to acquire data for extrapolation to lower temperatures. As a rule of thumb, by raising the temperature 10°C (18°F) the reaction rate doubles. If more than one environmental factor contributes to the service life of the plastic, then aging tests under the combined conditions must be considered, such as temperature and radiation, temperature and stress, temperature and moisture, etc.

Information on the long-term aging of polymers exposed to various environments is generally not available from material suppliers, and the information is almost always generated in house by the user. Some of this information is provided in Chap. 10, Refs. 14 to 29. The effects of chemicals are another important factor to consider in the selection of a plastic because while plastics do not rust or corrode like metals, they do craze, crack, discolor, melt, and dissolve when exposed to certain chemicals and/or environmental conditions. Generally this kind of information is provided by the material supplier. A few general guidelines follow: Stress affects product performance; as stress increases, resistance to a particular environment decreases. Stresses in plastics can arise in the processing and fabrication of polymers. Temperature combined with stress in plastics exposed to different chemical environments does not always result in premature part failure. Temperature sometimes can reduce the stress level, resulting in improved chemical resistance properties. Chemicals attack plastics in a variety of ways. Chemicals can attack the polymer chain directly, producing a progressive lowering of molecular weight (hydrolysis, chemical oxidation). Evidence of these changes is usually manifested by a decrease in short-term mechanical properties. Solvation effects of chemicals on plastics are usually manifested by swelling, weight, and dimensional changes. The degree of solvation depends on the nature of the plastic, its molecular weight, and the nature of the chemical. Another chemical effect is plasticization. If a chemical is miscible with a polymer, absorption and plasticization may occur which are usually evidenced by a decrease in stiffness, strength, creep resistance, and impact resistance. The polymer will also tend to swell and warp as a result of stress relaxation.

Assembly[6,7]

The ideal situation in plastic product design is the design of a one-piece item because it eliminates assembly operations. However, most applications for plastics require some type of assembly operation, and there are many assembly methods to choose from. These assembly methods can be grouped into the four following general categories:

Molded-in assembly systems

These methods are usually more economical because no additional assembly methods (adhesive, solvent, special equipment, or fasteners) are required. Examples of the molded-in assembly are the snap-fit, press-fit, pop-on, and molded-in threads. The advantages are fast assembly, low cost, and no need for additional parts. Disadvantages are that tooling can be complex and that not all plastics are suitable to this bonding method.

Fasteners

Many of the mechanical fasteners used for metals can be satisfactorily used with plastics as well. In fact, there are some specially designed fasteners for use with particular plastics. Typical fasteners include bolted, self-tapping screws, rivets, spring clips, and threaded inserts. The advantages of these assembly methods are ease of use, ready availability, and no need for complex tooling or part preparation. Disadvantages include the need to carry an inventory of extra parts and the overstressing of parts.

Chemical bonding methods[6,7]

The chemical bonding methods include the use of adhesives and/or solvent to join similar or dissimilar materials. These methods generally do not create assembly stresses; however, good joining of plastic parts requires properly matched materials as well as surface preparation and cure conditions. Solvent bonding involves applying a liquid solvent to the surfaces of the plastics, a process which causes some dissolution of the plastic surface. The parts are clamped together until the bond takes hold and the solvent evaporates. The solvent bonding technique is limited to materials which are compatible and will dissolve and/or soften in the same solvent. Certain solvents, however, can adversely affect some plastics, particularly stress plastics; they can cause crazing, cracking, and even shattering of the plastic if the stress is great enough. The supplier of the plastic or the materials engineer should be consulted to ensure that the solvent is compatible with the plastic. Solvent bonding applies mainly to thermoplastic amorphous resins. It cannot be used with polyolefins. Adhesive bonding differs from solvent bonding in that two surfaces are bonded together with a third substance connecting the two at the interface. This third substance, called an *adhesive,* can join plastics with other plastics, metals, elastomers, ceramics, glass, wood, etc.

Typical adhesives frequently recommended for use with plastics are epoxies, acrylics, polyurethanes, phenolics, elastomers, polyesters, hot melts, bismaleimides, cyanate esters, and others. Cyanoacrylate adhesives can also be used in plastic assemblies. Specific recommendations can be obtained from the plastic and adhesive manufacturers. Care should be exercised in using adhesives containing solvents because the solvent can attack the plastic surface, causing deterioration of both the bond and plastic. Surface preparation is often critical in ensuring a strong, lasting adhesive bond. A more detailed description of polymer adhesives is given in Chap. 12.

Thermal bonding methods

Thermal bonding methods do not generally require additional joining materials, chemicals, or ancillary equipment for coating and curing. Assembly with thermal bonding is fast, economical, and environmentally compatible. The thermal bonding methods available for plastics are ultrasonic, vibration, spin, radio frequency, and induction. Ultrasonic welding is economical for joining small to medium-size parts of similar plastics. High-frequency (20- to 40-kHz) vibrational energy is directed to the interface between two plastic parts, which creates localized molecular excitation, causing the plastic to melt. Pressure is maintained after the energy source is removed, and the plastic solidifies. The process takes less than 2 s, and the resulting weld has high strength, even approaching that of the base polymer. This method is useful with the following polymers: acetals, ABS, acrylics, nylons, polycarbonate, polyimide, polystyrene, styrene-acrylonitrile, and phenoxy. Vibration welding is the joining of plastic materials caused by frictional heat. In this process, two parts are rubbed together, producing frictional heat which results in melting at the interface. The parts move in a high-amplitude, low-frequency reciprocating motion which, when it stops and the melt cools, produces a very strong bond. Typical frequencies and amplitudes range from 120 to 240 Hz and 0.10 to 0.20 in of linear displacement, respectively. High-strength joints are produced in large parts and in irregular joint interfaces. Spin welding is used to join plastics with circular joint interfaces. Frictional heat is generated through rotation of one part against the other, causing melting. Upon cooling a strong weld joint forms. The process takes about 3 s to complete and is applicable to rigid thermoplastics. Radio-frequency welding or heat sealing is widely used with thermoplastic film and sheet material. A strong radio-frequency field is applied to the selected joint by a special metal die capable of applying clamping pressure to complete the weld. Some plastics are transparent to radio frequency and cannot be welded by this method. Induction welding uses a radio-frequency magnetic field to excite magnetic sensitive particles, which are incorporated into a gasket, preform, molded part, adhesive, etc. As heat increases, the parts are pressed together, causing the preform to melt and flow throughout the joint interface.

Machining, Finishing, and Decorating[6,7]

Both thermoplastics and thermoset materials are available in a variety of forms, e.g., rods, tubes, sheet, film, billets, laminates (paper or fabric), and cast blocks. Normally the fabrication of plastic parts from standard shapes is done only when the product cannot be molded.

Whether the part is molded or fabricated from a standard shape, it still can be shaped, machined, and finished with most of the equipment commonly used for machining metals. Proper tools and cutting conditions are mandatory and encompass tool geometry, cutting speeds, and coolants. The main problems encountered in machining thermoplastic materials are due to frictional heat generation. The plastic will distort and melt as the machining tool and plastic heat up. This can result in poor surface finish, tearing, localized melting, and jamming of the machining tool. All thermosets and thermoplastics will behave differently in machining operations depending on their physical and mechanical properties. Some are easier and faster to machine than others. Generally a high melting point, inherent lubricity, good hardness, and rigidity in a plastic are properties which tend to improve machinability. The presence of reinforcements in plastics significantly affects the machining characteristics; e.g., the cutting and machining of glass, organic fabric/fiber-filled plastic is quite different from short-fiber or filler-reinforced plastic. While standard drill equipment can be used in addition to special equipment for the drilling and reaming of thermoplastics, speeds and feed conditions must be controlled to prevent heat buildup. In addition, drill bit and reamer types and materials of construction must be carefully chosen (carbide, diamond tips, chrome-plated or nitrided surfaces). Cooling can include compressed air or a cooling liquid, but the former is preferred because no outside contamination is introduced.

> Many plastic parts use self-tapping screws, threaded metal inserts, or other fastener systems. When a machine thread must be added after molding, standard metal cutting taps and dies may be used, provided that the same precautions regarding heat, chip removal, tool maintenance, and lubrication for drilling are observed. For high production or with filled resins, carbide taps are recommended. Drilled or molded holes should generally be larger than those specified for steel, and threads finer than 28 threads per inch should be avoided. In cutting operations involving sawing, milling, turning, grinding, and routing standard end mills, the following can be used: standard end mills (two flutes), circular cutters, tool bits, wood saw blades, router bits, files, rasps, and sandpaper. Tools must be kept sharp and cool to avoid gumming, overheating, and poor finish quality.[8]

Not all processed plastics emerge from their processing operation in a perfect aesthetic state. Some plastics may require additional finishing operations including decorating depending on the end-use application. Plastics can be spray-painted, vacuum-metallized, hot-stamped, silk-screened, metal-plated, and printed. In some cases, the finish applied to the plastic may provide an extra level of protection from heat, radiation, chemicals, scratching, and abrasion. In other cases, conductive coatings are applied to the inside of the part for the dissipation of static electricity and for electromagnetic shielding applications. Such coatings are common in computer and other electronic equipment housings. An essential part of the coating and/or decoration of plastics is a clean surface to ensure good bonding. Cleaning with solvents, detergents, primers, etchants, abrasives, flaming, and corona has been used to enhance the bonding to the plastic surface. Most plastics can be painted, although some are a lot more difficult than others. Materials such as polyethylene, polypropylene, and acetal, which have waxy surfaces or other crystalline resins which are very solvent-resistant can be difficult to paint and require special primers or pretreatments for satisfactory adhesion. Many amorphous plastics easily accept a wide variety of coatings.

Although rolling and dipping are sometimes used, power spray painting is the usual method of paint application. Among the coatings used are polyurethane, epoxy, acrylic, alkyd, and vinyl-based paints. Since many paints are oven-cured, materials must have sufficient heat resistance to survive without distortion.

In the vacuum metallizing process, a special base coat is applied to the polymer surface. The part is placed in a vacuum chamber, and a metallic vapor is created, usually by resistance heating of a metal such as aluminum, silver, gold, or copper, which is deposited on the plastic part. A protective clear top coating is applied over the metallized surface for both abrasion resistance and environmental protection. In sputter-coating, a plasma or an electron beam is used to vaporize the metal and subsequently coat the plastic surface. These coatings are quite thin (<1 μm). Electroplating involves the electrolytic deposition of metals such as chrome, nickel, copper, silver, and brass onto pretreated plastic surfaces. Electroplated plastics are quite durable. In flame spraying and arc spraying, special equipment is used to deposit a fine spray of a molten metal on a plastic surface. The coating is relatively thick and rough and is used mainly in non-appearance-type applications such as in electromagnetic shielding and static electricity dissipation. Hot stamping is a process of transferring a high-quality image to a plastic part. A heated die transfers a pattern from a transfer tape or foil to a flat plastic surface. Lettering and/or decorative designs can be transferred. Almost all conventional printing processes can be used to decorate plastics. They include offset printing, pad printing, and silk screening.

As with any material, the proper selection, fabrication, and use should always be chosen after careful consultation with a knowledgeable materials engineer and the specific material manufacturer, to ensure optimum performance of the material in a given application.

References

1. E. Baer, ed., *Engineering Design for Plastics,* Reinhold Publishing, New York, 1964.
2. *Engineered Materials Handbook,* vol. 2: *Engineering Plastics,* ASM International, Metals Park, OH, 1988.
3. S. S. Schwartz and S. H. Goodman, *Plastics Materials and Processes,* Van Nostrand Reinhold Co., New York, 1982.
4. C. A. Harper, ed., *Handbook of Plastics, Elastomers, and Composites,* 2d ed., McGraw-Hill, New York, 1992.
5. T. W. Dakin, "Electrical Insulation Deterioration Treated as a Chemical Rate Phenomenon," *AIEE Transactions,* vol. 67, 1948, p. 113.
6. R. Greene, ed., *Modern Plastics Encyclopedia,* McGraw-Hill, New York, 1992.
7. M. W. Hunt, ed., *Materials Engineering,* Penton Publishers, Cleveland, OH, 1992.
8. *Designing with Plastic (The Fundamentals),* Design Manual TDM-1, Hoechst Celanese Corp., Chatham, NJ, 1989.

Part 3

Utilization of Plastics in Electronics

Chapter

12

Forms, Applications, and Uses

Polymers both natural and synthetic have had a tremendous positive impact on the growth of the electrical and electronics industry. These materials are used in an almost endless variety of forms for a multitude of applications. Table 12.1 lists some of these areas, and a more complete list of all types of resins and compounds, their forms, and their manufacturers can be found in Refs. 1 and 2. This chapter focuses on some of the major areas within the electrical and electronic industry where polymers play a significant part in the component to be manufactured. Some of these areas such as encapsulants and coatings have been discussed in earlier chapters but will be reviewed briefly in this chapter to emphasize the importance of these types of materials for specific applications. The purpose of this chapter, like the others before it, is to make the reader aware of the subject matter by providing information regarding the types of polymeric materials and their properties used in the fabrication of electrical and electronic products.

Testing and Characterization

ASTM section 10, volumes 10.01 to 10.05, covers most of the material forms discussed in this chapter. A summary of the topics covered is given in Table 12.2, and the reader is advised to review Ref. 4 for testing details.

TABLE 12.1 Forms, Uses, and Applications of Plastics in the Electronics Industry

Form	Use and application
Dielectric films	Coil wrap. Layer and interphase insulation. Wire insulation. Capacitor insulation. Magnetic tape. Pressure-sensitive tapes. Flexible circuit substrate. Slot insulation. Barrier insulation. Integrated circuits
Dielectric fabrics and papers	Wire and cable coil wraps. Slot insulation. Capacitor dielectrics. Electrical laminates
Tapes (adhesive and nonadhesive)	Wire and cable support. Busbar insulation; mechanical protection coil. Wire and cable wrap for phase and layer insulation. Slot insulation
Tubing and sleeving	Identification and binding of wires. Splices and termination insulation. Protective packaging for electronic parts. Outer jacket covering for wire harnesses. Mechanical protection
Molding compounds (thermoplastic, thermoset, and elastomers)	Electrical connectors, impregnants for coils, transformers, bushings, integrated circuits, coil forms, knobs, circuit breakers, insulator blocks, terminal bases, circuit boards, bobbin housings, switch plates, encapsulants, battery cases, distributor caps, sockets
Extrusion and pultrusion	Insulation covering, parts for control and switchgear devices, wire and cable insulation
Coatings and encapsulants	Integrated circuits, printed wiring boards, wire and cable, coil insulation, transformers, conformal coatings, laminates, mechanical and environmental protection, fiber-optic cables
Adhesives	All forms of bonding in electrical and electronic equipment
Cellular plastics	Coaxial-cable insulation; high-frequency cable insulation

Films and Sheets

Materials are classified as films and sheets if they are less than 0.015 in (film) or more than 0.015 in (sheet) (ASTM-D2305).[3] Films are prepared by processes called extrusion, casting, calendering, and skiving. The majority of films are thermoplastics, but virtually all polymers can be converted to films, although not all are useful in film form. Extruded films are basically produced from melting plastic powders or granules, forcing the melt through a die onto a continuously moving drum or belt, cooling the deposited film, and stripping it from the substrate. Casting is a solution-based process whereby the polymer is dissolved in a suitable solvent and the solution is cast or deposited on a moving belt,

TABLE 12.2 ASTM Electrical Insulation and Electronic Categories

Volumes 10.01 and 10.02	Electrical Insulation (I and II, D69 to D2484)
	■ Electrical insulating materials, composites, flexible sheet, tape, tubing, rods, molded materials, textiles
	■ Electrical heating unit insulation
	■ Electrical tests
	■ Flammability
	■ Hookup wire insulation
	■ Insulating papers
	■ Insulated wire and cable
	■ Magnet wire insulation
	■ Mica products
	■ Rubber tape
	■ Sampling and conditioning
	■ Solid filling, treating, encapsulating, and embedding compounds
	■ Varnishes and coatings for electrical insulation
	■ Terminology
Volume 10.03	Electrical Insulating Liquids and Gases; Electrical Protective Equipment
Volumes 10.04 and 10.05	Electronics (I and II)
	■ Microelectronic packaging

which is sent through ovens to remove the solvent. Calendering is a process whereby a compounded resin (exclusively PVC) is passed through a series of heated rolls which are successively placed more and more closely together. Skiving is a method of cutting thin films from a round stock of material. Films can be oriented, coated, and laminated to produce products with enhanced properties. Orientation can be monoaxial or biaxial with the latter being preferred. Orientation improves tensile strength, flexibility, and toughness. Coatings can be applied to films to act as adhesives in a subsequent bonding process, to improve the barrier properties of the film, to produce heat-sealable coatings, or to change some handling characteristic of the film. Coatings can be organic, inorganic, or metallic.

Films can also be laminated to each other or to a variety of other substrates such as metals, fabrics, and paper, to produce a composite material with properties far superior to those of the single-layer film. The most common use of polymer films is as a packaging material. They are used as tapes, laminates, and coated films for many applications in the electrical and electronics area, such as in capacitors, wire insulation, thermal insulation, and phase and layer insulation in transformers and rotating electrical equipment.

TABLE 12.3 Characteristics of Major Polymer Film Groups

Material	Characteristics
Acrylics	Weather and uv resistance
Cellulosics*: CA, CAB, CAP, CA_3	Good scuff, oil, and grease resistance. High gas permeability, medium MVTR,† medium clarity. CAB: good outdoor weathering; major uses: skin and blister packaging
Fluoroplastics	Excellent electrical, chemical, and thermal properties. Very low MVTR of any transparent film, LOX (liquid oxygen) compatible and flexible at cryogenic temperatures; uses: packaging of hygroscopic pharmaceuticals
Ionomers	Very tough, oil- and grease-resistant, heat-sealable through oil or grease
Polyamides	Good general-purpose, tough, strong material. Useful from −37 to +150°C
Polycarbonates	Strong, flexible material. Attacked by many chemicals. Must be dried thoroughly before processing or else bubbles result
Polyesters	Tensile strengths up to 40,000 lb/in² in tensilized (oriented) condition. Good high-temperature properties. Very good electrical properties. Metallized film used for thermal insulation on spacecraft
Polyethylenes	Tough, low-cost, versatile. Used for all types of packaging: trash bags, bread wraps, shrink film, antistatic film, etc.
Polyimides	Highest temperature resistance of all films. Excellent chemical and electrical properties. Heat-sealable using FEP coating
Polypropylenes	Strong, tough, low gas transmission, good chemical and grease resistance, best when oriented. Poor low-temperature properties
Polystyrenes	High gas permeability, good for produce packaging, not recommended for outdoor use. Many grades available. Good thermoforming capabilities for cups, blister packs, etc.
Polyurethanes	Very tough, good tear and abrasion resistance. Can be extruded and calendered onto fabric
Polyvinyl chlorides	Very versatile, strong, tough. Used for fresh meat and produce packaging
Polyvinylidene chlorides	Extremely low moisture and gas transmission rates. Good chemical resistance. Good shrink characteristics
Polyvinyl fluoride	Excellent weather and uv resistance, good chemical resistance
Polyvinylidene fluoride	Good uv, nuclear, weather, and chemical resistance; good electrical properties

*CA = cellulose acetate; CAB = cellulose acetate butyrate; CAP = cellulose acetate propionate; CA_3 = cellulose triacetate.

†MVTR = moisture vapor transmission rate, g/(100 in² · mil · 24 h) at 37°C, 90 to 100% RH.

SOURCE: Ref. 4.

A description of the major characteristics of polymer films is given in Table 12.3. The most useful films for electronics are polyester, polyimide, fluorocarbons, polysulfone, polycarbonate, and polyethylene. Their properties are given in Table 12.4.

Flexible Circuits

A flexible circuit is an arrangement of conductors on a flexible insulating base. Flexible circuits are used to make interconnections in systems that have limited space for the rigid-type printed-circuit board. Three polymeric components comprise the flexible circuit: the polymer film substrate material, the adhesive, and the coating. The metallic component is the copper cladding. Flexible circuits can be formed and twisted and bent without concern for fracturing either the metal conductor or the plastic substrate. The substrate material is selected from one of the flexible films listed in Table 12.4 but is typically polyester or polyimide.

In addition to the polymeric films listed in Table 12.4, some fibrous reinforcements can be used. These include Nomex paper and an epoxy-impregnated dacron fabric. Some of the properties of Nomex paper and polyester film were given in Chap. 3, and the properties of the substrate materials will not be reviewed here. The properties of the substrate are critical for optimum circuit performance. The important properties of a material for use as a substrate in flexible circuits include the tensile strength, coefficient of thermal expansion, elongation, tensile modulus, tear strength, flammability, use temperature, chemical resistance and dielectric strength, dissipation factor, dielectric constant, volume resistivity, thermal conductivity, and moisture absorption.

Adhesives are the next component of the flexible circuit, and they consist of epoxy, acrylic, polyester, polyimide, fluorocarbon, phenolic, and modifications thereof. The important properties of the adhesive are the compatibility with the film substrate, chemical resistance, electrical properties, adhesion, flexibility, and resistance to printed wiring board chemicals used in the manufacturing process. A comparison of the properties of various polymeric adhesives is given in Table 12.5.

The third nonmetallic component of the flexible circuit is the coating. Its function is to provide a protective covering for the metal circuit on the flexible substrate to protect it against moisture, contamination, damage, and stress. The coatings used typically are solventless resins such as acrylated epoxy or urethane and are cured by uv radiation. More detailed information on the materials and testing of flexible circuits is available from the Institute for Interconnecting and Packaging Electronic Circuits (IPC) (see Chap. 14).

TABLE 12.4 Properties of Polymer Films Used in the Electrical Industry

Properties	ASTM test method	Sulfone polymers	Ethylene-chlorotri-fluoro-ethylene copolymer (ECTFE)	Ethylene-tetra-fluoro-ethylene copolymer (ECTFE)	Ethylene-fluorinated ethylene propylene copolymer (FEP)	Per-fluoro-alkoxy (PFA)	Poly-chloro-trifluoro-ethylene copolymers (PCTFE)	Poly-tetra-fluoro-ethylene (PTFE)	Poly-vinyl fluoride (PVF)	Poly-vinyl-idene fluoride (PVDF)
Production method	—	Casting, extrusion	Extrusion	Extrusion	Extrusion	Extrusion	Extrusion	Skiving, casting	Extrusion	Extrusion
Forms available	—	Rolls, sheets	Rolls, sheets, tubes	Rolls, sheets, tubes	Rolls, sheets, tubes	Rolls, tubes	Sheets, tapes	Rolls, tubing	Rolls, sheets	—
Thickness range, in	—	0.0001–0.030	0.0005–0.090	—	0.0005–0.030	0.0005–0.090	0.00075–0.010	0.0005–0.030	0.0005–0.004	0.002–0.020
Maximum width, in	—	52	54	48	48	48	48	48	138	50
Area factor, in^2/(lb · mil)	—	20,000–22,355	16,300	12,900	12,900	13,000	12,800	17,200–20,000	16,000	—
Specific gravity	D1505	1.24–1.37	1.66–1.68	1.7	2.15	2.15	2.08–2.15	2.1–2.2	1.38–1.57	1.76
Tensile strength, lb/in^2	D882	8,400–12,000	8,000–10,000	7,000–8,000	2,500–3,000	4,000–7,000	5,000–10,000	1,500–4,000	7,000–18,000	5,500–6,500
Elongation, %	D882	20–150	150–250	300	300	200–600	50–150	100–400	115–250	300–500
Bursting strength, 1 mil	D774	60 lb/in^2	—	—	11	—	23–31	—	19–70	—
Tearing strength, g/mil propagating	D1922	7–16	900–1,300	600–900	125	40–70	2.5–40.0	10–100	12–100	50
Tearing strength, lb/in initial	D1004	236–950	—	—	600	650	330–900	—	997–1,400	—
Folding endurance	D2176	—	250,000	(1-mil)	—	—	—	—	5,000–47,000	>26,000
Water absorption 24 h, %	D570	0.3–2.1	<0.02	<0.02	<0.01	<0.03	Nil	Nil	<0.5	0.04
Rate of water vapor trans., g/(mil · 100 in^2 · 24 h) at 37.8°C	E96(E)	18	0.6	1.65	0.40	—	0.025–0.055	—	3.24	—
Permeability to gases, cm^3 · mil/ (100 in^2 · 24 h · atm) at 25°C CO_2	D1434	405–950	110	250	1,670	—	16–40	—	11.1	5.5
H_2		—	—	—	2,200	—	220–330	—	58	—
N_2		40	10	30	320	—	2.5	—	0.25	9
O_2		90–230	25	100	750	—	7–15	—	3	14

Property	ASTM								
Strong acids	D543	G	G	G	G	G	G	G	G
Strong alkalies	D543	G	G	G	G	G	G	G	G
Grease and oils	D722	G	G	G	G	G	G	G	G
Organic solvents	D543	P	G	G	G	G	G	G	G
Water	E96	G	G	G	G	G	G	G	G
High relative humidity	D756	G	G	G	G	G	G	G	G
Sunlight	D1435	P	—	G	G	G	E	G	G
Change in linear dimensions at 100°C for 30 min, %	D1204	30 min at 180°C 0.06%	—	<1	—	+2 – −2	+12 – −12	1	0.2
Dielectric constant, 1 kHz	D150	3.07–3.5	2.6	2.0–2.50	2.0–2.1	2.5–2.7	2.0–2.1	8.5	8.4
Dielectric constant, 1 MHz	D150	3.03–3.5	2.6	2.0–2.50	2.0–2.1	2.3–2.4	2.0–2.1	7.4	—
Dielectric constant, 1 GHz	D150	3.0–3.4	2.4	2.05	2.0–2.1	2.3	2.0–2.1	—	—
Dissipation factor, 1 kHz	D150	0.0008–0.0035	0.0008	<0.0002	0.022–0.024	<0.0001	1.6	0.019	—
Dissipation factor, 1 MHz	D150	0.0034–0.006	0.005	0.0003	0.0002	0.009–0.017	<0.0001	—	—
Dissipation factor, 1 GHz	D150	0.0041–0.005	0.0005	0.0015	0.00045	0.004	<0.0005	—	—
Dielectric strength, V/mil 1-mil	D149	5,800–7,500	5,000	6,500	4,000–5,000	3,000–3,900	2,200–4,400	3,500	—
5-mil		2,400 (4 mil)	—	2,500	2,000	—	2,700–3,300	1,000–2,000	1,700 (4-mil)
Volume resistivity, Ω · cm	D257	10^{16}–10^{18}	10^{16}	10^{19}	10^{18}	10^{18}	>10^{18}	10^{13}	10^{14}
Heat-sealing temp. range, °C	—	260–288	246–260	277–316	282–372	316–399	232–260	205–218	205–218

G = good, P = poor, E = excellent.

TABLE 12.4 Properties of Polymer Films Used in the Electrical Industry (Continued)

Properties	ASTM test method	Poly-carbonate	Polyester thermo-plastic polyethylene terephthalate (PET)	Low-density PE	Medium-density PE	High-density PE	Ultra-high-molecular weight PE	Ethylene-vinyl acetate copolymer	Polyimide
Production method	—	Casting, extrusion	Extrusion biaxially oriented	Extrusion	Extrusion	Extrusion	Skiving from molded billets	Extrusion	Polycondensation reaction
Forms available	—	Rolls (cast and extr.), sheets (extr.)	Rolls, sheets, tapes, tubes	Rolls, sheets, tapes, tubes	Rolls, sheets, tapes, tubes	Rolls, sheets, tubes	Rolls, tapes	Rolls, sheets, tubes	Rolls, tapes
Thickness range, in	—	0.00025–0.030	0.00008–0.014	≥0.0003	≥0.0003	≥0.0004	≥0.002	≥0.00075	
Maximum width, in	—	45 cast 54 extr.	67–120	480	240	60	24–60	480	60
Area factor, in²/(lb · mil)	—	23,100	19,800–22,600	30,000	29,500	29,000	29,600	29,000	19,500
Specific gravity	D1505	1.20	1.380–1.410	0.910–0.925	0.926–0.940	0.941–0.965	0.94	0.924–0.940	1.42
Tensile strength, lb/in²	D882	8,400–11,000	20,000–35,000	1,500–4,000	2,400–6,100	3,000–5,500	1,000–3,500	25,000	
Elongation, %	D882	40–105	60–165, 50 (tensilized)	100–700	50–650	10–650	300	400–800	70
Bursting strength, 1-mil thickness, Mullen points	D774 4 mil. 25–35	No break	55–80	10–12	—	—	—	10–12	75
Tearing strength, g/mil	D1922	20–25	50–300	50–300	50–300	15–300	—	50–300	8
Tearing strength, lb/in initial	D1004	700–1,600	1,000–3,000	65–575	—	—	—	80–500	1,120
Folding endurance	D2176	250–400	>100,000	—	—	—	—	—	10,000
Water absorption 24 h, %	D570	<0.8	0.25	<0.01	D570	Nil	Nil	<0.01	2.9
Rate of water vapor trans., g/(mil · 100 in² · 24 h) at 37.8°C	E96(E)	11.0	1.0–1.3	1.0–1.5	0.7	0.3	—	2.3	5.4
Permeability to gases, cm³ · mil/ (100 in² · 24 h · atm) at 25°C CO₂	D1434	1,075	15–25	2,700	1,000–2,500	580	—	6,000	45
H₂		1,600	100	1,950	1,950	—	—	—	250
N₂		50	0.7–1.0	180	85–315	42	—	400	6
O₂		300	3.0–6.0	500	250–535	185	—	840	25

280

Strong acids	D543	G	G	G	G	G	G	F	G
Strong alkalies	D543	P	P	G	G	F	G	F	P
Grease and oils	D722	G	G	P	F	G	G	P	G
Organic solvents	D543	G-P	G	F	F	G	G	F	G
Water	E96	G	G	G	G	G	G	G	G
High relative humidity	D756	G	G	G	G	G	G	G	G
Sunlight	D1435	F	F	F	F	F	F	F	G
Change in linear	D1204	Nil	<0.5	−2	0—−0.7	−0.7—−3.0	—	<1	30 min. at 250°C 0.3%
Dielectric constant, 1 kHz	D150	2.99	3.2	2.2	2.2	2.3	2.3	2.7–2.9	3.5
Dielectric constant, 1 MHz	D150	2.93	3.0	2.2	2.2	2.3	—	2.7–2.9	3.4
Dielectric constant, 1 GHz	D150	2.89	2.8	2.2	2.2	2.3	—	—	3.3
Dissipation factor, 1 kHz	D150	0.0015	0.005	0.0003	0.0003	0.0005	2.3×10^{-4}	0.01–0.02	0.0025
Dissipation factor, 1 MHz	D150	0.010	0.016	0.0003	0.0003	0.0005	—	0.01–0.02	0.010
Dissipation factor, 1 GHz	D150	0.012	0.003–0.008	0.0003	0.0003	0.0005	—	—	0.004
Dielectric strength, V/mil 1-mil	D149	6,300	7,500	5,000	5,000	5,000	1,300	5,000	7,000
Volume resistivity, Ω · cm 5-mil	D257	2,000	3,400	3,000	3,000	3,000	—	3,000	3,600
		10^{16}	10^{18}	10^{16}	10^{16}	10^{16}	10^{18}	10^{9}	10^{18}
Heat-sealing temp. range, °C	—	205–221	219–232	122–205	122–205	135–205	135–205	65–150	—

G = good, P = poor, E = excellent.

SOURCE: Ref. 1.

TABLE 12.5 Properties of Flexible Circuit Adhesives

Type	Temperature resistance	Chemical resistance	Electrical properties	Adhesion	Flexibility	Moisture absorption
Polyester	Fair	Good	Excellent	Excellent	Excellent	Fair
Acrylic	Very good	Good	Good	Excellent	Good	Poor
Modified epoxy	Good	Fair	Good–excellent	Excellent	Fair	Good
Polyimide	Excellent	Very good	Good	Very good	Fair	Poor
Fluorocarbon	Very good	Excellent	Good	Very good	Excellent	Excellent
Butyral phenolic	Good	Good	Good	Good	Good	Fair

SOURCE: Adapted from Ref. 5.

Coatings and Encapsulants

Polymeric materials used as coatings in the electrical industry have been described in Chap. 9, and they will not be repeated here except to mention those areas in the electrical industry where polymer coatings are used. Polymers are used as conformal coatings on circuit boards, as wire coatings, and as impregnants for fibers, fabrics, papers, encapsulants, varnishes, and adhesives.

Polymers that are or can be used as coatings and encapsulants include the polyimides, silicones, epoxies, urethanes, acrylics, silicones, paraxylylenes, fluorocarbons, benzocyclobutenes, and silicon-carbon resins.

Adhesives

The adhesives and sealants industry represented about $13 billion in 1992 sales to the electronics market, which grew by 9 percent to $235 million per year in 1992. In the die attachment adhesives which are used to hold semiconductor devices to chip carriers or other substrates, sales reached $40 million, which represents about 40 percent of the world market for electronics adhesives.[6] Epoxy resins play a significant role in die attachment adhesives with silicone-based adhesive products making some inroads. The latter adhesives are less rigid than the epoxies and have better heat resistance. Silicones are also ozone-resistant.

These two examples represent only a part of the uses of polymers as adhesives in the electrical and electronic industries. Polymers function in a dual capacity as adhesives: structural and electrical. As such, the adhesive must maintain the properties required to ensure both mechanical and electrical integrity of the product in which it is used. Polymeric adhesives provide a very useful assembly method and in many cases the only practical method of fastening components. Adhesives provide for better stress distribution than screws, rivets, or welding; and this stress distribution results in greater strength, rigidity, and fatigue resistance under vibration loads. In addition, adhesives do not break through, deform, or mar the surface of the assembly as mechanical fasteners do. Adhesives also help to provide a seal against the ingress of harmful liquids and provide electrical insulation as well as galvanic corrosion protection.

Some typical applications of adhesives in the electrical industry are the following:

1. Die attachment adhesives to join integrated circuits (ICs) and other chips in packages to provide heat conduction and mechanical stability
2. Surface-mounted technology where the encapsulated device is bonded directly to the circuit board
3. Joining of dissimilar materials in an electrical assembly
4. Joining thin materials such as films and foils
5. Joining of electrical laminates
6. Bonding of formed small wire coils and random packages of wire in rotating equipment
7. Bonding of steel laminations
8. Bonding metal cladding to printed wiring board substrates
9. As sealants

Polymers do indeed have a variety of applications as adhesives in the electrical and electronics industry.

Adhesive classes

Polymeric adhesives can be placed into four broad material classes: thermoplastic, thermosetting, elastomeric, and polymer alloys and blends. Thermoplastic adhesives have a useful temperature range up to about 80°C and are limited at the low-temperature end by their individual brittleness characteristics. They have poor creep resistance but good peel strength. Solvent resistance ranges from poor to excellent depending on the polymer. Thermosetting adhesives can operate at temperatures of 300°C, are more rigid than thermoplastics, and generally offer better chemical resistance. They have good creep resistance but only fair peel strength. The elastomeric adhesives can operate over a broad temperature range up to about 260°C. They have high peel strength, low overall strength, and high flexibility. Chemical resistance is variable depending on the elastomer. Alloys and blends can have the properties of all the other classes of materials but with a more balanced combination. Tables 12.6 to 12.9 detail the various types and properties of the four adhesive classes. Polymeric adhesives come in a variety of forms, and they are listed in Table 12.10.

Adhesive selection

No single factor controls the choice of an adhesive, but the following important considerations must be addressed to ensure a chance of success: (1) the joint type and materials to be bonded, i.e., adhesive-adherend combinations and their compatibility; (2) the service requirement, i.e., the effects of environmental conditions (temperature range, chemical exposure, radiation); (3) strength, i.e., the kind of stress involved (shear, tension, or peel) and the rate at which the stress is applied (various types of adhesives respond differently under impact, constant or cyclical loads, and vibration); (4) production requirements, i.e., application and bonding methods, adhesive state (tacky, dry, liquid, solid, drape, etc.), and cost, although it should be considered after other considerations are satisfied.

Microelectronic adhesive types

Not all the adhesives listed in Tables 12.6 to 12.9 are useful for all applications. There are some adhesives that have had a dominant influence in the electronic industry, and a brief review follows. These adhesives include phenolics, polyesters, modified epoxy, acrylic, urethanes, polyamides, silicones, polyimides, cyanoacrylates, and fluorocarbons.

Phenolics. Phenolic adhesives are used mostly for structural applications and offer very high bond strength. They are used in single-side laminates only because of the evolved volatiles during cure. Phenolic adhesives can be made in a range of flexibilities and are used also in tape bonding of electrical components.

Polyimides. Polyimides have one of the highest temperature capabilities of the high-temperature plastics (up to 370°C). They also require high cure temperatures and difficult process conditions, although some polyimides are much easier to process than others.

Polyurethanes. Polyurethanes offer easy rework but have low thermal stability (≤120°C maximum temperature limit). Polyurethanes have high outgassing.

(Text continues on p. 294.)

TABLE 12.6 Thermoplastic Adhesives

Adhesive	Description	Cure	Characteristics	Adherends
Cellulose acetate, cellulose acetate butyrate	Solvent solutions	Solvent evaporation	Water-clear, more heat-resistant but less water-resistant than cellulose acetate butyrate; has better heat and water resistance than cellulose acetate and is compatible with a wider range of plasticizers	Plastics, leather, paper, wood, glass, fabrics
Cellulose nitrate	Solvent solutions	Evaporation of solvent	Tough, develops strength rapidly, water-resistant; bonds to many surfaces; discolors in sunlight; dried adhesive is flammable	Glass, metal, cloth, plastics
Polyvinyl acetate	Solvent solutions and water emulsions, plasticized or unplasticized, often containing fillers and pigments. Also dried film which is light-stable, water-white, transparent	On evaporation of solvent or water; film by heat and pressure	Bond strength of several thousand pounds per square inch but not under continuous loading. The most versatile in terms of formulations and uses. Tasteless, odorless; good resistance to oil, grease, acid; fair water resistance	Emulsions particularly useful with porous materials like wood and paper. Solutions used with plastic films, mica, glass, metal, ceramics
Vinyl vinylidene	Solutions in solvents like methyl ethyl ketone	Evaporation of solvent	Tough, strong, transparent, and colorless. Resistant to hydrocarbon solvents	Particularly useful with textiles; also porous materials, plastics
Polyvinyl acetals	Solvent solutions, film, and solids	Evaporation of solvent; film and solid by heat and pressure	Flexible bond; modified with phenolics for structural use; good resistance to chemicals and oils; includes polyvinyl formal and polyvinyl butyral types	Metals, mica, glass, rubber, wood, paper

TABLE 12.6 Thermoplastic Adhesives (*Continued*)

Adhesive	Description	Cure	Characteristics	Adherends
Polyvinyl alcohol	Water solutions, often extended with starch or clay	Evaporation of water	Odorless, tasteless, and fungus-resistant (if desired). Excellent resistance to grease and oils; water-soluble	Porous materials such as fiberboard, paper, cloth
Polyamide	Solid hot-melt, film, solvent solutions	Heat and pressure	Good film flexibility; resistant to oil and water; used for heat-sealing compounds	Metals, paper, plastic films
Acrylic	Solvent solutions, emulsions, and mixtures requiring added catalysts	Evaporation of solvent; room temperature or elevated temp. (two-part)	Good low-temperature bonds; poor heat resistance; excellent resistance to ultraviolet; clear; colorless	Glass, metals, paper, textiles, metallic foils, plastics
Phenoxy	Solvent solutions, film, solid hot-melt	Heat and pressure	Retains high strength from 4 to 28°C; resists creep up to 82°C; suitable for structural use	Metals, wood, paper, plastic film

SOURCE: Ref. 7.

TABLE 12.7 Thermosetting Adhesives

Adhesive	Description	Cure	Characteristics	Adherends
Cyanoacrylate	One-part liquid	Rapidly at room temperature (RT) in absence of air	Fast-setting; good bond strength; low viscosity; high cost; poor heat and shock resistance; will not bond to acidic surfaces	Metals, plastics, glass
Polyester	Two-part liquid or paste	RT or higher	Resistant to chemicals, moisture, heat, weathering. Good electrical properties; wide range of strengths; some resins do not fully cure in presence of air; isocyanate-cured system bonds well to many plastic films	Metals, foils, plastics, plastic laminates, glass
Urea formaldehyde	Usually supplied as two-part resin and hardening agent. Extenders and fillers used	Under pressure	Not as durable as others but suitable for fair range of service conditions. Generally low cost and ease of application and cure. Pot life limited to 1 to 24 h	Plywood
Melamine formaldehyde	Powder to be mixed with hardening agent	Heat and pressure	Equivalent durability and water resistance (including boiling water) to phenolics and resorcinols. Often combined with ureas to lower cost. Higher service temperature than ureas	Plywood, other wood products
Resorcinol and phenol-resorcinol formaldehyde	Usually alcohol-water solutions to which formaldehyde must be added	RT or higher with moderate pressure	Suitable for exterior use; unaffected by boiling water, mold, fungus, grease, oil, most solvents. Bond strength equals or betters strength of wood; do not bond directly to metal	Wood, plastics, paper, textiles, fiberboard, plywood

TABLE 12.7 Thermosetting Adhesives (Continued)

Adhesive	Description	Cure	Characteristics	Adherends
Epoxy	Two-part liquid or paste; one-part liquid, paste, or solid; solutions	RT or higher	Most versatile adhesive available; excellent tensile shear strength; excellent resistance to moisture and solvents; low cure shrinkage, variety of curing agents and hardeners results in many variations	Metals, plastic, glass, rubber, wood, ceramics
Polyimide	Supported film, solvent solution	High temperature	Excellent thermal and oxidation resistance; suitable for continuous use at 288°C and short-term use to 480°C; expensive	Metals, metal foil, honeycomb core
Polybenzimidazole	Supported film	Long, high-temperature cure	Good strength at high temperatures; suitable for continuous use at 230°C and short-term use at 530°C; volatiles released during cure; deteriorates at high temperatures on exposure to air; expensive	Metals, metal foil, honeycomb
Acrylic	Two-part liquid or paste	RT	Excellent bond to many plastics, good weather resistance, fast cure; catalyst can be used as a substrate primer; poor peel and impact strength	Metals, many plastics, wood
Acrylate acid diester	One-part liquid or paste	RT or higher in absence of air	Chemically blocked, anaerobic type; excellent wetting ability; useful temperature range −54–150°C; withstands rapid thermal cycling; high-tensile-strength grade requires cure at 122°C, cures in minutes at 138°C	Metals, plastics, glass, wood

SOURCE: Ref. 7.

TABLE 12.8 Elastomeric Adhesives

Adhesive	Description	Curing method	Characteristics	Usual adherends
Natural rubber	Solvent solutions, latexes, and vulcanizing type	Solvent evaporation, vulcanizing type by heat or room temperature (RT) (two-part)	Excellent tack, good strength. Shear strength 30–180 lb/in^2; peel strength 0.56 lb/in width. Surface can be tack-free to touch and yet bond to similarly coated surface	Natural rubber, masonite, wood, felt, fabric, paper, metal
Reclaimed rubber	Solvent solutions, some water dispersions. Most are black, some gray and red	Evaporation of solvent	Low cost, widely used. Peel strength higher than that of natural rubber; failure occurs under relatively low constant loads	Rubber, sponge, rubber, fabric, leather, wood, metal, painted metal, building
Butyl	Solvent system, latex	Solvent evaporation, chemical cross-linking with curing agents and heat	Low permeability to gases, good resistance to water and chemicals, poor resistance to oils, low strength	Rubber, metals
Polyisobutylene	Solvent solution	Evaporation of solvent	Sticky, low-strength bonds; copolymers can be cured to improve adhesion, environmental paper resistance, and elasticity; good aging; poor thermal resistance; attacked by solvents	Plastic film, rubber, metal, metal foil, paper
Nitrile	Latexes and solvent solutions compounded with resins, metallic oxides, fillers, etc.	Evaporation of solvent and/or pressure	Most versatile rubber adhesive. Superior resistance to oil and hydrocarbon solvents. Inferior in tack range; but most dry tack-free an advantage in pre-coated assemblies. Shear strength of 150–2,000 lb/in^2 higher than neoprene, if cured.	Rubber (particularly nitrile), metal, vinyl plastics

TABLE 12.8 Elastomeric Adhesives (Continued)

Adhesive	Description	Curing method	Characteristics	Usual adherends
Styrene butadiene	Solvent solutions and latexes. Because tack is low, rubber resin is compounded with tackifier and plasticizing oils	Evaporation of solvent	Usually better aging properties than natural or reclaimed. Low dead-load strength; bond strength similar to reclaimed. Useful temperature range of −40–72°C	Fabrics, foils, plastics, film laminates, rubber and sponge rubber, wood
Polyurethane	Two-part liquid or paste	RT or higher	Excellent tensile-shear strength from −240 to +94°C; poor resistance to moisture before and after cure; good adhesion to plastics	Plastics, metals, rubber
Polysulfide	Two-part liquid or paste	RT or higher	Resistant to wide range of solvents, oils, and greases; good gas impermeability; resistant to weather, sunlight, ozone; retains flexibility over wide temperature range; not suitable for permanent load-bearing applications	Metals, wood, plastics
Silicone	Solvent solution; heat or RT curing pressure-sensitive; and RT vulcanizing solventless pastes	Solvent evaporation, RT or elevated temperature	Of primary interest is pressure-sensitive type used for tape. High strengths for other forms are reported at −73–260°C; limited service to 370°C. Excellent dielectric properties	Metals; glass; paper; plastics; and rubber, including silicone and butyl rubber and fluorocarbons
Neoprene	Latexes and solvent solutions, often compounded with resins, metallic oxides, fillers, etc.	Evaporation of solvent	Superior to other rubber adhesives in most respects—quickness; strength; max. temperature (to 94°C, sometimes 175°C); resistant to light, weathering, mold, acids, and oils	Metals, leather, fabric, plastics, rubber (particularly neoprene), wood, building materials

SOURCE: Ref. 7.

TABLE 12.9 Alloy and Blend Adhesives

Adhesive	Description	Curing method	Characteristics	Usual adherends
Epoxy-phenolic	Two-part paste, supported film	Heat and pressure	Good properties at moderate cures; volatiles released during cure; retains 50% of bond strength at 260°C; limited shelf life; low peel strength and shock resistance	Metals, honeycomb core, plastic laminates, ceramics
Epoxy-polysulfide	Two-part liquid or paste	RT or higher	Useful temperature range −55–94°C, greater resistance to impact, higher elongation, and less brittleness than epoxys	Metals, plastic, wood, concrete
Epoxy-nylon	Solvent solutions, supported and unsupported film	Heat and pressure	Excellent tensile-shear strength at cryogenic temperature range −250–82°C, limited shelf life	Metals, honeycomb core, plastics
Nitrile-phenolic	Solvent solutions, unsupported and supported film	Heat and pressure	Excellent shear strength; good peel strength; superior to vinyl and neoprene phenolics; good adhesion	Metals, plastics, glass, rubber
Neoprene-phenolic	Solvent solutions, supported and unsupported film	Heat and pressure	Good bonds to a variety of substrates: useful temperature range −55–94°C, excellent fatigue and impact strength	Metals, glass, plastics
Vinyl-phenolic	Solvent solutions and emulsions, tape, liquid, and coreacting powder	Heat and pressure	Good shear and peel strength; good heat resistance to weathering, humidity, oil, water, and solvents; vinyl formal and vinyl butyral forms available, vinyl formal-phenolic is strongest	Metals, paper, honeycomb core

SOURCE: Ref. 7.

TABLE 12.10 Adhesive Forms

Adhesive	Description	Advantages	Limitations	Processing considerations
Liquids: water base, anaerobics, solvent	Solvent- and water-based liquid adhesives, available in a wide number of bases (e.g., polyester, vinyl) in one- or two-part form fill bonding needs ranging from high-speed lamination to one-of-a-kind joining of dissimilar plastic parts. Solvents provide more bite, but cost much more than similar base water-type adhesive	Easy to apply; adhesives available to fit most applications	Shelf and pot life often limited. Solvents may cause pollution problems; water base not as strong	Application techniques range from simply brushing on to spraying and roller-coating lamination for very high production
	Anaerobics are a group of adhesives that cure in the absence of air, with a minimum amount of pressure required to effect the initial bond			Anaerobics are generally applied a drop at a time from a special bottle or dispenser
Mastics	Highly viscous single- or two-component materials which cure to a very hard or flexible joint depending on adhesive type. Includes epoxies, urethanes, rubber base, silicones, etc.	Does not run when applied	Shelf and pot life often limited. Can be used for structural bonds	Often applied via a trowel, knife of gun-type dispenser; one-component systems can be applied directly from a tube. Various types of roller coaters are also used

Type	Description	Advantages	Disadvantages	Application
Hot melts	100% solid adhesives that can flow when heat is applied. Often used to bond continuous flat surfaces. Made from polyethylene, saturated polyester, or polyamides, as granules, pellets, rope, or slugs	Fast application; clean operation	Virtually no structural hot melts for plastics	Hot melts are applied at high speeds via heating the adhesive, then extruding (actually squirting) it onto a substrate, roller coating, using a special dispenser or roll to apply dots, or simply dipping
Film	Available in several forms including hot melts, these are sheets of solid adhesive. Mostly used to bond film or sheet to a substrate. Includes epoxies, phenolics, polyamides, and elastomer base	Clean, efficient	High cost. Can be used for precision structural bonds	Film adhesive is reactivated by a heat source
Pressure-sensitive	Tacky adhesives used in a variety of commercial applications (e.g., cellophane). Often used with polyolefins. Usually rubber-base materials	Flexible	Bonds not very strong. Not used for structural bonds	Generally applied by spray with bonding effected by light pressure
Reinforcement	Impregnated fabric, paper. Used with a wide range of resins. Epoxies, imides, cyanate esters, silicones, bismaleimides, phenolic	Range of flexibilities (flexible to rigid). Strong bonds. Reinforcement	Process equipment is large	Applied in a prepregging operation

SOURCE: Ref. 4.

Polyamides. Like the polyurethanes, the polyamides are easy to rework. They also have high moisture absorption, high outgassing, and large variation in electrical properties because of their sensitivity to moisture.

Silicones. The silicone adhesives have high-temperature capability (≤250°C), are easy to rework, and have low outgassing and high purity. They also have low bond strength and a high coefficient of thermal expansion (CTE).

Epoxies. The epoxy and modified-epoxy resin adhesives are probably the most versatile. They offer a balanced combination of properties through modification. In general, the epoxy resins have good thermal properties, excellent adhesion, good electrical properties, good chemical resistance, and excellent process flexibility. They also can be filled to high loadings (60 percent). Depending on the curing agent and degree of cure, outgassing and leaching of corrosive chemicals could be a problem.

Acrylics. Acrylic adhesives are thermoplastic resins that can be hand- or wave-soldered. They have good electrical properties, chemical resistance, and flexibility, but strong alkalies will cause the adhesive to swell.

Fluorocarbons. The fluorocarbon adhesives are also thermoplastic and offer higher-temperature capability with a balanced combination of electrical, chemical resistance, thermal, and flexibility properties. They do have poor dimensional stability at high temperature. They are also self-extinguishing and have very low moisture absorption.

Adhesive bonding of plastics

In some instances it may be required that plastic be bonded to plastic, either the same material or another type of plastic. This section provides guidelines for such joining of materials.[8]

Acetal. Hexafluoroacetone sesquihydrate can be used for solvent bonding. Cyanoacrylate, polyester-urethane, and epoxy resins are other adhesives that can be used.

Acrylic. A solvent cement of methylene chloride plus methyl methacrylate monomer with a peroxide catalyst is a good bonding agent. For solvent bonding the following can be used: 1,1,2-trichloro-ethane, ethylene dichloride, and methylene dichloride.

ABS. For solvent bonding, methylethyl ketone, tetrahydrofuran, and methyl isobutyl ketone can be used. Other adhesives include epoxy, polyester, urethane, and hot melts.

Nylons. Aqueous solutions of phenol and formic acid as well as resorcinol and calcium chloride in alcohol can be used to bond nylon to itself.

Polycarbonate. Mixtures of methylene and ethylene dichloride can be used. Other adhesives that can be used include epoxy polyamides, silicon, and urethanes.

Polyolefins. Polyolefins cannot be solvent-cemented. The surface of the polyolefins must be pretreated by chemical, electrical, or thermal methods, but even then the bonds are not that strong but are acceptable. Epoxies have been used. Heat welding provides satisfactory bonds.

Polyphenylene oxide. Ethylene dichloride, toluene, and chloroform can be used for solvent cementing. Other adhesives include neoprene, nitrile, and urethane epoxy.

Polystyrene. For solvent cementing of polystyrene, the following solvents can be used. They are listed in order of decreasing evaporation rate because rapid evaporation of solvents may cause crazing and cracking of the joint. The solvents are methylene chloride, ethylacetate, methylethyl ketone, ethylene dichloride, trichloroethylene, toluene, and xylene. Adhesives that can be used are polyurethane, cyanoacrylate, and unsaturated polyester resins.

Polysulfone. A 5 percent solution of polysulfone resin in methylene chloride can be used to bond polysulfone to itself. For bonding to other substrates, epoxy resins are suitable.

Polyurethanes. Urethane-based adhesives provide an excellent choice for bonding polyurethane plastics. Resorcinol formaldehyde adhesives can be used, but they are more brittle. Nitrile adhesives are excellent materials for bonding polyurethanes.

Vinyls. The plasticizer in vinyls can migrate to the adhesive interface and degrade the adhesive. Consideration must be given to the compatibility between the adhesive and the plasticizer. Nitrile-based adhesives are excellent. Neoprene-based adhesives are suitable for cloth-backed vinyls. Cyanoacrylate adhesive gives good bonds to rigid vinyl and metals. Tetrahydrofuran and cyclohexanone can be used.

Thermosetting plastics. Most thermosetting plastics are not difficult to join together either to themselves or to other metals or thermosets. Solvent bonding cannot be used because thermoset plastics are not soluble. When a thermoset material is bonded to itself or to other substrates, an adhesive composition of the same thermoset that is to be bonded should almost always be tried first. If satisfactory bonds are not obtained, then the following adhesives can be used with all thermosets: Nitrile-based adhesives, phenolic-butyral, phenolic, polyester, epoxy, phenolic-neoprene, and cyanoacrylate adhesive compositions.

Fluoroplastics. To obtain reasonably good bonds to the fluoro polymers, their surfaces must be specially treated. These treatments include etching with a sodium naphthalene solution, electric discharge, or fusion.

Summary

The concept of a universal adhesive which will bond anything to anything is not so far-fetched. There are a multitude of materials—plastics and nonplastics—that can indeed be joined with the same adhesive system. However, the correct selection and optimum use of an adhesive still depend very much on the nature of the adherends and their functional requirements. Basically, one still has to match the adhesive with the application. References 7 and 8 provide an excellent resource for more detailed information on adhesives.

Tapes

Choosing the right tape for a particular electrical application can make a big difference in how the product performs and in its final design. In the electrical and electronic industry, plastic tapes are used in wire insulation, connector insulation, shielding, cable bundling, motor wrap coils, and wrapping and tabbing of lead wires. Tapes also provide abrasion and environmental protection, identification, reinforcement, and spacing and barrier protection. Electrical tapes are manufactured from a variety of backing and adhesive materials. The adhesive serves to anchor the tape to the substrate, and the backing or tape material provides the desired electrical function. The properties of both the adhesive and the backing material are important for the successful performance of the tape. Tapes are rated for a number of conditions, i.e., service temperature, conductor gauge, environmental conditions, operating voltage, moisture protection, fungus protection, etc. There are a number tapes to choose from that contain different backings and adhesives. The backing materials are paper, cloth, mat, plastic, and rubber; and the adhesive can be acrylic, silicone, elastomeric, and hot melt. They can be thermoplastic or thermoset.

Paper-backed tape

Paper tapes are economical and are used where the physical and electrical requirements are not excessive. They provide moderate service capabilities in low-humidity environments. Disadvantages are susceptibility to humidity, which in turn reduces electric strength, and degradation of the paper in high humidity under electric stress, forming acidic products which can be corrosive to metals. Types of paper tapes are impregnated creped kraft, micro-creped paper, creped cellulose fiber, impregnated flat-back rope, and aramid paper tapes. The decompositon of the paper under electrical stress and humidity is a compelling reason why cotton and paper tapes should be used with caution on fine-wire applications.

Woven-back tapes

Woven tapes include the polymer impregnated into cotton, cloth, glass, polyester, and aramid fabrics. Polymers include epoxy, silicone, polytetrafluoroethylene, polyethylene, polyimide, vinyl, and polyester. These tapes are used primarily for their excellent mechanical strength in addition to their electrical and thermal properties.

Vinyl-back tapes

These tapes are economical and abrasion-resistant and have excellent resistance to acids, oils, moisture, sunlight, and alkalies. These tapes have excellent conformability and are ideal for use in wrapping applications.

Acetate-back tapes

Acetate tapes are chemically modified cellulosics containing a plasticizer. Films and cloth are less hygroscopic than cellulose and yield tape products with good electrical properties for less severe environments.

Polyester-back tapes

Polyester tapes are usually specified for applications which require thinners, durability, and higher dielectric strength than the cellulosic tapes. Polyester tapes have excellent resistance to chemicals and moisture, are very tough, and conform well to a variety of shapes.

Fluoroplastic-back tapes

The fluoropolymers provide excellent high-temperature resistance (150 to 260°C) and chemical resistance. In addition, these tapes have excellent moisture resistance and low friction properties and do not support combustion.

Polyimide-back tapes

Polyimide-back tapes are the most thermally stable of all the polymer tapes available. They operate in environments where extreme temperature variations are encountered. The electrical and mechanical properties of polyimide tapes are essentially unchanged at high temperatures.

Silicone-back tapes

These tapes are very soft and conformable and are used as a high-temperature splicing tape. These tapes have an excellent balance of electrical, mechanical, and thermal properties.

Metallic-back tapes

Tapes with copper and aluminum foil backs are used in a number of electrostatic shielding operations in transformers, cables, and electronic apparatus such as in wound coils and toroids.

Adhesive systems

In addition to the tape backing material there is the anchoring system, and that is the adhesive. Adhesives are derived from such polymers as acrylics, silicones, and natural and synthetic rubber compositions and can be either thermosetting or nonthermosetting. The former adhesives provide greater adhesive strength, chemical resistance, and heat resistance than the latter.

Selection

When you are choosing a tape for a given application (high-performance or not), list the characteristics you want the tape to have, i.e., mechanical, electrical, radiation, and chemical resistance properties plus frictional and temperature requirements. Then look at thickness, length, width, and the permanence of adhesive, and finally match the available tape system with the application requirements. ASTM defines a number of specifications for tape products:

TABLE 12.11 Typical Properties of Plastic Tapes

Polymer	Adhesive	Polymer thickness, mils	Break strength, lb/in	Elongation, %	Adhesion, oz/in	Dielectric strength, V	Service temp., °C
PTFE	Acrylic	3	12	275	40	10.2	155
PTFE	Silicone	3	12	275	30	10.2	180
FEP	Silicone	2	6	275	20	8.4	180
Polyimide	Acrylic	2	50	50	32	10	155
Polyimide	Silicone	2	50	50	25	10	180
UHMW-PE	Acrylic	5	45	200	90	—	—
PTFE/glass cloth	Silicone	3	50	5	30	4	180
Silicone, rubber, glass cloth	Silicone	11	150	<5	20	—	180
Polyester	Acrylic	1	25	100	45	5.6	130
Polyester	Silicone	1	25	100	20	6	130
PVDF	Acrylic	2	35	300	35	3.1	150

ASTM D 119	Low-voltage rubber tape
ASTM D 1373	Medium-voltage rubber tape
ASTM D 3391	High-voltage rubber tape
ASTM D 1458	Tests for silicone and glass tape
ASTM D 1931	Silicone and glass tape
ASTM D 3390	Semiconducting rubber tape
ASTM D 4325	Tests for rubber splicing and repair tapes
ASTM D 4388	Tests for rubber splicing and repair tapes, high-voltage
MIL I 46852	Pressure-sensitive silicone tape
HH I 553	Rubber insulating tape

Some typical properties of tape products are given in Table 12.11. Some tape suppliers are Norton Performance Plastics, 3M Co., CHR Co., Permacel Co., Dewal Inc., Taconic Co., Chemfab Co., and Oak.

Sleeving and Tubing

Plastic sleeving and tubing are used for a variety of purposes in the electrical and electronic industries: shaft insulators, separators, coil forms, supports, identification and binding of wires together, protective outer jacket, and the insulation of lead wires. Plastic sleeving and tubing can be rigid and/or flexible, and many plastics and elastomers can be converted to these forms. The products include rigid insulating

TABLE 12.12 Generic Tubing and Sleeving Materials

Material	Useful voltage range, V	Maximum useful temp. range, °C
Vinyl	300–600	−55–+105
Polyolefin	300–600	−75–+135
Neoprene	150–600	−70–+120
PTFE	150–600	−55–+260
FEP	300–600	−55–+200
PVDF	600	−55–+175
Silicone	400–660	−75–+200
PFA	600	−55–+260
PCTFE	620	−55–+150

tubes, flexible tubes of plastic, paper, impregnated paper, composite sleeving, heat-shrinkable tubing, coated sleeves, and braided sleeving. In the United States, *tubing* generally refers to extruded plastics and elastomers that do not contain any fabric reinforcement, and *sleeving* refers to a reinforced tubular product. The reinforcement is selected from braided yarns of cotton, rayon, fiberglass, nylon, aramid, quartz, ceramics, etc. Some typical tubing and sleeving products are generically identified in Table 12.12. All these products are available in a number for forms, i.e., heat-shrinkable or non-heat-shrinkable, cross-linked or non-cross-linked.

Types

Rigid tubing is made from almost any type of plastic and most fabrication methods, such as molding, extrusion, filament winding, and machining. In addition to the fabrication of tubing from the neat plastic, impregnated materials lend themselves to be converted to tubular forms such as epoxy-impregnated paper or fabric.

Composite tubing and sleeving can be made from sheet materials by bonding longitudinally with adhesives. Materials include Nomex (a high-temperature aramid paper from the du Pont Co.), kraft paper, polyimide and polyester film, and combinations of these films with various papers.

Heat-shrinkable tubing products are available that utilize the following polymers: polyolefin, polyvinyl chloride, fluorocarbon polymers, neoprene, butyl, silicone rubber, ethylene propylene rubber, and polyester. Heat-shrinkable products, as the name implies, shrink in diameter when exposed to a heat source. After shrinking the part is encapsulated by the shrinkable tubing.

Shrinkable products with a broad range of materials, shrink ratios (as high as 10 to 1), and shrink temperatures (70 to 300°C) are available. There are two broad classes of heat-shrinkable polymers, based on their shrink mechanism. One mechanism is based on the stress relieving of frozen-in strains during processing; the second mechanism is based on the elastic memory of cross-linked polymers which behave as thermoplastics at room temperature and as elastomers at elevated temperatures. Because of the cross-linkages these polymers cannot melt, and precise control of shrink dimensions can be obtained from this mechanism. Heat-shrinkable products are available for operating temperatures from 80 to 260°C. Several products are available that are shrinkable without the use of heat: polyvinyl chloride, polyurethane, chlorosulfonated polyethylene, silicone, and neoprene elastomers. These products are covered in MIL Specification I-85080.[9]

Coated sleeving products are another form of insulation, and they are prepared by impregnating or coating various types of reinforcement with a polymer. Typical coating resins are the oleoresin varnishes, acrylics, epoxy, urethane, polyester, silicone, imide, fluorocarbon, and vinyl. The main reason for the use of reinforcement in tubing and sleeving products is to provide support to those coating materials that would be impossible to use in extruded form because of insufficient physical strength. The American Society for Testing and Materials classifies seven types of flexible treated sleeving products for use on lead wires and in connections in motors, transformers, and switchgear equipment.[10]

1. A flexible, treated, organic-based sleeving material such as cotton, rayon, or nylon which is impregnated and coated with an insulating resin and has an operating service temperature of 105°C.

2. A flexible, treated, inorganic-based sleeving material such as glass fibers impregnated and coated with an insulating resin and has an operating service temperature of 130°C.

3. A flexible, treated, inorganic-based sleeving such as fiberglass which is impregnated and coated with an insulating material (polyvinyl chloride) and has an operating service temperature of 105°C.

4. A flexible, treated, inorganic-based sleeving such as fiberglass which is impregnated and coated with an insulating resin such as silicone or a fluorocarbon resin and has an operating service temperature of 200°C.

5. A flexible, treated, inorganic-based sleeving such as fiberglass which is impregnated and coated with a silicone elastomer and has an operating service temperature of 200°C.

6. A flexible, treated, inorganic-based sleeving such as fiberglass which is impregnated and coated with epoxy, polyester, polyurethane, or acrylic and has an operating service temperature of 155°C.

7. A flexible, treated sleeving made from an inorganic-based material such as fiberglass which is impregnated and coated with an insulating resin to provide an operating service temperature of 180°C.

In addition, high-temperature organic-based reinforcement (aramid braided sleevings) can also be impregnated and coated with silicone, fluorocarbons, and imide resins to provide a product capable of operating at 200°C.

Specifications and test methods

The American Society for Testing and Materials (ASTM), federal specifications (MIL), and the National Electrical Manufacturers Association (NEMA) all have specifications and test methods for tubing and sleeving material. The ASTM specifications are listed in Table 12.13. Additional information on federal specifications and NEMA standards can be found in Ref. 9.

TABLE 12.13 Specifications and Test Methods for Sleeving and Tubing

ASTM D350	Standard Test Methods for Flexible Treated Sleeving Used for Electrical Insulation
ASTM D372	Standard Specification for Flexible Treated Sleeving Used for Electrical Insulation
ASTM D2902	Standard Specification for Fluoropolymer Resin Heat-Shrinkable Tubing
ASTM D2671	Standard Specification for Heat-Shrinkable Tubing for Electrical Use
ASTM D2903	Standard Specification for Neoprene Heat-Shrinkable Tubing
ASTM D876	Standard Test Method for Nonrigid PVC Tubing Used for Electrical Insulation
ASTM D3149	Standard Specification for Cross-linked Polyolefin Heat-Shrinkable Tubing
ASTM D3150	Standard for Cross-linked and Non-cross-linked PVC Heat-Shrinkable Tubing
ASTM D3144	Standard Specification for PVDF Heat-Shrinkable Tubing
ASTM D348	Standard Test Methods for Rigid Tubes Used for Electrical Insulation
ASTM D619	Standard Test Methods for Vulcanized Fiber Used for Electrical Insulation
ASTM D710	Standard Specification for Vulcanized Fiber Sheets, Rods, and Tubes Used for Electrical Insulation

SOURCE: Ref. 11.

Manufacturers

The following are a few companies which supply tubing and sleeving products: Bentley-Harris, Lionville, Pennsylvania; Brand-Fex, Willimantic, Connecticut; Markel, Norristown, Pennsylvania; Varflex, Rome, New York; Highlander, Chicago, Illinois.

Reinforced Electrical Insulation Materials

The products discussed in this section are limited to forms used for electrical applications even though the reinforced material may also be used in structural applications where electrical properties are not a requirement. The reinforced materials described here consist of a thermoplastic or thermosetting polymer combined with a reinforcing member such as fibers, fabric, mat, or paper. This combination produces a variety of forms for use in the electrical industry. These forms are identified as laminates (clad and unclad), sheet molding compounds, bulk molding compounds, and fabric tapes.

Reinforcements

Although glass fiber is the most common reinforcement material, many other types are used. Table 12.14 lists the reinforcements and some of their properties. Fibrous glass is used as woven fabrics, nonwoven mat, chopped and milled fibers, and unidirectional yarns and rovings. The fabrics are woven from inorganic or organic fibers into a variety of patterns. The one most commonly used in the electrical industry is called

TABLE 12.14 Approximate Properties of Fiber and Paper Reinforcements Used in Plastic Laminates

Reinforcement	Dielectric constant	Dissipation factor	Specific gravity	CTE, ppm/°C	Strength, $\times 10^3$ lb/in^2	Modulus, $\times 10^6$ lb/in^2
E-glass	6.3	0.0037	2.54	5.0	500	10.5
S-glass	6.0	0.002	2.49	2.9	675	12.5
D-glass	4.6	0.0015	2.16	3.0	350	7.5
Quartz	3.7	0.0002	2.2	0.55	130	10
Teflon	2.0	0.0002	2.2	—	50	1.3
Kevlar	3.7	0.002	1.44	−5.0	525	18.5
Polybenzoxazole	2.4	0.002	1.6	−8	575	40
Aramid paper	1.6–3.7	0.004	0.72	−5−−7.5	25*	—
Kraft paper	2.1	0.13	0.95	—	5.0	—
PBT	—	—	—	−10	—	55

*lb/in of width.

plain weave. Nonwoven mats are either composed of chopped strands laid down in a random manner or made of swirled continuous yarns. The fibers are held together by binders or by stitching. Bulk chopped fiber is used in resin premixes for molding compounds, and milled fibers are small modules of glass filaments that are used mostly in casting resins. Some synthetic and natural organic fibers that are used as reinforcements include rayon, nylon, polyester acrylic, cotton, and paper. Cotton and paper are widely used with melamine and phenolic resins in the electrical industry. Paper is generally used for its electrical qualities rather than mechanical properties; and cotton is used because it has high mechanical strength, bulks well, and is resistant to stretching and heat. A general guide to matching the reinforcement with the laminate property that it affects is given in Table 12.15. More detailed information about reinforcements can be obtained in Ref. 12.

Resins

Although almost any polymer (thermoplastic, thermoset, or elastomer) can be combined with a reinforcement material, the principal resins have been epoxy, silicone, phenolic, melamine, polyester, bismal-

TABLE 12.15 Laminate Property and Reinforcement Fiber

Reinforcing fiber	Mechanical strength	Electrical properties	Impact resistance	Chemical resistance	Machining and punching	Heat resistance	Moisture resistance	Abrasion resistance	Low cost	Stiffness
Glass strands	X	X	X			X	X		X	X
Glass fabric	X	X	X	X		X	X			X
Glass mat			X	X		X	X		X	X
Asbestos		X	X			X				
Paper		X		X					X	
Cotton and linen	X	X	X		X				X	
Nylon		X	X	X				X		
Short inorganic fibers	X		X						X	
Organic fibers	X	X	X	X	X			X		
Ribbons		X					X			
Metals	X		X			X				X
Polyethylene	X		X		X					X
Aramid	X	X	X			X				X
Boron	X		X			X				X
Carbon and graphite	X		X			X				X
Ceramic	X		X					X		

SOURCE: Ref. 13.

eimide, cyanate ester, imide, allyl phthalates, and polytetrafluoroethylene. The chemistry and basic properties of the neat resins have been reviewed in Chaps. 3 and 4. These resins are used in the preparation of laminates, SMC, BMC, and fabric-based tapes. Not all resins are useful in each category.

Silicone resin. The silicone resins as a class of materials have an excellent balance of physical, chemical, thermal, and electrical properties and find wide use in electrical applications. These resins have high arc resistance, low dielectric constant and dissipation factor, resistance to corona, and excellent dielectric strength even in moist environments. They also have good weathering properties.

Phenolic resin. The phenolic resins are also widely used in electrical applications. These resins are available as liquids or solids and can be formulated to produce a variety of molding and impregnating compositions. The phenolics are poor heat conductors and have excellent thermal properties, chemical resistance, and flame resistance.

Melamine resin. The melamine resins have excellent arc and track resistance which makes these reinforced materials good choices for use in high-power arcing applications such as circuit breakers. The reinforced melamines have poor dimensional stability when exposed to cyclic humidity conditions.

Polyester. The unsaturated polyester-reinforced materials have excellent chemical resistance and can be formulated to produce flame-retardant compositions that have high arc and track resistance. These materials are easy to process and fabricate and are used in SMC and BMC.

Polyimides. The polyimide-reinforced materials are more difficult to process than the epoxy, silicone, phenolic, and polyester materials but offer temperature performance levels much higher than those of the traditional materials mentioned above. These reinforced resins have exceptional radiation resistance, dimensional stability, and flame resistance.

Bismaleimides. The bismaleimide-reinforced resins are similar in behavior to the polyimides. They process much more easily than the polyimides with some sacrifice in thermal stability. Polyimides can operate at about 260°C while bismaleimides can operate at about 200°C.

Cyanate ester. These resins also have epoxylike processing capabilities and offer operating temperature capability between the epoxy- and bis-maleimide-reinforced resins.

Allyl phthalate. The allyl phthalate resins are easy to process and can operate in moist environments at elevated temperatures (up to 210°C) while maintaining electrical integrity.

Polytetrafluoroethylene. This fluorocarbon polymer is used as a circuit board material in high-frequency applications. It has extremely stable dielectric properties, low moisture absorption, chemical inertness, and flame resistance. It can operate at temperatures up to 260°C. On the other hand, polytetrafluoroethylene is soft, is subject to creep, and has a high CTE.

Types of reinforced products

All the resins which were briefly described in the previous section can be converted to a variety of reinforced products. For printed wiring board applications, Table 12.16 shows the general range of properties of the different neat resins available. These resins are formulated and combined with various reinforcements such as in E-glass laminates to yield the range of laminate properties shown in Table 12.17. In addition to printed wiring board materials, the electrical industry uses a host of other reinforced and laminated plastic products. These products are defined by National Electrical Manufacturers Association NEMA

TABLE 12.16 Approximate Neat Resin Properties for Polymers Used in Printed Wiring Boards

Resin	T_g, °C	Dielectric constant, 1 MHz	Dissipation factor, 1 MHz
FR-4 epoxy	125–130	3.5–3.6	0.03
HP FR-4 epoxy	135–150	3.5–3.6	0.03
HP polyfunctional epoxy	155–180	3.7–4.0	0.03
Polyimide	240–300	3.3–3.5	0.02
Bismaleimide and triazine	230–260	2.9–3.4	0.004
Cyanate ester	220–260	2.7–3.1	0.005
PTFE	327	2.0–2.1	0.00002
Fluorinated polyimide	260	2.9	0.0004
Polysulfone	192	3.0	0.004
Polyethersulfone	230	3.5	0.003
Polyetherimide	212	3.1	0.006

TABLE 12.17 Approximate Property Range of E-Glass: Laminates and Other Laminate Systems

Resin	Reinforcement	Dielectric constant at 1 MHz	Dissipation factor at 1 MHz	Flammability, UL-94
FR-4 epoxy	E-glass	4.1–4.3	0.020	V-0
HP, FR-4 epoxy	E-glass	4.1–4.3	0.020	V-0
HP polyfunctional epoxy	E-glass	4.3–4.6	0.020	V-0
Polyimide	E-glass	3.9–4.6	0.010	V-1
Bismaleimide/triazine	E-glass	4.0–4.2	0.013	V-0
Cyanate ester	E-glass	3.5–3.6	0.005	V-0
PTFE	E-glass	2.4–2.6	0.003	V-0
Cyanate ester blends*	Aramid	2.9–3.1	0.009	V-0/V-1
	S-glass	3.0–3.2	0.009	V-0/V-1
	PBO	2.9	0.003	V-0
	D-glass	3.5	0.008	V-0
Bismaleimide blends*	Aramid	2.9–3.0	0.005–0.008	V-0 to V-1
	S-glass			
	PBO			

*From Ref. 14.

and are listed in Table 12.18. More detailed property information on the materials listed in Table 12.18 can be obtained from the NEMA specifications.[15] Although the NEMA standards cover a variety of resins and reinforcements, there are many more single resins and resin combinations that an experienced chemist or materials scientist could synthesize or formulate to yield special properties in combination with various reinforcements. The point is that the design engineer is not limited to those products covered by existing specifications. New products can and will be developed as new polymers or combinations of polymers are synthesized and formulated.

Sheet molding and bulk molding compounds. Sheet and bulk molding compounds consist basically of unsaturated polyester resin formulations mixed with stranded fiberglass reinforcement and processed into a sheet form suitable for matched die molding. In addition to the unsaturated polyester resin, the formulations contain fillers, thickeners, catalysts, and reactive diluents. In contrast to sheet molding compounds (SMCs) which are actually formed from relatively low-viscosity formulations, the bulk molding compounds (BMCs) are puttylike in consistency and contain short-fiber-length reinforcement. Both SMC and BMC materials have good electrical properties, are arc- and track-resistant and low-cost, and are used in many electrical housings such as outlet boxes and switches, circuit breakers, and slot wedges. The

308 Utilization of Plastics in Electronics

TABLE 12.18 NEMA Grades of Electrical Laminates

Grade	Resin	Reinforcement	Comments
X	Phenolic	Paper	Used in nonelectrical applications. Affected by humidity
XP	Phenolic	Paper	Hot-punchable. Better than X grade
XPC	Phenolic	Paper	Cold-punchable
XX	Phenolic	Paper	Easy to machine. Some electrical applications
XXP	Phenolic	Paper	Punchable. Some electrical applications
XXX	Phenolic	Paper	Excellent electrical laminate. Radio-frequency applications
XXXP	Phenolic	Paper	Punchable hot. Excellent electrical laminate
XXXPC	Phenolic	Paper	Punchable cold. Excellent electrical laminate
C	Phenolic	Cotton	Not for electrical applications
CE	Phenolic	Cotton	Not recommended for use in excess of 600 V
L	Phenolic	Cotton	Not for electrical applications
G-3	Phenolic	Glass fabric	Not for electrical applications
G-5	Melamine	Glass fabric	Excellent electrical properties. Heat-resistant. Used in arcing applications
G-7	Silicone	Glass fabric	Excellent electrical properties. Heat-resistant. Used in arcing applications
G-9	Melamine	Glass fabric	Good wet electrical properties. Used in arcing applications
G-10	Epoxy	Glass fabric	Excellent mechanical properties and electrical properties
G-11	Epoxy	Glass fabric	Same as G-10 but higher heat resistance
N-1	Phenolic	Mylar cloth	Excellent electrical properties under humid conditions
FR-1	Phenolic	Paper	Flame-resistant grade of XP
FR-2	Phenolic	Paper	Flame-resistant grade of XXXPC
FR-3	Epoxy	Paper	Flame resistant. High flexural strength. Good electrical properties
FR-4	Epoxy	Glass fabric	Flame-resistant. Printed-wiring board grade
FR-5	Epoxy	Glass fabric	Similar to G-11 but flame-resistant
CEM-1	Epoxy	Cotton and glass	Punchable. Similar to FR-2 and FR-3 with electrical properties approaching FR-4
CEM-3	Epoxy	Woven glass and glass mat	Good electrical properties. Suitable for plated-through holes. Used in computers, automobiles, and printed wiring boards
GPO-1	Polyester	Glass mat	General-purpose electrical laminate
GPO-2	Polyester	Glass mat	Flame-resistant electrical laminates
GPO-3	Polyester	Glass mat	Arc- and track-resistant laminates
GPO-1P, 2P, 3P			Punchable grades of GPO-1,2,3 used in switches, circuit breakers, switch gears, motors, and transformers

sheet and bulk polyester molding compounds are covered in Federal Specification MIL M14G. SMCs are compression-molded whereas BMCs can be injection-molded and transfer-molded as well as compression-molded.

Extruded forms. Plastic extruded forms are produced by the processes of pultrusion, melt, and liquid extrusion. These processes are the principal methods used in the manufacture of wires, cords, jackets, cables, and profiled pultruded shapes.

Pultrusion. Pultrusion as a plastic molding fabrication process was described in Chap. 7, but briefly it is a way to make profiled shapes in a continuous process. The resins used are polyesters, epoxy, silicone resins, and vinyl esters. Pultruded products have very high strength in the longitudinal direction, excellent chemical resistance, and adequate electrical properties and are easily fabricated. Reinforcements include E- and S-glass, graphite, and Kevlar. Electrical applications of pultruded forms include component parts for control and switchgear devices, electric bus duct, and third-rail conversion railroads.

Melt and liquid extrusion. These processes were also described in Chap. 7, but briefly melt extrusion is a process in which a plastic melt is forced under pressure through a die. The liquid extrusion process is a process in which a wire is passed through a liquid resin bath and subsequently passes through a die to remove excess liquid resin and then passes into an oven to cure the liquid coating to a hard, infusible solid. This wire enameling process was described in Chap. 9. The principal polymers used for extrusion are polyethylene; polypropylene; polyvinyl chloride; the fluoropolymers—polytetrafluoroethylene (PTFE), fluorinated ethylene-propylene (FEP) copolymer, ethylene-tetrafluoroethylene (ETFE), polyvinylidene fluoride (PVF, KYNAR), perfluoroalkoxy (PFA), ethylene-chlorotrifluoroethylene (E-CTFE) copolymer; the elastomers, including silicone rubber, natural rubber, styrene-butadiene rubber, butyl rubber, ethylene-propylene rubber, chloroprene, and chlorosulfonated polyethylene; and the thermoplastic elastomers.

Batteries

Another electrical area where plastics play a critical role in product performance is in electrochemical energy storage cells, commonly known as batteries. Here plastics are used as the housing material in which the battery components including the electrolyte are contained and as a separator within the electrochemical cell. Plastics used for the housing include polyolefins, polysulfone, acrylonitrile-butadinestyrene, and hard rubber (ebonite). The properties of these plastics are reviewed in Chap. 3 and will not be repeated here. Of particular importance in the operation of a battery is the separator material, and its properties are discussed in this section.

A *separator* may be defined as a porous component in an electrochemical cell that separates the positive from the negative electrode. It has two primary functions. First, it provides electronic separation of both electrodes to prevent short circuits, by not allowing the anode and cathode to come into physical contact with each other. Second, it allows flow of ionic current between the electrodes immersed in a suitable electrolyte. Separators are important components in electrochemical systems such as batteries and fuel cells. Their effectiveness is a major factor in the electrochemical energy efficiency, performance, and cost of these systems. Separators are not an active component of the battery; i.e., they do not participate in the electrochemical reactions of the cell. They are, however, a necessary and critical component of a battery because the wrong separator can prevent the battery from operating at all or can shorten its life cycle considerably. A wide variety of separators are used in batteries because of the different electrochemical environments (acid, alkaline, nonaqueous). These environments are aggressive in themselves, and when coupled with the redox reactions of the electrochemical cell, they become even more active and can seriously degrade the separator material. Because of these environments, a wide variety of properties are needed in the separator to allow the battery to reach optimum performance levels. There is no universal separator that can give optimum performance in all battery systems.

The various properties required of separator materials are summarized as follows:

Chemical stability and degradation resistance. The material must have the necessary chemical resistance to degradation by either the electrolyte or the active materials in the anode and cathode as oxidation and reduction reactions occur. Separator degradation in any battery can occur in a number of different ways, e.g., alkaline/acid hydrolysis, thermal/oxidative depolymerization, dehydrochlorination, free-radical oxidation. Each of these processes is enhanced by increasing temperature, a necessary operational component of batteries. One or more of these processes may occur if the separator is susceptible to degradation. Clearly, there are many elements that pose a serious threat to separator performance over the life of the battery, and this is why it is so important to choose the proper separator material.

Mechanical strength and dimensional stability. Separators must be strong enough to withstand mechanical handling during assembly and to resist the operational effects of the cell such as shock and vibration. Changes on wetting with electrolyte are important in the proper design and assembly of cells. Subsequent dimensional changes during cell operation may be responsible for defects leading to premature failure or inadequate life. Expansion pressure due to swelling may contribute to crazing of cell containers, and shrinkage may lead to tearing and reacting of the separator, especially at bends and at plate edges. Changes may also occur as a consequence of chemical instability. Degradation by either hydrolysis or oxidation will influence the polarity of the molecule, which in turn influences the shrink and swell of the separator.

Electrolyte absorption, retention, and permeability. Cells are usually assembled with a minimum amount of electrolyte, and wetting and wicking of the separator are of paramount importance. This is especially true in sealed cells where no free electrolyte is present. A maximum amount of retained electrolyte is desirable to achieve minimum cell internal resistance and to maximize capacity. A good separator must not only absorb electrolyte but also retain it if the cell is subjected to acceleration. If electrolyte is lost, serious degradation of the electrical performance of the cell will occur. The separator must allow diffusion of current-carrying species of electrolyte and retard diffusion of soluble metal oxides. It must also prevent the transport of colloidal particles or ions to prevent self-discharge and dendrite formation.

Separator resistance. Separators must be porous enough to allow for the flow of current-carrying species in order for the cell to operate properly. Ideally, the separator must not significantly contribute to heating and/or cell resistance in the battery. The placement of the separator between the electrodes constricts the area of the electrolyte path and renders it tortuous. It is the resistance of the constricted, tortuous electrolyte path which is measured. The resistance is a valuable indicator of the performance of separators in working cells because the voltage loss can be calculated from separator resistance values and as such is a useful design criterion. Separator resistance can also be used to check the statistical variation in total porosity over the entire surface of the separator. This is useful because uniformity of porosity is highly desirable. Nonuniform porosity leads to nonuniform current density distribution, which in turn may lead to reduced activity of the active materials and eventual failure because some parts of the electrodes will be worked harder than others.

Homogeneous. The separator should have a uniform pore distribution to ensure good current efficiency and uniform current distribution.

High insulation resistance. While the electrolytic resistance of the separator in the electrolyte should be minimal, the separator must provide a barrier to prevent short-circuiting the electrodes during assembly.

Minimum thickness. This might seem an obvious requirement, but nevertheless it is extremely important because it affects the battery in several ways. First, one needs to minimize both space and weight in the battery; second, resistance voltage and heating must be kept to a minimum; third, cost is reduced; fourth, high-energy density batteries require the use of thin separators.

Additives. The separator must not contain any substances that could adversely affect battery performance.

Separator properties. Table 12.19 presents some comparative data on the electrolytic resistance properties of various separators in an alkaline electrolyte.[16] Additional information on battery separators can be found in Refs. 17 to 25.

High-voltage uses. As electrical insulation, plastics are used in many different applications where the applied voltage can range from a few millivolts to more than several thousand kilovolts. The electrical behavior of plastics is very much influenced by a number of factors; and in addition to the plastic composition and its uniformity, those other factors include temperature, time, moisture, contaminants in the bulk of the polymer as well as on its surface, part geometry, frequency, and magnitude of the applied voltage and stress. All these, while affecting the electrical behavior of a given type of plastic, also significantly affect the life of the plastic; and in applications requiring high-voltage use, these effects are magnified. Electric power losses become significant at high voltages and high frequencies such as near 100 GHz. Deterioration of plastics in high-voltage areas is usually associated with the interface, as in solid to gas, solid to liquid, or solid to solid. These interfaces can be looked at as potential electrical weaknesses in the polymer. Some examples of those interfaces are the presence of voids in the plastic insulation, fissures, conductor surface asperities, surface contamination on the insulation, and surface charge entrapment. In high-voltage applications, plastic insulation must have a high-voltage breakdown strength as well as mechanical strength. The following are typical requirements of an insulating material for all electrical applications and are particularly important in high-voltage applications: They have a high dielectric strength to withstand an electric field between the conductors; a high insulation resistance to prevent leakage current across the conductors; a high arc resistance to prevent damage during arcing incidents; a high mechanical strength;

TABLE 12.19 Electrolyte −30% KOH + 15 g/L LiOH Electrolytic Resistance as Function of Temperature

Separator	Resistance* at: −17.7°C	22°C	32°C	Polymer type	Manufacturer
Celgard 5501	48	17	15	Polypropylene	Hoechst Celanese Co.
Powersep, ribbed	96	34	28	Polyolefin	W. R. Grace Co.
PVC	179	52	44	Polyvinyl chloride	Jungfer Separators, Inc.
SM-517	8	6	5	Nylon-polypropylene	International Paper Co.
Cellophane	17	8	6	Cellophane	Pall RAI Inc.
P-2193 40/20	167	36	29	Polyethylene-acrylic	Pall RAI Inc.
PKHE	243	49	40	Polyethylene-acrylic	Pall RAI Inc.
PKE	224	46	37	Polyethylene-acrylic	Pall RAI Inc.
PK-3060	588	89	71	Polyethylene-acrylic	Pall RAI Inc.
Zamm-3	80	28	18	Polyethylene-cellophane-polyethylene	Pall RAI Inc.
Zamm-0	32	9	7	Polyethylene-acrylic	Pall RAI Inc.
Powersep gray, 2 mil	32	11	9	Polyolefin	W. R. Grace Co.
PHDC	20	7	6	Polypropylene	Pall RAI Inc.
Nomex E88C-309C	8	5	5	Aromatic polyamide	Du Pont
Powersep gray, 8 mil	84	33	29	Polyolefin	W. R. Grace Co.
Powersep white, 5 mil	106	40	35	Polyolefin	W. R. Grace Co.
Celgard K-932	11	9	8	Polyethylene	Hoechst Celanese Co.
Celgard 3401	34	17	15	Polypropylene	Hoechst Celanese Co.
Celgard 5401	66	29	25	Polypropylene	Hoechst Celanese Co.
NFWA	47	24	21	Carboxy-acrylic ion-exchange resin	Gelman
Veratec 1030	11	6	7	Nylon	International Paper Co.
Veratec 1488	43	20	19	Polypropylene	International Paper Co.
Veratec SM-450	7	5	5	Polypropylene	International Paper Co.
Veratec SM-516	9	6	6	Polypropylene	International Paper Co.

*Resistance readings are in milliohms times square inches (m$\Omega \cdot$ in^2).
SOURCE: Ref. 16.

and the ability to resist the deterioration and degradative effects of environmental conditions such as temperature, humidity, and radiation. The key properties one should be examining relative to these requirements are the dielectric constant, dielectric strength, dissipation factor, volume, and surface resistivity.[26]

TABLE 12.20 Typical Properties of XLPE versus EPR

Property	XLPE*	EPR†
Density, $g \cdot cm^{-3}$	0.92	1.2–1.4
Tensile strength, MPa	19	9–12
Elongation, %	500	250–350
Modulus of elasticity, MPa	121	5–14
Heat distortion, %	20	5–8
Thermal conductivity, W/(m · °C) at 90°C	0.27	0.27–0.35
Dielectric constant	2.3	2.5–3.0
Dissipation factor, % at		
23°C	<0.03	0.16–0.30
90°C	<0.03	0.30–1.0
Volume resistivity, $\Omega \cdot m$ at 23°C	10^{16}	10^{13}
Short-term ac breakdown on miniature cable, kV/mm	48	30–40

*Cross-linked polyethylene.
†Ethylene propylene rubber.
SOURCE: Ref. 27.

There are two dielectric materials that are excellent choices for insulation in high-voltage extruded cables rated up to 230 kV: cross-linked polyethylene and ethylene propylene rubber. The properties of these two materials are shown in Table 12.20. Cross-linked polyethylene was introduced in the late 1950s for use in medium-voltage cables up to 35 kV, and today it is being used in Japan and Europe for voltages up to 500 kV.[27]

References

1. R. Greene, ed., *Modern Plastics Encyclopedia,* McGraw-Hill, New York, 1992.
2. M. W. Hunt, ed., *Materials Engineering—Materials Selector 1993,* Penton Publication, Cleveland, OH, 1992.
3. ASTM D-2305, *Method of Testing Polymeric Films Used for Electrical Insulation,* American Society for Testing and Materials, Philadelphia, 1992.
4. S. S. Schwartz and S. H. Goodman, *Plastics Materials and Processes,* Van Nostrand Reinhold, New York, 1982.
5. T. H. Shepler and K. L. Casson, "Flexible Printed Boards," in *Electronic Materials Handbook,* vol. 1, ASM International, Materials Park, OH, 1989.
6. P. Kemezis, "Electronics Segment Continues Growth Trajectory," *Chemical Week,* March 10, 1993, p. 33.
7. C. A. Harper, ed., *Handbook of Plastics, Elastomers and Composites,* McGraw-Hill, New York, 1992, chap. 9 (E. Petrie).
8. I. Skeist, ed., *Handbook of Adhesives,* Van Nostrand Reinhold, New York, 1977.
9. W. T. Shugg, *Handbook of Electrical and Electronic Insulating Materials,* Van Nostrand Reinhold, New York, 1986.
10. ASTM D-372, *Standard Specification for Flexible Treated Sleeving Used for Electrical Insulation,* American Society for Testing and Materials, Philadelphia, 1992.

11. American Society for Testing and Materials, *Electrical Insulation I and II*, vols. 10.01 and 10.02, Philadelphia, 1992.
12. J. V. Milewski in H. S. Katz, ed., *Handbook of Reinforcements for Plastics*, Van Nostrand Reinhold, New York, 1987, Secs. I–IV, pp. 1–419.
13. C. A. Harper, ed., *Handbook of Plastics, Elastomers and Composites*, McGraw-Hill, New York, 1992, chap. 4 (R. Sampson).
14. W. M. Alvino and Z. N. Sanjana, unpublished results, Westinghouse Electric Corp., 1988–1990.
15. NEMA Publ. LI1, *Standards Publication for Industrial Laminated Thermosetting Products*, Washington, DC, 1989.
16. W. M. Alvino, unpublished results, Westinghouse Electric Corp., Science & Technology Center, Pittsburgh, PA, 1992.
17. J. W. Van Zee, R. White, K. Kinoshita, and H. Burney, eds., *Symposium Proceedings on "Diaphragms, Separators and Ion Exchange Membranes,"* The Electrochemical Society, vol. 86-13, 1986.
18. *Proceedings of the Symposium on Battery Separators*, Columbus Section of the Electrochemical Society, February 18, 1970, Columbus, Ohio.
19. S. Uno Falk and A. J. Salkind, *Alkaline Storage Batteries*, Wiley, New York, 1969.
20. H. V. Venkatasetty, *Lithium Battery Technology*, Wiley, New York, 1984.
21. V. D'Agnostino, J. Lee, and R. Coyle, "Low Temperature Alkaline Battery Separators," *Proceedings of the 27th Power Sources Conference*, June 1976.
22. G. Hsiue and W. K. Huang, "Preirradiation Grafting of Acrylic and Methacrylic Acid onto Polyethylene Films," *J. Appl. Polym. Sci.*, vol. 30, 1985, p. 1023.
23. L. C. Christensen and Q. Kampf, "Non-Woven Battery Separators: Applications and Requirements," INDA-TEC, International Non-Woven Technological Conference, Hilton Head, SC, May 18, 1987.
24. J. J. Lander, "Battery Separator Materials," *SAMPE Quarterly*, vol. 3, no. 1, October 1971, p. 27.
25. K. Ruhling and A. Winsel, "Separators for Acidic and Alkaline Batteries," *J. Appl. Electrochemistry*, vol. 19, 1989, p. 553.
26. R. Liepins and C. C. Ku, *Electrical Properties of Polymers*, Hanser Publishers, New York, 1987.
27. C. J. Chan, M. D. Hartley, and L. J. Hiivala, "Performance Characteristics of XLPE versus EPR as Insulation for High Voltage Cables," *Electrical Insulation Magazine*, vol. 9, no. 3, May/June 1993, p. 8.

Chapter

13

Plastics in Microelectronics

Since the development of the transistor in 1948, the electronics industry has enjoyed an almost exponential growth. However, the size of the electronic components and devices has undergone a tremendous reduction. Table 13.1 shows comparative wire densities for various technologies. This size reduction has placed new demands on the materials needed to fabricate and process these devices. The materials whether inorganic or organic must provide high reliability, durability, and manufacturing ease for the electronic product. This size reduction, however, has also significantly advanced integrated-circuit (IC) technology from small-scale integration (SSI) through medium-scale integration (MSI) and large-scale integration (LSI) to very large-scale integration (VLSI) and eventually to ultralarge-scale integration (ULSI) with over a million components per chip. This miniaturizing is due in part to new developments in polymer materials which are critical components in microelectronic packages. These critical polymer components include photoresists, IC encapsulants, die attachment adhesives, interlayer dielectrics, and packaging materials. Also included in this list are the

TABLE 13.1 Comparative Wire Densities for Various Technologies

Technology	Wire density, in/in^2
House wiring	0.2
Printed wiring board (PWB)	50.
Multichip module (MCM)	500.
Integrated circuit (IC)	5000.

SOURCE: Ref. 2.

polymeric materials that are needed to support the development and growth in new microelectronic substrate and packaging technologies such as multichip modules (MCMs), advanced printed wiring boards (PWBs), and ball grid arrays (BGAs) for packaging. For all these electronic applications, polymers have indeed played a critical role. All these applications of polymers in IC devices can be divided into three major areas: resists, interlayer dielectrics, and packaging. While the polymers used in these applications do not take an active part in the functioning of the device, even though the polymers can affect the device's performance, there are new emerging areas of technology within the microelectronic arena where polymers can and do take an active part in the functioning of the device. These areas include conducting polymers, polymer electrolytes, Langmuir-Blodgett films, and polymeric optical waveguides.

The emphasis in this chapter, as it has been throughout this book, will be on the nature and properties of the polymeric materials used in these technologies. For detailed information on the various technologies mentioned above, see Refs. 1 to 10.

Resists

The integrated circuit is a three-dimensional structure consisting of patterned layers of metals, dielectric material, and semiconductors on a silicon wafer substrate. These patterned layers are delineated with a polymeric imaging material called a *resist*. A resist is a highly radiation-sensitive material capable of undergoing alteration when exposed to such an energy source. Radiation can be ultraviolet (uv), electron beam (EB), *x* rays, and ion beam; and the corresponding resist materials are designed as photographic or uv resists, electron-beam resists, *x*-ray resists, and ion-beam resists. When a polymer is exposed to these radiation sources, changes occur in the polymer which affect its solubility. Those changes can be cross-linking, chain scission, and/or molecular rearrangement. Those polymers that become more soluble after exposure are called *positive resists,* and the increased solubility results from chain scission. Those polymers that become less soluble after exposure to radiation are called *negative resists,* and the decreased solubility results from cross-linking. After exposure through a mask, the resist-coated substrate is immersed in a developing solution which washes away the soluble areas of the resist. This procedure can be repeated to produce many layers with an array of circuit patterns. A pictorial representation of this process is shown in Fig. 13.1. The polymers used for resists, aside from being sensitive to the particular radiation source, must possess the following:

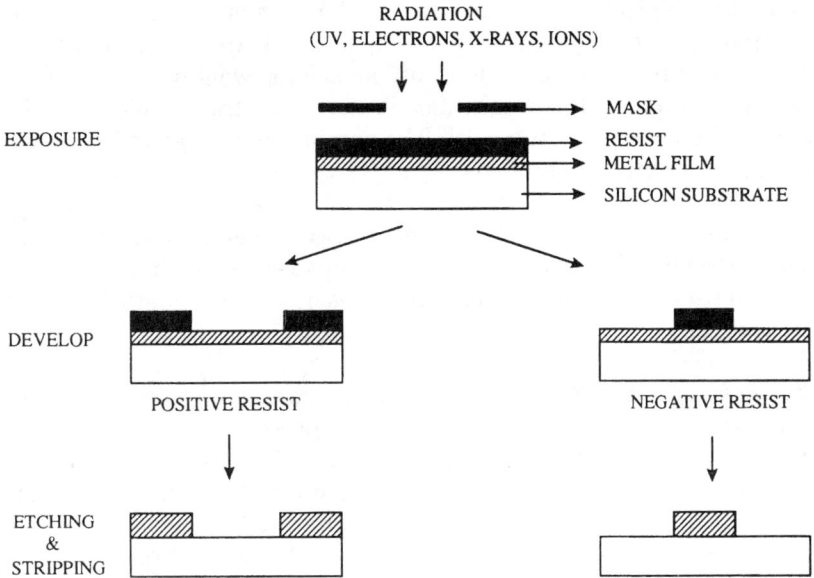

Figure 13.1 Illustration of lithographic process or polymer resists.

Film-forming capability. The polymers must have sufficient molecular weight and possess good rheological properties so that they can be used and processed easily as coatings.

Sensitivity. The sensitivity of a resist is a measure of how well it responds to a given amount of radiation and of the exposure required to delineate a resist pattern. It is inversely proportional to the absorbed dose, so that smaller values represent higher sensitivity. The sensitivity of a polymer is a function of its chemical composition, i.e., what types of bonds are present that might undergo scission, cross-linking, or cleavage and what types of chemical groups or species are attached to the polymer to make it susceptible to absorbing a specific wavelength of radiation. With optical resists, polymers generally do not absorb radiation with wavelengths greater than 300 nm, and these polymers require the incorporation of chemical sensitizers to form either positive or negative resists. As such, the sensitivity of photoresists is usually more dependent on the photosensitizer than on the base polymer itself. With electron, x-ray, and ion-beam resists, the sensitivity of the resist is much more dependent on the structure, molecular weight, and its distribution in the polymer.

Contrast and resolution. The contrast of a resist is a measure of how fast a chemical change takes place upon exposure to the radiation source. Contrast is independent of molecular weight but strongly dependent on molecular weight distribution; i.e., the broader the distribution the lower the contrast.[10] The resolution of a resist is directly related to the contrast of a resist.

Optical density. The amount of light absorbed per unit thickness of resist material determines whether the desired photochemical reactions occur and the homogeneity of the deposited radiation through the resist thickness.

Etch resistance. The dry and wet etch resistances of a resist refer to the ability of the resist to withstand the process conditions during transfer of the resist pattern to the substrate. In this regard, another resist property comes into play, i.e., adhesion of the resist to the substrate, which is more critical in wet than dry etching. Polymers that have strong bonds in the polymer chain as well as polymers containing aromatic and/or polar functional groups are more resistant to plasma etching. Polymers containing chlorine etch faster.[7]

Purity. The contaminant level in any resist material must be extremely low (down to parts per billion) because of the sensitivity of semiconductors to contaminants.

Types of resists

There are basically five categories of polymer resists: photo, deep uv, electron-beam, x ray, and ion-beam. All these essentially operate by creating a differential solubility in the exposed and unexposed areas of the polymer film after exposure to a radiation source. Examples are presented in Table 13.2.

Photoresists. Polymers classified as photoresists contain a photosensitizer called a *photoinitiator* which is sensitive to uv in the range of 300 to 450 nm. Polymers in this class contain chromophoric groups to enhance the scission or cross-linking process. Deep uv resists can also be included because these materials are photon-activated in the range of 200 to 300 nm. At this wavelength the energy is sufficient to break carbon—carbon bonds in the polymer, causing either scission or cross-linking. Photoresists can be either positive or negative depending on their solubility after exposure. If the exposed area is more soluble, it is a positive resist; if less soluble, it is a negative resist. Negative resists have much poorer resolution than positive resists. In addition, positive photoresists have better etch resistance and thermal stability.

TABLE 13.2 Examples of Polymer Resist Materials*

Class	Type	Polymer
Photoresist	Positive	Novolac resin
(300–450 nm)	Negative	Polyisoprene
Electron-	Negative	Siloxane-benzocyclobutene
beam	Negative	Polyimide, siloxane-BCB
(10–100 nm)	Positive	Polymethyl methacrylate and copolymers
	Positive	Polyolefin sulfones
	Negative	Polystyrene and chlorostyrenes
	Negative	Polyglycidyl methacrylates
Deep uv	Positive	Polymethyl methacrylate and copolymers
(200–300 nm)	Positive	Novolac resin
	Negative	PMMA with poly n-butyl α-chloroacrylate
	Negative	Chlorinated polystyrenes
x ray	Positive	PMMA and copolymers
(0.5–5 nm)	Positive	Polybutene sulfone
	Negative	Polychloromethyl styrene
	Negative	Chlorinated polyvinyl ethers

*Adapted from Ref. 7.

Electron-beam resists. Polymers are exposed to a high-energy focused beam of electrons (10 to 100 nm) where the polymer molecules are excited and ionized. This focused beam is scanned over the resist-coated surface of the wafer to produce a pattern without the need for a mask, as required with photoresist materials. The electron beam enters the polymer, causing chain scission and/or cross-linking or both, leading to either positive or negative resists. Negative resists are generally more sensitive than positive electron-beam resists, but their resolution is lower than that of positive resists. Electron backscattering can make these polymer resists difficult to remove.

Dry-film resists. Dry-film resists are polymeric photosensitive materials that are laminated as dry films to a printed-wiring board (PWB). The photopolymer (an acrylate) is sandwiched between a polyester (Mylar) and polyolefin separator sheet. The latter is removed prior to laminating it to the PWB. The photopolymer is processed in the same manner as a normal liquid photoresist polymer. The dry films can be either positive or negative working resists. Dry-film resists are easier to apply when the covered area is large, and these resists eliminate prebaking as well as solvents.

Of all the processes briefly described above, only electron-beam and optical lithography are in commercial use for semiconductor processing. The three most important properties of a polymer resist are its sensitivity, resolution, and contrast. Shaw has presented an excellent summary of these two important polymer resist properties.[11] This summary is repeated here.

The sensitivity of resist materials to various forms of radiation is defined as the minimum incident exposure dose (not absorbed dose) required to differentiate between the exposed and unexposed regions. These sensitivity values are obtained by plotting the resist thickness removed (for positive resists) or remaining (for negative resists) vs. the log of incident exposure dose, as is shown in Fig. 13.2. The sensitivity value for positive resists is the exposure dose necessary to remove 100% of the exposed area in a given development time (D_o) and is reported in mJ/cm² for uv and X-ray resists, and in coulombs per centimeter squared (C/cm²) for e-beam resists. For negative systems, D_g is the dose at which the polymer begins to crosslink, and D_i is the dose needed to totally crosslink the resist film. The slope of the linear portion of the sensitivity curve determines the contrast (γ) of the system, or the exposure range over which the material will respond; it is given by the equations shown in Fig. 13.2. It is desirable to have a high-contrast system in order to minimize resist response to low-level radiation from diffraction effects or backscattered secondary electrons.

The resolution capability of the polymer is defined as the minimum line width that can be obtained in a given resist thickness. It is determined, in the final analysis, by (1) the type of radiation used to expose the resist and the design of the exposure system; (2) the type of radiation-sensitive material used and its exposure response, as seen in the sensitivity and contrast curve, and material properties, such as molecular weight, absorptivity, thermal stability, and development response; and (3) process control such as bake temperatures, thickness control, and developer pH and temperature control. In general, it is difficult to achieve greater than a 1:1 aspect

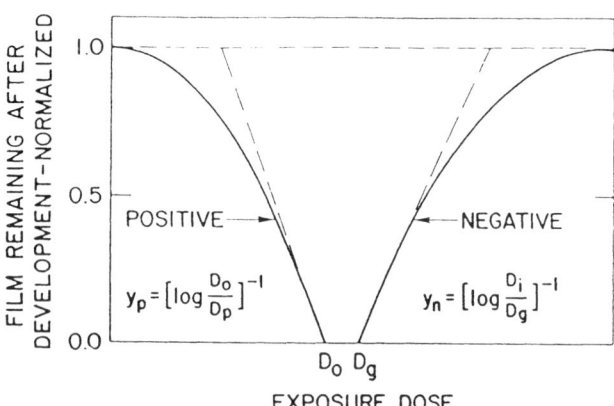

Figure 13.2 Sensitivity and contrast for positive and negative resists. (*Source:* Ref. 11. Reprinted with permission.)

ratio where the width of the line is the same dimension as the thickness of the resist. The choice of a radiation-sensitive polymer depends on resolution and processing requirements. Resists used for chip applications are applied from a liquid solution (10–30 wt. % solids in a solvent) to achieve the thin films needed to meet high-resolution requirements. Dry film resists are laminated as thick films and are used throughout the industry to pattern the thicker metal wiring (50 microns) needed for printed circuit boards.

Other Developments

Two high-performance polymer materials—the polyimides (PIs) and the benzocyclobutenes (BCBs)—have been converted to photoimageable versions that could be useful as resists. While both have excellent high-temperature stability with T_g values in excess of 290°C, the BCB resins have lower dielectric constant, lower moisture absorption, and better processing characteristics than the PI materials. The photoimageable BCB material is based on a vinyl siloxane modification.[12,13] It is a negative resist material that is photosensitive to 365-nm wavelength (i-line) or 436-nm wavelength (g-line). It reportedly gave resolution down to 16-μm lines and spaces on a 5.6-μm film with good sidewall angles. This material is available from Dow Chemical Co. The polyimide materials are modified to create positive or negative imaging materials[4] and are used also as interlayer dielectrics, protective coatings, and alpha-ray shielding materials in semiconductor devices.

New chemically amplified deep uv resists have been developed[14] based on acrylate polymer technology which show promise as positive resists that can deliver high resolution.

Suppliers[15]

A summary of the various types of resists is given below, along with supplier details.

Positive optical photoresists: Novolac resins with diazonaphthaquiones as photoactive compounds.

Negative optical photoresists: Cyclized synthetic rubber resin with a *bis*-arylazide as photoactive compound.

Deep uv photoresists: Includes PMMA, modified PMMA, and modified novolacs.

Planarizing resists (used in multilevel resists): The planarizing layer can be conventional positive photoresist, polyimide, or polysulfone.

Electron-beam negative resists:

Poly(glycidyl) methacrylate monopolymers made by Tokyo Okha

PGMA Copoly(ethylacrylate) made by Mead

Poly(chloromethyl styrene) made by Hewlett-Packard

Poly(iodostyrene) made by Hitachi

Poly(siloxane) made by IBM

Electron-beam positive resists:

Poly(methyl methacrylate) made by numerous manufacturers

PMMA Copoly(acrylonitrile) made by Tokyo Okha

Poly(trifluoro-alpha-chloroacrylate)

Poly(fluoroalkylmethacrylate) made by Daikin Kogyo

PMMA Xlinked made by Microimage

Poly(butene sulfone) made by Mead

Novolac-diazoquinone resist made by Az

Novolac-poly(methylpentene sulfone) made by Hitachi

Negative x-ray resists: Poly(2,3-dichloro-1-propyl-acrylate) mixed with COP for adhesion, and called DCOPA.

Ion-beam resists: Substituted polystyrenes.

ACSI, Milpitas, CA, (408) 262-8000: Electron-beam, photooptical, resist developers only

Convac-APT, Fremont, CA, (510) 659-8370: Electron-beam, photooptical, x-ray, processing equipment only

Cyantek Corp., Fremont, CA, (510) 651-3341: Electron-beam, photooptical, developers only

ESSCHEM, Essington, PA, (215) 521-3800: Electron-beam, x-ray, polymer house only

Genesis, Santa Clara, CA, (408) 986-1636: Silylated and surface imaging systems

> The PAC in novolac resins is also a silylation inhibitor. When it is exposed, the resin becomes silylatable. The dry development of this resist results in images that are better than those obtained even with trilayer resists. Chemically amplified resists are used in both electron-beam and deep uv. In electron-beam, this can help cure proximity effects.

Hoechst Celanese, Somerville, NJ, (908) 231-5800

> DUV, polyhydroxy styrenes and substituted polyhydroxy styrenes
> Positive electron-beam, PMMA, novolacs, and polyhydroxy styrenes
> Negative electron-beam, PN114 (a novolac resin)
> Negative optical, cyclized rubber (polyisoprenes)
> Positive optical, novolac resins
> X-ray, same as positive electron-beam (novolacs)

Jelight Co. Inc., Laguna Hills, CA, (714) 380-8774: UV/DUV, resist strippers only

KTI, Sunnyvale, CA, (408) 733-3500; Wallingford, CT, (203) 265-9242: I-line

Laser Energetics, Orlando, FL, (407) 658-7525: Laser machining tool systems only

Microsi, Inc., Phoenix, AZ, (602) 893-8898: Contrast enhancement, nitrone dye (a bleachable dye)

OCG Microelectronic Materials, West Paterson, NJ, (201) 977-6002: DUV, electron-beam, negative, optical, plasma-resistant, planarizing, positive, silylated, uv/DUV, x ray

OHKRA America, Inc., Milpitas, CA, (408) 956-9901: DUV, negative, positive, two- and three-level processing, uv/DUV, x ray

Semiconductor Systems, Inc., Fremont, CA, (510) 683-8858: electron-beam

Shipley, Marlborough, MA, (508) 481-7950; Whitehall, PA, (215) 820-9777

Deep uv	Acrylics
Positive	Optical
Electron-beam negative	Novolac, chemically amplified
Electron-beam positive	PMMA
X-ray	PMMA and novolac resins
Planarizing	Polydimethy glutarimide
Ion-beam	Novolac and polyvinyl phenols
Surface imaging	Siloxanes, polysiloxanes, novolacs
Silylated	Polysiloxanes

Technics Plasma US, Florence, KY, (606) 283-0200: Ion-beam, strippers only

The Light House, West Bridgewater, MA, (508) 584-4666: DUV, electron-beam, ion-beam, uv/DUV, x-ray, exposure systems, and lamps only

UCB-JSR Electronics, Inc., Sunnyvale, CA, (408) 736-5990

Deep uv	Contain chemical amplifiers (acid-releasing, catalyzing material that unzips the polymer) *tert*-butyl ester of *p*-hydroxy benzoic acid. There are many different polymers for DUV.
Negative optical	The old systems were rubber. There are new ones now.
Positive optical	Novolacs.

In addition, Ref. 16 provides a list of companies supplying materials to the printed wiring board industry.

Materials for Electronic Packages

The rapid development of the microelectronics industry created a need for polymers to be used in the manufacture of IC devices. Polymers are used as interlayer dielectrics on the IC device itself, as die attachment adhesives, encapsulants, conformal coatings, passivation layers within the device, and molding compounds as the final protective layer to form the body of the package. A pictorial representation of one type of microelectronic device illustrating the use of polymer materials is shown in Fig. 13.3.

The main functions of the materials in this package are to provide environmental and mechanical protection to the IC contained there while still maintaining electrical interconnections and adequate thermal paths during operation to minimize overheating of the chip.

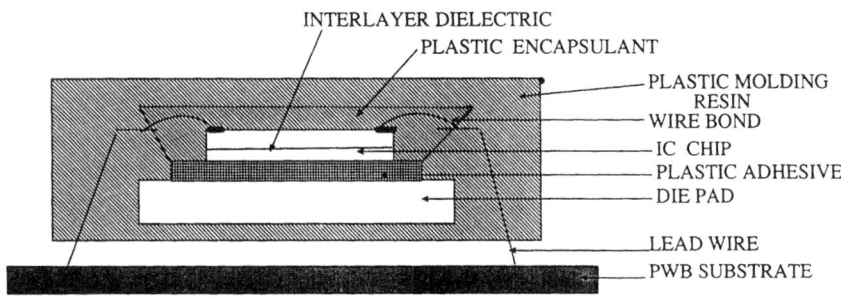

Figure 13.3 Illustration of plastic use in a microelectronic device.

Die attachment materials

Die attachment is the bonding of the silicon wafer containing the IC to the center pad of the lead frame (this is called the *die attachment pad*); see Fig. 13.3. Solder bonding alloys,[17] silver-filled glass,[6] and polymeric adhesives[18] are used. The advantages of the polymer-based adhesives are the lower temperature of attachment of the adhesive to the pad (≤150°C), rework capability, and lower residual stress. Performance requirements for die attachment adhesives include ease of processing [i.e., good dispensability, short cure time and low cure temperature (<300°C)], solventless adhesive, no outgassing during cure or in postcure operations, high adhesive strength, negligible ionic contaminants, high thermal stability, thermal conductivity, and electrical conductivity. The polymer adhesives contain metal fillers (silver) and include epoxy, polyimide, silicon-carbon (SYCAR), silicone, imide-siloxane, and bismaleimide resin). Some suppliers of die attachment adhesives include Johnson Matthey, ITK Inc., Occidental Chem. Co., National Starch, Amicon Co., and Hercules Inc.

It was reported[19] that a novel die attachment adhesive paste was developed that can be cured in 60 s at 200°C, that does not show any appreciable weight loss or outgassing, and that has a high T_g (240°C). The proprietary material is designated JM-7000 from Johnson Matthey, San Diego, California.

Interlayer dielectrics

Multilayer polymer metal structures were fabricated to meet the demands of VLSI and ULSI integrated-circuit fabrication. A key material in this technological development is the use of a thin-film polymer dielectric between the metal layers, called an *interlayer* or *intermetal dielectric*. The purpose of this layer is to isolate the metal conductors so as to prevent short circuits, cross-talk, and/or signal coupling. The ideal dielectric material must meet the following criteria: good dielectric properties (high dielectric strength) and low dielectric constant, hydrolytically stable, thermally stable to withstand the rigors of the assembly operations, low-temperature cure to protect the circuit components, ability to withstand thermal cycling, good adhesion, processible, good chemical resistance, and good planarization properties.[20] That covers a multitude of properties, but of critical importance are the electrical properties because the dielectric layer is quite thin. For a dielectric constant of 3.5 and dielectric strength of 10^6 V/cm and a typical controlled impedance and capacitance, the thickness of the dielectric layer that is required for memory chips is about 1 to 2 μm.[21] A number of polymeric materials were evaluated as interlayer dielectrics and they include epoxy, novolac, acrylic, vinylic, benzocyclobutene,

TABLE 13.3 General Properties of Interlayer Dielectrics

Material property	PI	BCB*	Ultem*	Hybrid triazine*	Quinox	SYCAR
T_g, °C	>300	>350	215	180	>300	160–200
Dielectric constant e	3.5	2.7	3.1	3.3	3.0	2.6
tan δ	0.001	0.0006	0.001	0.025	0.0005	0.002
Coefficient of thermal expansion, ppm/°C	20	27–70	—	65	—	40–80
H_2O, % absorption	0.24	0.25	0.25	—	<1%	0.04
Planarization	OK	OK	OK	OK	OK	OK
Thermal stability	300	300	170	180	300	160–180
Dielectric strength, V/mil	7700	—	—	900	—	750

*See Refs. 22 and 23.

quinoxaline, imide, triazine, and silicon-carbon (SYCAR resins).[7,17,20,22] Of these, the last five polymers offer the best overall combination of properties. Some of their properties are listed in Table 13.3. The properties in this table do not pertain to a specific composition but are more a reflection of the overall material properties. See Refs. 7 and 20 to 22 for additional information. Dow Chemical supplies the BCB resin; Hercules, the SYCAR resins; General Electric, the Ultem material; du Pont, the polyimides; and AT&T, the triazine resin technology.

Encapsulants

Microelectronic devices are protected by a variety of different polymeric encapsulants. These polymeric encapsulants provide only a nonhermetic seal.

The purpose of an encapsulant is to provide protection to the IC devices from moisture, mobile-ion contaminants, and adverse environmental conditions such as temperature, radiation, humidity, and mechanical and physical damage. The polymeric material types used for encapsulants include silicones, epoxy, polyurethane, polyimide, parylene, benzocyclobutene, and the silicon-carbon resins. The properties of these materials have been reviewed in Chaps. 4, 9, and 12. The use of these materials for IC applications is limited by the level of purity that the materials can attain. Ultrapure resin formulations are required for microelectronic applications. For dual in-line package (DIP) devices, epoxies and silicones tend to dominate in die attachment adhesives and wire bonding to the lead frame. For glob-topped encapsulants, the silicones and epoxies are used. On the IC chip itself, polyimides, silicones, epoxy, and BCB polymer coatings can be used as passivating top layers. The requirements of these encapsulants are excellent adhesion, easy processing, low dielectric loss, and low moisture absorption.

The properties of these polymers used as encapsulants were covered in Chap. 9 and will not be discussed again here. Those encapsulants essentially dealt with the surface application (dipping or film) of a coating material. In this section, the discussion is limited to the encapsulation of the entire chip and lead frame by using transfer molding. In the transfer molding process, a polymeric resin is preheated in a mold until it is liquefied. The resin is forced and transferred from this preheat chamber through sprues and runners, through a construction called a *gate,* and finally into a closed mold cavity containing the device to be encapsulated. The processes takes from 1 to 3 min to complete the cure, depending on the resin used as the molding compound. Mold pressures and temperatures range from 500 to 1000 lb/in^2 and 150 to 200°C, respectively. The molding compounds used for device encapsulation include phenolic, epoxy, silicone, epoxy-novolac, and polyimide resins with epoxy-novolac resin being the dominant material. The chemical structures of the epoxy resins are shown in Fig. 13.4. Besides the base resin, the molding compounds contain additives such as the cure agent, filler, catalyst, and extenders to enhance the properties of the molding compound. Some of these properties include adhesion, shrinkage, thermal conductivity, thermal expansion, and moisture absorption. Stress buildup and void formation within the molded resin can adversely affect both the properties of the resin and the device performance. It is best to work with the molder and material scientist to optimize the resin formulation for a specific application.

There have been many improvements in both the transfer molding process and the development of new resin materials for device encapsulation. Some of these improvements have been the reduction of ionic impurities in the molding resins, especially the epoxy resins, improved rheological properties of the molding resins to minimize stress buildup, and the development of new resins for use as molding compounds. Notable among these resins are the liquid-crystal polymers (LCPs). These polymers were discussed in Chap. 3. A key advantage of the liquid-crystal polymers is that their molding cycle time is much faster than that of the other molding resins, not to mention the fact that LCP rheology is more favorable for the molding process. Another material that could be a potential candidate for a transfer-molded resin is polyphenylene sulfide (PPS). This material has good melt flow characteristics and extremely low moisture absorption. Ionic impurities in PPS would prevent this material from being widely used as a microelectronic device encapsulant. For detailed information on plastic packaging of microelectronic devices, see Ref. 5. A general summary of transfer-molding materials and properties is given in Table 13.4.

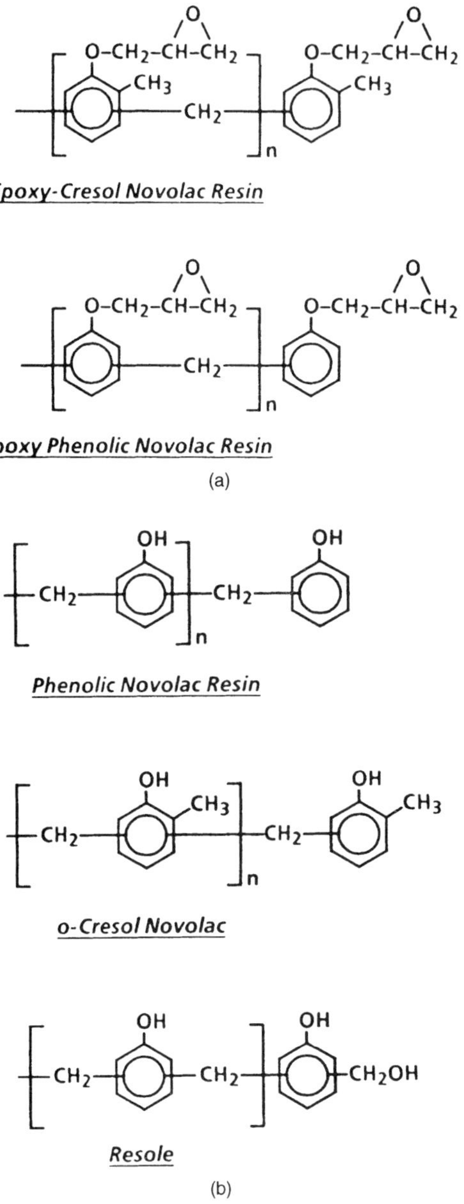

Figure 13.4 Generalized structures of (*a*) epoxy and (*b*) phenolic novolac resins used in transfer-molding processes.

TABLE 13.4 Transfer-Molding Resins

Molding resin	Pros	Cons
Epoxy	Excellent adhesion. Accepts fillers and modifiers well. Low level of ionic impurities. Low pressure molding	Further viscosity and CTE reductions may be difficult to achieve. Higher thermal conductivity needed
Phenolic	Excellent overall balance of properties	High-pressure molding. Electrical properties are fair. Color limited to brown and black
Allyl	Superior moisture barrier. Excellent dimensional properties, heat resistance, and electrical properties	
Imides	Superior heat resistance. High- and low-pressure types	High cost
Silicone	Excellent long-term high-temperature stability and electrical properties. Excellent moisture resistance	Soft. Low thermal conductivity. High CTE
Polyphenylene sulfide	Good rheological properties. Good high-temperature properties. Superior moisture resistance	Ionic impurities. Outgassing. High CTE. Low thermal conductivity
Liquid crystal	Superior rheological properties. High heat resistance	High cost. High-pressure molding. High molding temperatures

Conducting polymers

We can define electrical conductivity as the transport of charge carriers through a medium under the influence of a potential gradient. Conduction is either ionic or electronic. In general, polymers are insulating materials with conductivities in the range 10^{-10} $(\Omega \cdot cm)^{-1}$ for PVC to 10^{-18} $(\Omega \cdot cm)^{-1}$ for PTFE. Figure 13.5 shows the conductivities of various materials. The superior insulating properties of polymers, however, are one of the major reasons that this class of materials has found widespread use throughout the electrical and electronic industries as insulators, dielectrics, resists, etc. Nevertheless, during the 1970s research began on the development of a new class of polymer materials that exhibited levels of conductivity heretofore not obtainable in plastics unless metal fillers, such as silver, nickel, gold, or cop-

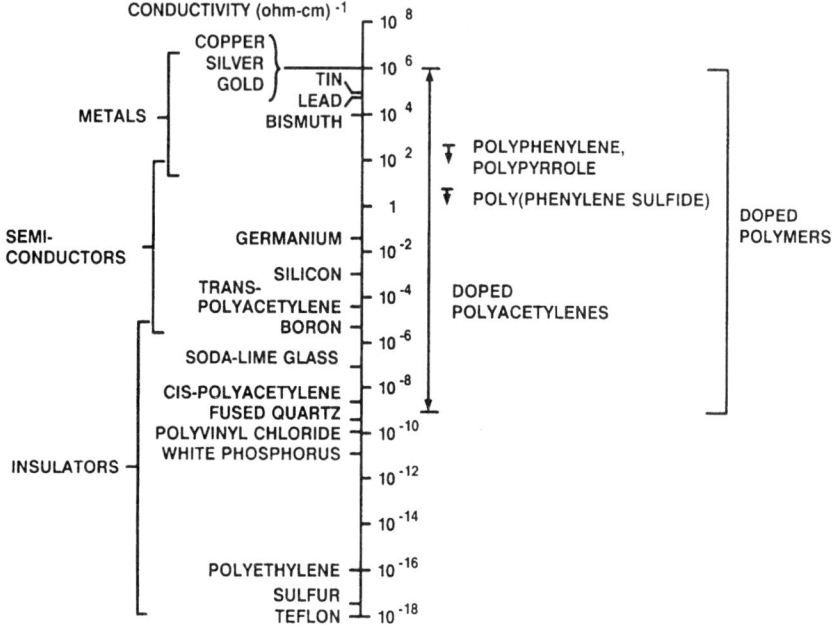

Figure 13.5 Conductivities of various materials.

per, were incorporated into the polymer. The addition of metallic fillers, while imparting electrical conduction, also enhanced the thermal conductivity of the material. The conductivity of filled polymers depended on the type of metal, particle size and shape of the filler, and filler loading. The characteristics of diversity of synthesis, high strength-to-weight ratio, excellent combination of mechanical and physical properties, ease of fabrication and processing into a variety of complex shapes; the potential for molecular engineering of the desired properties; and the low cost of polymers have made them extremely attractive materials for electrically conductive applications. This section is designed to provide the reader with general information about the kinds of organic polymers that conduct electricity, their properties, and where they are used in electrical and electronic applications. See Refs. 24 to 27 for detailed information on conductive polymers.

Conductive polymer types

A disparate number of polymer materials exhibit conductivity, and except for the structural similarity of π-electron conjugation, these polymers are indeed vastly different. Conductive polymers can have aliphatic, aromatic, or heterocyclic groups within their overall chain structure. Examples of conductive polymers are given in Table 13.5.

TABLE 13.5 Some Types of Conductive Polymers

Polymer	Structure	Conductivity,* $1/(\Omega \cdot cm)$ or S/cm
Polyacetylene		500–150,000
Polyphenylene		500
Polyphenylene sulfide		1
Polypyrrole		600
Polythiophene		100
Polyaniline		1–10
Polyquinoline		50
BBL		1–2

*See Refs. 8, 28, and 29.

The conductivity values shown in the table represent a particular composition and may or may not contain a dopant to enhance conductivity. In some cases, the dopant level may be as high as 50 percent of the final weight of the conducting polymer composition. Dopants function as charge-transfer agents and contribute to the conductivity primarily by increasing the charge (electron donors or acceptors) concentration.

Doping is carried out by chemical means by direct exposure of the dopant to the polymer in the gas or solution phase or by electrochemical means (oxidation and reduction). Radiation can also induce conductivity in polymers.[25] Although the list in Table 13.5 is quite short, there are numerous other polymers that have been reported to exhibit varying degrees of conductivity.[25,26] Conductivity is not limited to single polymers; and multicomponent systems such as blends and composites can be made conductive by dispersing the conductive polymer in a mixture of insulating polymers to form a new polymeric matrix with a range of conductivities.[24]

Applications

Although conductive polymers at present are used in a variety of applications, the total quantity of material is quite small. In 1986 the volume use of those materials was estimated at less than 20,000 lb/yr and decreasing.

Potential commercial applications of conductive polymers include power cable sheath (EMI shielding), batteries (as electrodes or electrolytes), signal processing, photovoltaic devices, chemical sensors (electrode coatings) and radiation detectors, fillers for special encapsulants and adhesive formulations, optoelectronic devices, catalysts, Schottky diodes, FETs,[30] and electrostatic agents.

In IC devices, conductive polymer compositions are or could be used as interconnect materials for die attachment adhesives and solder replacements to attachment leads of a component to the PWB.[5,10,31] Probably the single largest area of focused research utilizing conductive polymers is in the development of a high-power, high-energy density polymer battery for electric vehicle applications. The work is sponsored by the U.S. Advanced Battery Consortium (USABC), consisting of Chrysler, Ford, General Motors, the Electric Power Research Institute (EPRI), and the Department of Energy (DOE). The question of whether conductive polymers can be a significant commercial material still remains unanswered. Much work still needs to be done to develop polymers with higher conductivities as well as electrochemical and environmental stability. Despite this formidable task, research is continuing with conductive polymers in the areas of molecular switching, thin-film transistors, transparent electrodes, and photoconductors.[32,33]

Polymer electrolytes

Current is transported by electrons or ions. In the latter case, ions may be anions (negative) or cations (positive). If the ions within a polymer are attracted too strongly to each other, the resultant ion pair is neutral and no charge is conducted under the influence of an electric field. A considerable number of polymer compositions are known which conduct electricity by ion migration. Both organic- and inorganic-based polymers are known. Polymer electrolytes are those materials that allow conduction by ion transport. Typically these materials are polymer-salt complexes such as the polyethylene oxide (PEO)-alkaline salt complexes. Some of these complexes are PEO-LiClO$_4$, PEO-LiCF$_3$SO$_3$, PEO-LiAsF$_6$, PEO-NH$_4$SCN, PEO-LiBF$_4$, PEO-LiPF$_6$, and PEO-LiAlCl$_4$.

The ionic conductivity of these electrolytes is proportional to the ion concentration times its mobility. There are several ways to put ions into polymers. In the gel polymer electrolyte, the polymer is swollen with a salt-solvent solution. In the polymer salt electrolyte, also called *polyelectrolyte,* there are ionic groups (anionic or cationic) bound to the polymer chain which act as counter ions to the unbound groups and potentially mobile ions, thus maintaining the electroneutrality of the salt. In the dry state, these polyelectrolytes are very poor conductors; but when they are wet or swollen with solvent, significant conductivity results due to the increased mobility of the unbound ions. The third way of placing ions into polymers is to have the polymer act as the solvent for the salt. This is called the *solvating polymer electrolyte,* whereby the polymer dissolves the salt. This latter system appears to be the preferred polymer electrolyte because it allows fast ion conduction. Conductance in these polymeric electrolytes take place in the amorphous phase of the polymer and not from defects in the crystalline phase.[27] Typical conductivities are 10^{-6} to 10^{-8} S/cm at room temperature and 10^{-3} to 10^{-4} S/cm at 100°C. Recently, California Institute of Technology researchers working under a NASA contract have developed a gelled polymer electrolyte with a stable bulk room-temperature conductivity of 10^{-3} S/cm. That represents a big step up from other polymeric ion conductors, which generally show only 10^{-7} S/cm conductivities at room temperature. Users must heat conventional materials to around 100°C to achieve useful conductivities, and those temperatures soon degrade polymer performance.

At CalTech, Ganesan Nagasubramanian, Alan Attia, and Gerald Halpert make the electrolyte thin film by pouring a heated mixture of polyacrylonitrile, LiBF$_4$, and propylene carbonate into molds and holding them in a vacuum oven for 24 h.

The resulting 100-μm-thick films showed bulk resistivity of 1000 to 1200 $\Omega \cdot$ cm, a level comparable to liquid electrolytes. The interfacial charge-transfer resistivity eventually stabilizes at 1000 $\Omega \cdot$ cm^2, and double-layer capacitance reaches 3 μF. Tests show Li exchange-current density of 0.05 mA/cm^2. The material retained its electrochemical stability for at least 172 h.[34] The major application area for polymer electrolytes is in the development of high-energy density batteries. The greatest potential advantage of these polymer electrolytes is the fact they could be formed into thin films (with micrometer thickness) which is ideal in the fabrication of solid-state devices. However, much research is needed to improve these polymeric electrolytes so as to be useful in this area. This section on polymer electrolytes only scratched the surface of this subject area. It was intended to introduce the reader to a new area of polymer research and how these materials could affect the electrical and electronics industry. References 24, 27, and 35 provide more in-depth coverage.

Langmuir-Blodgett films

Ever since the fascination of Benjamin Franklin in the 18th century with the spreading of oil on water, interest in the preparation of thin films (1 molecule thick) has grown tremendously. Calculations based on this early experiment (2 cm^3 oil/acre of water) indicate a surface film thickness of approximately 1 nm. From this experiment and others came what is now referred to as the *Langmuir-Blodgett (L-B) film technology*. L-B film technology allows one to construct organic monolayer films with controlled molecular geometries. The physical technique of doing this involves spreading a suitable organic molecule onto the water surface, compressing the resultant film to form a compact monolayer, and transferring that film to a suitable substrate. Transference is accomplished by pulling the substrate through the film-water layer. References 36 to 40 provide considerable detail about L-B film technology. This section on L-B films will serve only as a brief introduction to the material types and applications. Polymeric materials can function as active or passive elements in electronic devices. The growth of the polymer and electronic industries has been almost synergistic; and as each industry advanced, new demands were placed on material properties and device performance. In the microelectronics area, devices are becoming smaller and more complex, and the level of control one needs over the material and device architecture is becoming exceedingly more difficult. Langmuir-Blodgett technology has made it possible to process and control polymeric materials to precise levels of thickness and architecture at the molecular level, and indeed this ability makes the L-B technique one of the primary methods of choice in the construction of thin-film microelectronic devices. In general, to form monolayers by using L-B techniques, organic molecules must possess appropriate structural characteristics, i.e., contain a hydrophilic group and a hydrophobic group. The former group promotes surface activity on the water (i.e., spreading), and the latter prevents the molecule from dissolving in the water medium and provides control over the molecule's orientation. The molecules used to build single or multiple monolayers can be small (nonpolymeric) or large (polymeric) so that theoretically any organic molecule that had the necessary hydrophilic and hydrophobic features could be converted to monolayers by using L-B technology. Some polymers that have been converted to L-B films are the polyimides,[41] polythiophenes,[4] poly(p-phenylenevinylene),[4] poly-pyrrole, polyvinyl stearate,[7] phthalocyanines,[4] and polyacetylene.[4,7] Some areas of application for L-B films include sensors, pyroelectric devices, electrooptics, resists, quantum mechanical tunneling devices, biofilms, and field effect transistors (FETs). L-B film technology has considerable potential for microelectronic device development on a molecular level.

Polymer waveguides

One area of study in the use of polymers for photonic applications is polymer waveguides for the interconnection of electronics and optics. In electronics, information is transmitted via electrons through copper conductors. In optics, that same information is transmitted via photons through a waveguide. An optical waveguide is simply a pathway to guide optical signals from one point to another. Polymers can provide the medium through which to send the guided wave. The integration of optics and electronics is becoming the technology of choice for the development of high-performance electronic processor systems, and waveguides provide a technology for achieving this. Optical waveguides and interconnections offer several key advantages over electrical interconnections, e.g., improved thermal management, reduced propagation delay, large bandwidth, capability of being fabricated into thin films, low crosstalk, and the fact that waveguides occupy negligible space within the system. A number of areas need to be addressed before successful integration can be accomplished, at least in multilayer printed wiring boards: The material must be able to be manufactured and fabricated on a high-volume level; the material must have low losses; the material must lend itself to complex structure fabrication (straights, cross-throughs, bends, fanouts); transmitter-receiver coupling must be improved; and the waveguide material must be stable over the long term. There are a number of optical pathway options for backplanes, but polymer waveguides offer considerable promise.[42,43] The class of polymers that seem the most appropriate for this application is the photopolymerizable resins because as a class they are highly efficient in terms of fabrication, processing, properties, and overall compatibility with the PWB manufacturing process. The characteristics of these resins are as follows: (1) solventless resins, (2) rapid cure (<3 s),

(3) complete resin utilitization, (4) compatibility with heat-sensitive substrates, (5) PWB manufacturing compatibility, (6) wide formulation latitude, and (7) balance of electrical chemical, mechanical, and thermal properties.

Several groups have fabricated polymeric optical waveguides. Salvaraj et al.[44] and Sullivan et al.[45,46] investigated polyimides; Kapoor et al.[47] looked at polyurethanes; Townsend et al.[49] studied polyacetylenes; Ives and Reichert[50] studied polystyrenes; du Pont manufactures a photopolymer called Polyguide; and Mitsubishi Gas Chemical Co. designed a polycarbonate resin for optical waveguide applications. Hartman et al.[48] used commercially available radiation-curable polymer adhesives to create channel optical waveguides on PWB such as FR-4, ceramic, and silicon. Westinghouse[51] is investigating a number of uv-curable acrylated epoxies, novolacs, urethanes, imides, esters, and amide-imide polymers. These resins were developed in house, and a number of waveguides were fabricated on PWB that contained straights, bends, and crossovers. Polymer waveguides up to 125 μm thick were selectively patterned with uv through a photomask. Propagation losses on selected waveguides up to 3 in long ranged from 0.07 to 0.3 dB/cm at 0.63- and 0.78-μm wavelengths. To ensure that the light waves are confined to the waveguide core, the refractive index of the core must be larger than that of its surroundings. Thus the uv-curable acrylated resins mentioned above could function as both the core and the cladding material depending on their refractive index. This technology is at an early stage of development, and much work needs to be done in the areas of material development, fabrication, processing, and long-term waveguide stability. It is difficult to predict when or even if polymer waveguides will become practical, but remarkable progress has been made in just a few years. More detailed information can be obtained in Refs. 52 and 53.

References

1. "Advanced Packaging and Interconnections," *IEEE MICRO,* vol. 13, no. 2, April 1993.
2. "Configuring for High Performance: Multichip Modules," *Computer,* vol. 26, no. 4, April 1993.
3. "Special MCM Issue," *International Journal of Microcircuits and Electronic Packaging,* vol. 15, no. 4, 4th quarter 1992.
4. C. P. Wong, ed., *Polymers for Electronic and Photonic Applications,* Academic Press, New York, 1993.
5. L. T. Mangione, *Plastic Packaging of Microelectronic Devices,* Van Nostrand Reinhold, New York, 1990.
6. D. S. Soane and Z. Martyneko, *Polymers in Microelectronics,* Elsevier, New York, 1989.
7. J. H. Lai, *Polymers for Electronic Applications,* CRC Press, Boca Raton, FL, 1989.
8. M. J. Bowden and S. R. Turner, eds., *Electronic and Photonic Applications of Polymers,* American Chemical Society, Washington, 1988.

9. *IEEE Transactions on Components, Hybrids and Manufacturing Technology*, vol. 15, nos. 1–6, February-December 1992.
10. M. T. Gossey, ed., *Plastics for Electronics*, Elsevier, New York, 1985.
11. J. M. Shaw, "Overview of Polymers for Electronic and Photonic Applications," in C. P. Wong, ed., *Polymers for Electronics and Photonic Applications*, Academic Press, New York, 1993, Chap. 1, pp. 10–11.
12. E. W. Rutter, E. Moyer, R. Harris, D. Frye, V. L. St. Jeor, and F. Oaks, "A Photodefinable Benzocyclobutene Resin for Thin Film Microelectronic Applications," *Proceedings of The First International Conference on MCM (ISHM and IEPS)*, Denver, CO, April 1992, p. 394.
13. E. S. Moyer, E. Rutter, M. Bernius, P. Towsend, R. Harris, H. Pranjoto, and D. Denton, "Photodefinable Benzocyclobutene Formulations for Thin Film Microelectronic Applications. Part II," *IEPS Proceedings*, Austin, TX, September 1992.
14. R. D. Allen, G. Wallraff, W. Hinsberg, W. Conley, and R. Kunz, "New Single Layer Positive Resists for 193 nm and 243 nm Lithography Using Methacrylate Polymers," *Solid State Tech.*, November 1993, p. 53.
15. S. Wolf and R. N. Tauber, "Silicon Processing for the VSLI Era," in *1993 Solid State Technology Resource Guide*, vol. 1, Lattice Press, Sunset Beach, CA, 1986, p. 102.
16. D. J. Esposito, ed., *Printed Circuit Fabrication*, Miller Freeman Publisher, San Francisco, 1993.
17. L. T. Manjione, *Plastic Packaging of Microelectronic Devices*, Van Nostrand Reinhold, New York, 1990, p. 82.
18. C. A. Dostal, ed., *Electronic Materials Handbook*, vol. 1: *Packaging*, ASM International, Materials Park, OH, 1989, p. 470.
19. M. N. Nguyen and M. B. Grosse, "Low Moisture Polymer Adhesive for Hermetic Packages," *IEEE Transactions on Components, Hybrids and Manufacturing Technology*, vol. 15, no. 9, December 1992, p. 964.
20. A. Schlitz, "A Review of Planar Techniques for Multichip Modules," *IEEE Transactions on Components, Hybrids and Manufacturing Technology*, vol. 15, no. 2, April 1992, p. 236.
21. D. S. Soane and Z. Martynenko, *Polymers in Micro-Electronics*, Elsevier, New York, 1989, p. 170.
22. *The International Journal of Microcircuits and Electronic Packaging*, vol. 15, no. 4, 4th quarter 1992.
23. J. R. Susko, R. W. Synder, and R. A. Susko, eds., *Proceedings of Symposium on Polymeric Materials for Electronic Packaging and High Technology Applications*, Electrochemical Society, vol. 88-17, 1988.
24. T. A. Skotheim, ed., *Electroresponsive Molecular and Polymeric Systems*, vols. 1 and 2, Marcel Dekker, New York, 1988.
25. D. B. Cotts and Z. Reyes, *Electrically Conductive Organic Polymers for Advanced Applications*, Noyes Data Corp., Park Ridge, NJ, 1986.
26. T. A. Skotheim, ed., *Handbook of Conducting Polymers*, vols. 1 and 2, Marcel Dekker, New York, 1986.
27. J. R. MacCullum and C. A. Vincent, eds., *Polymer Electrolyte Reviews I*, Elsevier, New York, 1987.
28. H. Mark, N. Bakales, C. Overberger, G. Menges, eds., *Encyclopedia of Polymer Science and Engineering*, 2d ed., vol. 5, Wiley, New York, 1986, p. 462.
29. Kim Oh-Kil, "Electrical Conductivity of Heteroaromatic Ladder Polymers," *J. Polym. Sci. Polym. Letters, Ed.*, vol. 20, 1982, p. 663.
30. P. Singer, ed., "Conductive Polymers Studied," in *Semiconductor International*, January 1993, p. 26.
31. G. P Nguyen, J. Williams, and E. Gibson, "Conductive Adhesives," *Circuits Assembly*, January 1993, p. 36.
32. V. Comello, "Conducting Polymers Finding Niche Uses," *Res. & Dev. Magazine*, July 1993, p. 63.
33. R. Dagani, "Polymer Film Effect Is Basis of Molecular Switch," *C&EN*, February 22, 1993, p. 24.

34. *High-Tech Materials Alert,* vol. 10, no. 12, Technical Insights Inc., Englewood, NJ, December 1993.
35. R. G. Linford, ed., *Electrochemical Science and Technology of Polymers,* vols. 1 and 2, Elsevier, New York, 1987–1988.
36. G. G. Roberts, "An Applied Science Perspective of Langmuir-Blodgett Films," *Advances in Physics,* vol. 34, no. 4, 1985, p. 475.
37. T. M. Ginnai, "Monomolecular Films: Trends and Materials for Electronic Applications," *Ind. Eng. Chem. Prod. Res. Dev.,* vol. 24, 1985, p. 188.
38. V. K. Agrawal, "Langmuir-Blodgett Films," *Physics Today,* vol. 40, 1988.
39. I. R. Peterson, "Langmuir-Blodgett Films: Structure and Application," *J. Molec. Elec.,* vol. 3, 1987, p. 103.
40. G. Roberts, ed., *Langmuir-Blodgett Films,* Plenum Press, New York, 1990.
41. H. Sotobayaski, T. Schilling, and B. Teschi, "Scanning Tunneling Microscopy of Polyimide Monolayers Prepared by the Langmuir-Blodgett Technique," vol. 6, 1990, p. 1246.
42. M. J. Goodwin, "Optical Interconnect Technologies for High Performance Electronic Processor Systems," *GEC J. of Res.,* vol. 10, no. 2, 1993, p. 81.
43. A. J. Williams and A. C. Carter, "Optoelectronic Integrated Circuits," *GEC J. of Res.,* vol. 10, no. 2, 1993, p. 91.
44. R. Salvaraj, H. T. Lin, and J. F. McDonald, "Integrated Optical Waveguides in Polyimide for Wafer Scale Integration," *J. Lightwave Tech.,* vol. 6, no. 6, 1988, p. 1034.
45. C. Sullivan, A. Guha, J. Bristow, and A. Husain, "Optical Interconnections for Massively Parallel Architectures," *Appl. Optics,* vol. 29, no. 8, 1990, p. 1077.
46. C. Sullivan and A. Husain, "Guided-Wave Optical Interconnects for VLSI Systems," *Opt. Computing and Nonlinear Materials,* vol. 881, 1988, p. 172.
47. S. K. Kapoor, C. D. Pandey, J. Joshi, A. Dawar, N. Tripathy, and V. Gupta, "Fabrication and Characterization of Polyester and Acrylic Polyurethane Optical Waveguides," *Appl. Optics,* vol. 28, no. 1, 1989, p. 37.
48. D. Hartmann, G. Lalk, J. Howse, and R. Krchnavek, "Radiant Cured Polymer Optical Waveguides on Printed Circuit Board for Photonic Interconnection Use," *Appl. Optics,* vol. 28, no. 1, 1989, p. 40.
49. S. D. Townsend, G. L. Baker, N. Schlotter, C. Klausner, and S. E. Temad, "Fabrication of Waveguide Structures from Soluble Polydiacetylenes," *Synth. Metals,* vol. 28, 1989, p. 639.
50. J. T. Ives and Reichert, "Polymer Thin Film Integrated Optics," *J. Appl. Poly. Sci.,* vol. 36, 1988, p. 429.
51. W. M. Alvino, L. Cargnel, T. Henningsen, I. Lieberman, H. Saunders, and E. Supertzi, "Polymeric Optical Waveguides," Paper presented at 25th central regional meeting of the ACS, October 4–6, 1993, Pittsburgh.
52. J. E. Midwinter, ed., *Photonics in Switching,* vol. 1, Academic Press, New York, 1993.
53. B. L. Booth, "Polymers for Integrated Optical Waveguides," in C. P. Wong, ed., *Polymers for Electronic and Photonic Applications,* Academic Press, New York, 1993.

Part 4

Reference Sources on Plastic Materials

Chapter 14

Information Sources

This chapter is intended to provide the reader with a directory of where to find more detailed information on plastics in general and on their involvement in the electrical and electronic industries. These resources will provide information on theory, properties, processing, applications, and manufacturers of plastics as well as electrical and electronic industry resources. This chapter lists technical societies, private industry sources, property databases, conferences and symposiums, periodicals, encyclopedias, and reference books that will be helpful to both experienced and inexperienced persons in the fields of plastics and electronics.

Technical Societies

American National Standards Institute (ANSI)[1, 2]
11 W. 42d St., New York, NY 10036-8002
(212) 642-4900

ANSI does not develop standards but provides the means for determining the need for standards; it ensures that qualified organizations develop standards, and it coordinates standards approval in the United States. ANSI represents U.S. interests in international standardization and provides information on and access to world standards such as the International Organization for Standardization (ISO) and the International Electrotechnical Commission (IEC). Its membership includes 1300 companies; 260 trade associations, professional societies, labor and consumer groups; and 30 government organizations. Publications include *ANSI Reporter,* a newsletter that updates members on major national and international standards activities; *Standards Action,* a newsletter that outlines all national draft standards currently under the approval process; *Annual Catalog and Supplement,* a complete list of all ANSI-approved standards.

Electrical Apparatus Service Association (EASA)[3]
1331 Baur Boulevard, St. Louis, MO 63132
(314) 993-2220

EASA is an international trade organization of more than 2800 companies that sell service and repair industrial electric motors, generators, transformers, and related electromechanical and electronic equipment. It provides engineering, consulting, technical publications, and computer engineering programs as well as seminars on all current aspects of the above-mentioned service and repair of electric equipment.

Publications include *AC Motor Redesign, Internal Connection Diagrams, EASA Technical Manual,* and the *Core Iron Study.*

Electronic Industries Association[4]
2001 Pennsylvania Ave. NW, Washington, DC 20006-1813
(202) 457-4900

EIA has expanded since its birth in 1924 and has absorbed many electronic industry associations over the years. It services all facets of the electronics industry and is composed of the following groups:

Industrial Electronics Group

Consumer Electronics Group

Government Division

Telecommunications Sector and the Electronic Industries Foundation

The EIA through its various groups serves the entire electronics industry in a variety of ways, all designed to enhance the electronics industry as a whole and its individual members.

Institute of Electrical and Electronics Engineers (IEEE)[5]
345 E. 47th St., New York, NY 10017
(212) 705-7900

The IEEE is the world's largest technical professional society. The objectives of the IEEE focus on advancing the theory and practice of electrical and electronics science through the sponsorship of conferences, symposia, and local meetings. Industry coverage of IEEE includes aerospace, communications, computers, electric power, consumer electronics, and biomedicine. A multitude of publications are available from IEEE covering various aspects of the industries mentioned above. Of particular interest to plastics users is the EI/DEIC group within this organization and the Electrical Insulation Conference sponsored by this organization.

American Society for Testing and Materials (ASTM)[6]
1916 Race St., Philadelphia, PA 19103
(215) 299-5585

The ASTM develops consensus standards on specifications, tests, practices, guides, and definitions for a variety of materials, products, systems, and services. ASTM also publishes information on state-of-the-art testing techniques. The 1993 ASTM publications of specific interest to plastics and electronics are as follows:

ASTM vols. 08.01 to 08.04, *Plastics*

ASTM vols. 09.01 and 09.02, *Rubber*

ASTM vols. 10.01 to 10.05, *Electrical Insulation and Electronics*

ASTM ISO Handbook 21 on Plastics

ASTM also has an International Directory of Testing Laboratories and a number of publications on a variety of subjects from activated carbon to wood.

The Institute for Interconnnecting and Packaging Electronic Circuits (IPC)[7]
7380 N. Lincoln Ave., Lincolnwood, IL 60646
(708) 677-2850

The IPC is a U.S.-based trade association with international membership that brings together all the players in the interconnection industry. Its membership comprises manufacturers and users of printed wiring boards, flat cable, discrete wiring hybrid circuits, and printed wiring board assemblers. Also included in its membership are suppliers to the industry, government, and educational agencies. The IPC develops standards that cover all facets of the electronic interconnection industry and include engineering and design, materials, components, printed board fabrication, interconnection structure performance, quality, and testing. The IPC also provides training programs, technical conferences, publications, marketing, and environmental information about the interconnection industry.

National Electrical Manufacturers Association (NEMA)[8]
2101 L Street, NW, Suite 300, Washington, DC 20037
(202) 457-8490

NEMA serves the electrical and electronics industries by improving the competitiveness of its member companies by providing services to promote the best there is in safe manufacture and use of electrical products, by development of codes and standards, by representing industry in government, and by conducting educational forums. Each NEMA subdivision is responsible for its product(s), and they are as follows: industrial automation, lighting equipment, electronics, industrial equipment, insulating materials, wire and cable, building equipment, power equipment, diagnostic imaging, and therapy systems division.

Society of Plastics Engineers (SPE)[9]
14 Fairfield Dr., Brookfield, CT 06804
(203) 775-0471

The SPE is an international professional society dedicated to promoting scientific and engineering knowledge relative to plastics. It accomplishes this through publications, technical conferences, and seminars on plastics-related issues. It publishes a list of books on various aspects of the fundamental properties and processing of plastics and provides video cassettes and computer software on plastics. It also holds regional and annual technical conferences.

Society of the Plastics Industry, Inc. (SPI)
1275 K St., NW, Suite 400, Washington, DC 20005
(202) 371-5200

The SPI was established in 1937 to promote the development of the plastics industry and to enhance public understanding of the nature and uses of plastics. It has a wide range of activities that addresses many facets of the plastics industry through its operating units, industry trade expositions, regional and sectional meetings, and special service groups. The SPI also provides extensive publications on plastics.

Underwriters Laboratories (UL)
333 Pfingsten Rd., Northbrooks, IL 60062
(708) 272-8800

Underwriters Laboratories is chartered under the laws of the state of Delaware as a not-for-profit organization to establish, maintain, and operate a laboratory to investigate the safety aspects of devices, systems, and materials. Its principal business is the safety investigation of (1) electrical and electronic equipment, products, and components; (2) mechanical products; (3) building materials and construction systems; (4) fire protection equipment; (5) burglary protector systems; and (6) marine products. The UL mark is the most widely recognized mark of compliance with safety requirements. UL offers the following services to manufacturers and inspection authorities: Listing and identification of UL-approved products, classification of product safety issues, certification services, field evaluation, international compliance and inspection, and research. It informs its members and the industry at large with publications, meetings, and conferences.

Plastics Information Databases[10]

Material selection is often an ordeal for anyone who is responsible for choosing the right material for a given application. A number of materials databases are available to make the materials engineer's choice a little easier, and a database directory is provided in Table 14.1.

Conferences

All the organizations listed in this chapter support national, regional, and local conferences, meetings, and seminars. Because these events are varied, the reader is advised to contact the organization for an up-to-date program of events for the year.

Publications

Numerous sources of information are available on plastics and electronics in the form of encyclopedias, handbooks, and periodicals. A list of some of these is given here.

General polymer information

Encyclopedia of Chemical Technology, Wiley

Encyclopedia of Polymer Science and Technology, Wiley

Encyclopedia of Science and Technology, McGraw-Hill

Modern Plastics, McGraw-Hill

Polymer Handbook, Wiley-Interscience

Electrical and electronic and polymers

IEEE Electrical Insulation Magazine

International Journal of Microcircuits and Electronic Packaging

IEEE Transactions on Electrical Insulation

TABLE 14.1 Plastic Materials Database Directory

Company	Product	Materials	Data types	Formats	Cost, $
ASM International Materials Park, OH (216) 338-5151	Mat. DB	Over 8,000 plastics, also other materials	Properties	PC, publication	1,064 with plastics
BASF Corp. Parsippany, NJ (800) 227-3746	CAMPUS	BASF materials	Properties, ratings, chemical, resistance	PC	F.Q.C. (free to qualified customers)
D.A.T.A. Business Publishing Englewood, CO (800) 447-4666	D.A.T.A Plastics Digest	Over 14,000 thermoplastics, thermosets, elastomers	Properties, ratings	Publication	220
	Adhesive Digest	Over 7,800 adhesives, sealants, primers	Properties, ratings	Publication	180
Dow Plastics Midland, MI (800) 258-CHEM	591 Ways to Succeed	Dow materials	Properties, ratings, chemical resistance	PC	F.Q.C.
GE Plastics Pittsfield, MA (800) 845-0600	Engineering Design Database (EDD)	GE thermoplastics and foams	Properties, ratings, design data	On-line	F.Q.C.
Hoechst Celanese Corp. Los Angeles, CA (800) 235-2637	Fast Focus	Hoechst materials and equivalents	Properties, ratings, chemical resistance	PC	F.Q.C.
IDES, Inc. Laramie, WY (307) 742-9227	Prospector*	Over 10,000 thermoplastics, thermosets, elastomers, films	Properties, ratings, design data, chemical resistance	PC, Macintosh	495–695
Information Indexing Garden Grove, CA (800) 888-0608	CenBase Materials	Over 10,000 plastics, also other materials	Properties, design data, chemical resistance	PC, workstation	1,450–3,500

348

Company	Product	Description	Type	Price
LNP Engineering Plastics, Exton, PA (215) 363-4500	EPOS	LNP materials	Properties, ratings, design data, chemical resistance	PC	F.Q.C.
McGraw-Hill Inc./Polydata, New York, NY (800) 845-5056	DataPlus	7,000 engineering thermoplastics	Properties, ratings, chemical resistance	PC	995–3,000
Miles, Pittsburgh, PA (412) 777-3800	CAMPUS	Miles materials	Properties, ratings, chemical resistance	PC	F.Q.C.
Plaspec, Yardley, PA (800) 743-1060	Plaspec	Over 10,000 thermoplastics, thermosets, elastomers	Properties, ratings, chemical resistance	On-line	$100–160
Plastics Design Library, New York, NY (212) 838-2817	Chemical Compatibility and ESCR†	More than 60 families of plastics	Chemical resistance	Publication	285
Prime Alliance, Des Moines, IA (800) 247-8038	Prime Alliance Database	Over 700: BASF, Miles, Mobil, Monsanto, OxyChem, Rexene	Properties, ratings	PC	F.Q.C.
Rapra Technology Ltd. Shropshire, England Telex 35134	Plascams	Contains generic materials	Properties publication	PC	990

*Also available: Locator with properties and ratings, $199; Chemplas with ratings and chemical resistance, $395.
†Also available: *Effect of Temperature on Plastics*, $285; *Effect of Creep on Plastics*, $285.
SOURCE: Ref. 10. Reprinted with permission from *Plastics Design Forum*.

Electronic Packaging and Production, Cahners

Printed Circuit Fabrication, Miller Freeman, Inc.

Circuits Assembly, Miller Freeman, Inc.

Electronics, McGraw-Hill

Electrical World, McGraw-Hill

Newspapers[11]

Chemical Marketing Reporter, 100 Church St., New York, NY 10007, (212) 732-9820

Journal of Commerce, 110 Wall St., New York, NY 10005, (212) 425-1616

Plastics News, 1725 Merriman Rd., Suite 300, Akron, OH 44313, (216) 836-9180

Trade Associations and Professional Organizations[11]

American Chemical Society, 1155 16th St., NW, Washington, DC 20036, (202) 872-4600

Chemical Management and Resource Association, 60 Bay St., Suite 702, Staten Island, NY 10301, (718) 876-8800

Chemical Manufacturers Association, 2501 M St., NW, Washington, DC 20037, (202) 887-1100

Chemical Specialties Manufacturers Association, 1001 Connecticut Ave., NW, Washington, DC 20036, (202) 223-8891

Composite Can and Tube Institute, 1818 N St., NW, Suite T-10, Washington, DC 20036, (202) 223-4840

Drug, Chemical and Allied Trades Association, 42-40 Bell Blvd., Suite 604, Bayside, NY 11361, (718) 229-8891

Dry Color Manufacturers Association, 206 N. Washington St., Suite 202, P.O. Box 20839, Alexandria, VA 22314, (703) 684-4044

Flexible Packaging Association, 1090 Vermont Ave., NW, Suite 500, Washington, DC 20005, (202) 842-3880

INDA, 1700 Broadway, New York, NY 10019, (212) 582-8401

International Institute of Synthetic Rubber Producers, 2077 S. Gessner Rd., Suite 133, Houston, TX 77063, (713) 783-7511

International Isocyanate Institute, 119 Cherry Hill Rd., Parsippany, NJ 07054, (201) 263-7517

National Paint and Coatings Association, 1500 Rhode Island Ave., NW, Washington, DC 20005, (202) 462-6272

Polyurethane Manufacturers Association, 1500 K St., NW, Washington, DC 20005, (202) 682-1338

Society of Plastics Engineers, 14 Fairfield Dr., Brookfield Center, CT 06805, (203) 775-0471

Society of the Plastics Industry-Canada, 1262 Don Mills Rd., Suite 104, Don Mills, Ontario M3B2W7, Canada, (416) 449-3444

Synthetic Organic Chemical Manufacturers Association, 1330 Connecticut Ave., NW, Suite 300, Washington, DC 20036, (202) 659-0060

Trade Publications[11]

Canadian Plastics, 1450 Don Mills Rd., Don Mills, Ontario M3B2X7, Canada, (416) 445-6641

Chemical Engineering, 1221 Ave. of the Americas, New York, NY 10020, (212) 512-3198

Chemical and Engineering News, 1155 16th St., NW, Washington, DC 20036, (202) 872-4600

Chemical Week, 810 7th Ave., New York, NY 10019, (212) 586-3430

Material Engineering, 1100 Superior Ave., Cleveland, OH 44114, (216) 696-7000

Modern Plastics, 1221 Ave. of the Americas, New York, NY 10020, (212) 512-6242

Packaging Digest, 122 E. 42d St., New York, NY 10168, (212) 867-9191

Plastics Compounding, 1129 E. 17th Ave., Denver, CO 80118, (303) 832-1022

Plastics Connection—Plastics Focus, P.O. Box 814, Amherst, MA 01004, (413) 549-5020

Plastics Design Forum, 1129 E. 17th Ave., Denver, CO 80218, (303) 834-1022

Plastics Design and Processing, 17730 Peterson Rd., Libertyville, IL 60048, (312) 362-8711

Plastics Engineering, 14 Fairfield Dr., Brookfield Center, CT 06805, (203) 775-0471

Plastics Industry News, 1129 E. 17th Ave., Denver, CO 80218, (303) 832-1022

Plastics Machinery and Equipment, 1129 E. 17th Ave., Denver, CO 80218, (303) 832-1022

Plastics News, Crain Communications Inc., 1725 Merriman Rd., Akron, OH 44313, (216) 836-9180

Plastics Packaging, 1129 E. 17th Ave., Denver, CO 80218, (303) 832-1022

Plastics Technology, 633 Third Ave., New York, NY 10017, (212) 984-2283

Plastics Trends, P.O. Box 3640, Culver City, CA 90231-3640, (213) 337-9717

Plastics World, 275 Washington St., Newton, MA 02158, (617) 964-3030

Business Consulting Groups and Publishers[11]

Battelle Columbus Laboratories, 505 King Ave., Columbus, OH 43201, (614) 424-5579

Bonner & Moore Marketing Consultants, 2727 Allen Pkwy., Houston, TX 77019, (713) 522-6800

Business Communications Co., Inc., 25 Van Zant St., Norwalk, CT 06855, (203) 853-4266

CMAI, 11757 Katy Fwy., Suite 750, Houston, TX 77079, (713) 531-4660

CPI Consulting Associates, Inc., 244 W. Chester, White Plains, NY 10604, (914) 949-4402

Charles River Associates, John Hancock Tower, 200 Clarendon St., Boston, MA 02116, (617) 266-0500

Chem Systems Inc., 303 S. Broadway, Tarrytown, NY 10591, (914) 631-2828

Chemical Data, Inc., 2900 N. Loop West, Suite 830, Houston, TX 77092, (713) 683-3900

Composite Services Corp., 56 Glenwood Ave., Demarest, NJ 07627, (201) 767-4164

DRI/McGraw-Hill, 24 Hartwell Ave., Lexington, MA 02173, (617) 863-5100

DeWitt & Company, Inc., 120 N. Atrium, 16800 Greenspoint Park, Houston, TX 77060, (713) 875-5525

Eldib Engineering & Research, Inc., 613 Springfield Ave., Berkeley Heights, NJ 07922, (201) 464-2244

Franklin Associates, Ltd., 4121 W. 83d St., Prairie Village, KA 66208, (913) 649-2225

Frost & Sullivan, Inc., 106 Fulton St., New York, NY 10038, (212) 233-1080

Hull & Company, 5 Oak St., P.O. Box 4520, Greenwich, CT 06830, (203) 622-9120

International PC, Inc., P.O. Box 79224, Houston, TX 77279, (713) 497-6053

International Plastics Consultants Corp., 1492 High Ridge Rd., Stamford, CT 06903, (203) 968-1233

Kline & Co., Inc., 165 Passaic Ave., Fairfield, NJ 07006, (201) 227-6262

R. M. Kossoff & Associates, Inc., 10 Rockefeller Plaza, New York, NY 10020, (212) 246-4035

Arthur D. Little, Inc., 25 Acorn Park, Cambridge, MA 02140, (617) 864-5770

Lovell Associates, 421 Burd St., Pennington, NJ 08534, (609) 737-3622

Margolis Marketing & Research Co., 232 Madison Ave., Suite 905, New York, NY 10016, (212) 685-1278

Marilyn Bakker Technology Forecast, 9 Drumlin Rd., Westport, CT 06880, (203) 226-9929

Market Search, Inc., 2727 Holland Sylvania Rd., Suite A, Toledo, OH 43615, (419) 535-7899

Mastio & Company, 11263 Wright St., Omaha, NE 68144, (402) 334-7899

Morton Research Corporation, 28 Merrick Ave., Merrick, NY 11566, (516) 378-1066

Noyes Data Corp., Mill Road at Grand Ave., Park Ridge, NJ 07656, (201) 391-8484

Peter Sherwood Associates, Inc., 20 Harlem Ave., White Plains, NY 10603, (914) 761-3033

Phillip Townsend Associates, P.O. Box 90327, Houston, TX 77090, (713) 873-8733

Predicasts, Inc., 11001 Cedar Ave., Cleveland, OH 44106, (216) 795-3000

Skeist Laboratories, Inc., 375 Route 10, Whippany, NJ 07981, (201) 515-2020

Springborn Laboratories, Inc., 20 Springborn Ctr., Enfield, CT 06082, (203) 749-8371

SRI International, 333 Ravenswood Ave., Menlo Park, CA 94025, (415) 326-6200

Strategic Analysis Inc., 11 Fairlane Rd., Reading, PA 19606, (215) 779-9080

The Target Group, 1000 Harston Lane, Philadelphia, PA 19118, (215) 233-4083

Technomic Consultants, Inc., 300 S. Riverside Plaza, Chicago, IL 60606, (312) 876-0004

Technomic Publishing Co., 851 New Holland Ave., Box 3535, Lancaster, PA 17604, (717) 291-5609

Government Publications[11]

Annual Survey of Manufacturers, Bureau of the Census

Census of Manufacturers, Bureau of the Census

Current Industrial Reports, Bureau of the Census

Foreign Trade Reports—1M 146 (imports), EM 545 (exports), Bureau of the Census

Guide to Census Bureau Data Files and Special Tabulations, Bureau of the Census

Guide to Foreign Trade Statistics, Bureau of the Census

Producer Price Indexes, Bureau of Labor Statistics

Standard Industrial Classification Manual, Office of Management and Budget

Statistical Abstract of the United States, Department of Commerce

Statistical Services of the U.S. Government, Office of Management and Budget

Synthetic Organic Chemicals, International Trade Commission

U.S. Industrial Outlook, Department of Commerce

Plastics Suppliers

There are literally hundreds of companies that supply plastic materials, and rather than list them, it seems more appropriate to list seven references that contain the information to locate the supplier of a particular plastic.

Modern Plastics Encyclopedia, McGraw-Hill

Machine Design—Materials Selector Issue, Penton Publication

ChemCycloPedia, American Chemical Society, publishers

Adhesives Age—Directory Issue, Communication Channels Inc., publishers

Plastics World Yellow Pages, A. Cahners

Advanced Composites Directory of Products and Services, Bluebook, Advanstar Publishers

Plastics Compounding Redbook, Advanstar Publishers

These references are revised yearly and contain up-to-date information on names, addresses, and phone numbers of plastic suppliers.

References

1. American National Standards Institute, *Annual Report 1992.*
2. ANSI bulletin *Questions and Answers,* 1992.
3. EASA bulletin *A Partnership for Progress,* 1992.
4. Electronic Industries Association, *1992 Annual Report.*
5. *IEEE Membership Information Booklet, 1993.*
6. *ASTM Publication Catalog 1993.*
7. *IPC Information Package 1993.*
8. *NEMA Annual Report 1992.*
9. Society of Plastics Engineers Catalog 1993.
10. LaVerne Leonard, ed., "Comparable Data for Plastic Materials—Help Is on the Way," *Plastics Design Forum,* January/February 1993, p. 37.
11. *Facts and Figures of the U.S. Plastics Industry,* Society of Plastics Institute, October 1992.

Index

Abrasion resistance, 30
Accelerator, 30
Acetal resins, 54
Acrylics:
    elastomers, 114
    plastics, 52–53
Acrylonitrile butadiene styrene (ABS), 40
Addition polymer, definition of, 13–14
Adhesive:
    alloys and blends, 283, 291
    classes of, 283
    definition of, 30
    die attach, 327
    elastomer, 283, 285–286
    in flex circuits, 277–282
    forms, 292–293
    in microelectronics, 284, 294
    in plastic bonding, 294
    selection of, 284
    in tapes, 298
    thermoplastic, 283, 285–286
    thermoset, 283, 287–288
Aging:
    Arrhenius equation in, 237
    definition of, 30
    in design considerations, 263
    nature of, 238
    service temperature of plastics, 240–241
    test methods, 239
Alkyds, 86
Alloys, 129
    adhesives, 283, 291
    characterization, 133
    chemistry of, 129, 131
    definition of, 129–130
    differential scanning calorimetry of, 133

Alloys (*Cont.*):
    electrical uses of, 136
    glass transition temperature, 133
    morphology, 130
    preparation of, 131
    processing of, 131
    properties of, 133, 136
    trade names of, 134–136
    types, 134–136
        block and graft copolymers, 132
        IPEN, 132
        IPN, 132
        liquid crystal, 132
Allyl resins, 87
American National Standards Institute (ANSI) (*see* Technical societies for plastics)
American Society for Testing and Materials (ASTM) (*see* Technical societies for plastics)
Amorphous polymers, definition of, 15–20
Applications, of plastics, 5
Arc and track resistance, 29–30, 221
Autoclave molding, 140

B-stage, 30
Batteries:
    properties of, 310–313
    separation in, 310
Benzocyclobutene, 103
Bismaleimide, 91
Blends (*see* Alloys)
Blow molding, 140
Blowing agent, 30
Bond strength, 30
Branched polymer, 15, 18–19
Bulk polymerization, 26

357

358 Index

Business consulting groups for plastics, 352
Butadienes, 115
Butyl rubber, 110

Calendering, 142
Capacitance, 30
Casting, 141
  centrifugal, 142
Catalyst, 30
Cellulose:
  acetate butyrates, 55
  acetates, 55
  proprionates, 55
Chemistry:
  of alloys, 129–131
  of blends, 129–131
Chips, integration of, 10
Chlorinated polyethylene, 115
Chloroprene, 112
Chlorosulfonated polyethylene, 110
Coatings, 282
  advantages and disadvantages, 198
  application methods, 196–197
    brush, 197
    casting, 199
    dipping, 197
    encapsulation, 196–199
    flow, 197
    glob-top, 200
    molding, 200
    potting, 199
    spraying, 197
  conformal, 189–195
    acrylic, 190
    allyl phthalate, 194
    benzocyclobutene, 194
    characteristics of, 189, 191
    epoxy, 190
    fluorocarbon, 193
    parylene, 192
    polyimide, 193
    polyurethane, 192
    silicon-carbon, 194
    silicone, 192
  extruded forms of, 309
  overview, 177–178
  properties, 180–185
    abrasion resistance, 187
    adhesion, 187
    bonding, 189

Coatings, properties (*Cont.*):
  cut-through, 187
  dielectric strength, 188
  flexibility, 187
  heat shock, 187
  high-voltage continuity, 188
  low-voltage continuity, 188
  selection of, 195
  solderability, 189
  spring-back, 187
  types, 179
    acrylics, 179
    alkyds, 179
    epoxy, 179
    diallyl phthalate, 179
    phenolic, 185
    plain enamels, 179
    polyester, 186
    silicone, 186
    urethane, 186
    wire enamels, 182–185
Coefficient of expansion, 30, 225
Cold flow, 30
Cold forming, 143
Composite, electrical and electronic market, 7
Compression molding, 142
Condensation polymers, 13–14
Conducting polymers, 331–334
  applications of, 334
Conductivities of materials, 332
Contact bonding, 30
Copolyester elastomers, 127
Copolymers, 15
Cross-linking 18–19, 30
  radiation, 79
  thermoplastics, 79–84
  (*See also* Vulcanization)
Crystalline melt point, 31
Cure agent, 31
Cure temperature, 31
Cure time, 31
Cyanate ester, 99–103

Design:
  aging considerations, 263
  approach, 259
  assembly, 264
  bonding:
    chemical, 265
    thermal, 266

Index

Design (*Cont.*):
  decorating, 266
  electrical properties, 261
  engineering properties, 260
  environmental properties, 262
  finishing, 266
  general concepts, 259
  machining, 266
  mechanical property considerations, 262
  selection criteria, 259
Die attach materials, 327
Dielectric constant, 29, 31
Dielectric loss, 31
Dielectric loss angle, 31
Dielectric loss factor, 31
Dielectric phase angle, 31
Dielectric power factor, 31
Dielectric strength, 28, 31
Differential scanning calorimetry, of blends, 133
Dissipation factor, 29, 31

Elastomers, 16, 31
  abrasion resistance, 118
  characteristics of, 109
  compression set, 119
  electrical properties of, 119–124
  flexural resistance of, 119
  hardness, 118
  hysteresis, 119
  low-temperature properties, 119
  properties of, 111, 117–124
  tear resistance of, 118
  tensile properties of, 117
  types, 110–117
Electric Industries Association (EIA) (*see* Technical societies for plastics)
Electrical Apparatus Service Association (EASA) (*see* Technical societies for plastics)
Electrical properties, 217–222
  arc and track resistance, 28–30, 221
  corona, 221
  dielectric constant, 29, 31
  dissipation factor, 29, 31, 219
  insulation resistance, 220
  resistivity, 34, 220
  surface resistivity, 34, 220
  thermoplastics, 34, 52
  volume resistivity, 220
Electrolytes, polymer, 334–336

Electronic packages, 326
Embedding, 149
Encapsulation, 149, 282, 328
Environmental properties, 227
  elastomers, 120–123, 235–237
  thermoplastics, 228–233
  thermosets, 233–235
Epichlorohydrin, 112
Epoxy resin, 91–94
Ethyl cellulose, 55
Ethylene acrylic, 74, 114
Ethylene propylene, 112
Ethylene vinyl acetate, 41, 74
Exotherm, 32
Extrusion, 144

Fibers, 17
Filament winding, 144
Fillers:
  definition of, 32
  effect on property, 3, 37–39
  types, 37–39
Film casting, 142
Film and sheet materials, 274–276
  characteristics of, 276
  properties of, 278–281
Flammability, 225
  properties of polymers, 227
  tests, 226
Flex resistance, of elastomers, 119
Flexibilizer, 32
Flexible circuit, 277
Flexural modulus, 32
Flexural strength, 32
Fluorinated elastomers, 113
Fluorocarbons, 32
Fluoropolymers, 55–57
  fluorinated ethylene propylene (FEP), 55–57
  polychlorotrifluoroethylene vinylidene chloride (CTFE), 55–57
  polyethylene-chlorotrifluoroethylene (ECTFE), 55–57
  polytetrafluoroethylene (PTFE), 55–57
  polyvinyl fluoride (PUF), 55–57
Fluorosilicone, 113
Fungus resistance, of plastics, 240–241

Gel, definition of, 32
Glass transition temperature, 25, 32, 224

360 Index

Glob-top (*see* Coatings, conformal)
Government publications on plastics, 354

Hand layup, 145
Hardener, 32
Heat distortion temperature, 32, 225
Heat sealing, 32
Heat-shrinkable tubing, 300
High-temperature polymers, 153–176
　advantages, 153
　applications, 153
　business outlook, 9
　classification, 156
　demographics, 153–154
　distribution of, 153–154
　fibers, 170–171
　heterocyclic, 167–170
　hybrid, 164
　properties, 174
　thermal stability, definition of, 154
　types, 155
　　acetylene-terminated polyimides (ATPI), 163
　　addition polyimides (API), 160–162
　　bismaleimide-triazine (BT), 163
　　bismaleimides (BMI), 163
　　condensation polyimides (CI), 155–160
　　cyanate esters (CE), 166
　　epoxy imide (EI), 162
　　liquid crystal polymers (LCP), 172
　　matrimid, 161
　　polyamide-imide, 172
　　polyarylate, 172
　　polybenzimidazoles (PBI), 164
　　polyether-imide, 172
　　polyether sulfone, 172
　　polyimide, 155–159
　　polyketone, 172
　　polyphenyl sulfide, 172
　　polyphenylene ether, 172
　　polyphenylene sulfone, 173
　　polyphenylenes, 167
　　polyphthalamide, 173
　　polysulfone, 173
　　silicon-carbon (SYCAR), 166
　　thermoplastic polyimides (TPI), 159, 171–173
　　usage of, 154

High voltage:
　continuity in, 188
　plastic use in, 312
Hot-melt adhesive, 32
Hydrocarbon, 32
Hydrolysis, 33
Hysteresis, 119

Impact strength (*see* Mechanical properties)
Impregnate, 33, 149
Impregnating resins, 200
　application methods, 200
　　trickle impregnation, 201
　　vacuum impregnation, 201
　reinforcement in, 201
Inhibitor, 33
Injection molding, 146
Institute of Electrical and Electronics Engineers (IEEE) (*see* Technical societies for plastics)
Institute for Interconnecting and Packaging Electronic Circuits (IPC) (*see* Technical societies for plastics)
Insulation:
　primary function of, 5
　resistance of, 33
Insulators, propagation delay in, 219
Integrated circuit:
　driving force in, 9
　plastics in, 317–331
Interlayer dielectrics, 327
Ionomer, 57
Isoprene, 114

Ketone plastics, 58–59

Laminates, 303–304, 307–308
Laminating, 148
Langmuir-Blodgett films, 336
Liquid-crystal blends, 132
Liquid-crystal polymers, 20, 60–63
Low-voltage continuity, 188

Mechanical properties:
　compressive strength, 216
　creep, 217
　elongation, 215

Mechanical properties (*Cont.*):
  flex strength, 216
  hardness, 217
  impact, 216
  modulus, 33, 215
  of plastics and elastomers, 206–214
  tensile strength, 215
  thermoplastics, 50–51
Melamines, 105
Microelectronics:
  adhesives in, 284
  overview, 317
  plastics in, 317–331
Mixtures (*see* Alloys)
Modulus of elasticity, 33
Molecular parameters, influence on properties, 75
Molecular weight, 21–24
Molecular weight distribution, 21–34

National Electrical Manufacturers Association (NEMA) (*see* Technical societies for plastics)
Natural rubber, 114
Newspapers dealing with plastics, 350
Nitrile rubber, 114
Nuclear radiation effects, 241
  on elastomers, 243–246
  on organic fluids, 25
  on plastics, 246–255
Nylons, types, 64–66
  copolymers, 64
  high-temperature, 65–66
    Kevlar, 65
    Nomex, 65
  nylon 6, 64
  nylon 6/6, 64
  nylon 6/10, 64
  nylon 8, 64
  nylon 11, 64
  nylon 12, 64

Olefins (*see* Polyolefins)
Organic fluids, effect of nuclear radiation on, 257

Packaging:
  in electronics, 326
  trends in, 11

Parylene, 43
Permittivity, 33
Phenolics, 94
Phenoxy, 43
Phenylene oxide, 75
Piezoelectric, 58
Plasticizer, 33
Plastics:
  applications, 5
  consumption outlook, 9
  demographics, 7
  distribution, in electronics, 6–8
  forms, applications, and uses, 273–314
  growth, 8
  importance of, 3–4
  information databases, 347–348
  organizations, 4
  outlook, 9
  primary function of, 5
  processing, 139–151
    autoclave molding, 140
    blow molding, 140
    calendering, 142
    casting, 141
      centrifugal, 142
    cold forming, 143
    compression molding, 142
    embedding, 149
    encapsulation, 149
    extrusion, 144
    filament winding, 144
    hand layup, 145
    impregnation, 149
    injection molding, 146
    laminating, 148
    potting, 150
    pulforming, 146
    pultrusion, 146
    radiation, 150–151
    reaction injection molding, 147
    rotational molding, 147
    spray-up, 145
    thermoforming, 147
    transfer molding, 144
  propagation delay, in insulators, 219
  properties:
    overview, 205
    variation of polymer, 21
  publications, 347
  statistics, 6–9
  suppliers, 355
  thermoplastic, 37–84

Plastics (*Cont.*):
  thermoset, 6–7, 85–108
  use of, 8
  (*See also* Polymers)
Polymerization (*see* Synthesis, of polymers)
Polymers, 33
  comparison with fibers, plastics, and elastomers, 18
  classification, 17
  definition of, 13
  glass transition temperature, 25, 32
  liquid-crystal polymer, 20
  molecular weight, 21–24
  molecular weight distribution, 21–24
  thermoplastic, 6–7, 15, 34, 37–84
  thermoset, 6–7, 15, 34, 85–108
  types, 13
    acetals, 54
    acetylene-terminated polyimides (ATPI), 163
    acrylics, 52–53
    acrylonitirile-butadiene-styrene, 40
    addition polyimides, 160
    alkyds, 86
    allyl resins, 87
    benzocyclobutenes, 103
    bismaleimides, 91
    butadienes, 115
    butyl, 110
    cellulose acetate butyrate, 55
    cellulose acetates, 55
    cellulose proprionate, 55
    chlorinated polyethylene, 115
    chloroprene, 112
    chlorosulfonated polyethylene, 110
    cyanate esters, 99–103
    diallyl phthalate, 87
    epichlorohydrin, 112
    epoxy resin, 91–94
    ethycellulose, 55
    ethylene acrylic, 74
    ethylene propylene, 112
    ethylene vinyl acetate, 41, 74
    fluorinated elastomers, 113
    fluoropolymers (*see* Fluoropolymers)
    fluorosilicones, 113
    ionomers, 57
    liquid-crystal polymers, 60–63
    melamines, 105
    nylons, 64–66
    parylene, 43

Polymers, types (*Cont.*):
    phenolic resin, 94
    phenoxy, 43
    polyamide (*see* Nylons, types)
    polyamide-imide, 66
    polyarylate, 70
    polyarylether, 44
    polybenzimidazole, 164
    polybutadiene, 89
    polybutylene, 44
    polycarbonate, 72–73
    polyester, 70–71, 95–96
    polyetherimide, 48, 69–70
    polyethersulfone, 77
    polyethylene, 72–75
    polyimide, 67–69
    polyketone, 58–59
    polymethylpentene, 44
    polyphenyl sulfone, 77–78
    polyphenylene, 167
    polyphenylene ether, 75
    polyphenylene sulfide, 76
    polyphthalamide, 48
    polysulfide, 115
    polyurethane, 50, 52, 126
    silicon-carbon, 106–107
    silicone, 98–99, 116
    styrene-butadiene, 115
    urea-formaldehyde, 105
    vinyls, 46–47, 78
Polyolefins, 72–75
Pot life, 34
Potting, 33, 150
Power factor, 34
Printed circuits, board evolution, 11
Processing (*see* Plastics)
Promoter (*see* Accelerator)
Pulforming, 146
Pultrusion, 146

Quinoline polymers, 169–170
Quinoxalone polymers, 169–170

Radiation processing, 150–151
Radiation resistance of polymers, 240–255
Reaction injection molding (RIM), 147
Reinforcement:
  in bulk molding compounds (BMC), 307, 309

Reinforcement (*Cont.*):
    fiber types, 303
    in laminates, 303–304, 307–308
    in printed wiring boards, 306
    in resins, 304–306
    in sheet molding compounds (SMC), 307, 309
Resistivity, 34
Resists:
    resolution of, 320
    types, 320–321
        dry film, 321
        electron beam (EB), 321
        photo, 320
        suppliers of, 323–326
Rockwell hardness number, 34
Rotational molding, 147
Rubber (*see* Elastomers)

Separators, in batteries, 309–310
Sheet molding compounds, 307, 309
Shore hardness, 34
Silicon-carbon resins, 106–107
Silicone elastomers, 116
Silicone resins, 98–99
Sleeving, 299
Solderability, coating property, 189
Solution polymerization, 26
Specific heat, 223
Spray-up, 145
Spring-back, coating property, 187
Storage life, 34
Strain, 16, 34
Stress, 16, 34
Structure, of polymers, 18
Styrene-butadiene resins, 40, 115
Surface resistivity, 28
Synthesis, of polymers, 26

Tapes, 296
    properties, 299
    selection, 298
    types, 297–298
Technical societies for plastics, 343
Terminology, 27–35
Thermal conductivity, 34, 222
Thermal properties of plastics, 222

Thermoforming, 147
Thermoplastic, 34
    applications, 40–48
    characteristics, 40–48
    processing, 40–48, 139–151
    properties, 49–52
    types, 37
Thermoplastic elastomers, 124
    olefinic, 126
    properties of, 125
    styrene, 126
    urethane, 126
Thermoset, 34, 85
    applications, 88–90
    characteristics, 88–90
    processing, 88–90
    properties, 86
    types, 88–90
Thixotropic, 34
Trade publications, 350–351
Transfer molding, 144
    resins, 329–331
Tubing:
    heat-shrinkable, 300
    test methods, 302
    types, 300

Ultraviolet radiation, resistance of plastics, 240
Urea-formaldehyde, 105
Urethanes, 50, 52

Vacuum impregnation, 201
Vicat softening temperature, 35
Vinyl polymers, 78
    flexible, 46
    rigid, 47
Viscosity, 34
Volume resistivity, 34
Vulcanization, 34

Waveguides, polymer, 337–338
Wetting, 34
Wire coatings, 182–189
Wire density, of chips, 317

ABOUT THE AUTHOR

William M. Alvino has more than 33 years of experience in the engineering application of plastics, and is presently a Fellow Scientist in the System, Processes, and Technologies Division of Westinghouse in Pittsburgh, Pennsylvania. He previously worked at the Bell Telephone Laboratories in Murray Hill, New Jersey. Mr. Alvino has written numerous technical papers and holds 18 patents in the area of organic/polymer chemistry. He is a member of the American Chemical Society.